T0318755

Time-Critical Cooperative Control of Autonomous Air Vehicles

Time-Critical Cooperative Control of Autonomous Air Vehicles

Isaac Kaminer, António Pascoal,
Enric Xargay, Naira Hovakimyan,
Venanzio Cichella, and
Vladimir Dobrokhodov

Butterworth-Heinemann
An imprint of Elsevier

Butterworth-Heinemann is an imprint of Elsevier
The Boulevard, Langford Lane, Kidlington, Oxford OX5 1GB, United Kingdom
50 Hampshire Street, 5th Floor, Cambridge, MA 02139, United States

Library of Congress Cataloging-in-Publication Data
A catalog record for this book is available from the Library of Congress

British Library Cataloguing-in-Publication Data
A catalogue record for this book is available from the British Library

ISBN: 978-0-12-809946-9

For information on all Butterworth-Heinemann publications
visit our website at https://www.elsevier.com/books-and-journals

Publisher: Joe Hayton
Acquisition Editor: Sonnini R. Yura
Editorial Project Manager: Ana Claudia Garcia
Production Project Manager: Kiruthika Govindaraju
Designer: Vicky Pearson

Typeset by VTeX

To the precious memory of my mother Olga and grandmother Leah,
to my father Isaac, to my sons Emanuel and Conor,
to Jenny, Cristof, and Shoshi and to Ekaterina, my love.

Isaac Kaminer

To the beloved memory of my parents Manuel and Idalina,
to my wife Stephanie and son Ricardo with boundless love and gratitude.

António Pascoal

To my parents, Elvira and Enric, with admiration, gratitude,
and unconditional love.

Enric Xargay

To the children: Arik, Arsen, and Maria with all my love, with all my heart.

Naira Hovakimyan

To my grandfather Ferruccio,
to the loving memory of my grandparents Gina, Giovanna, and Venanzio,
always present in my thoughts, in my heart, in my life.

Venanzio Cichella

To my parents, Valentina and Nicholai, with gratitude and admiration,
to my girls Elena and Anna who are the purpose of my life,
and to my brother Oleg who is always there for me.

Vladimir Dobrokhodov

Contents

List of Figures xi
Foreword xv
Preface xvii
Acknowledgments xxi
Notation and Symbols xxv

Part One Time-Critical Cooperative Control: An Overview 1

1 Introduction 3
1.1 General Description 3
1.2 Practical Motivation and Mission Scenarios 5
1.2.1 Cooperative Road Search 5
1.2.2 Sequential Auto-Landing 8
1.3 Literature Review 9
1.3.1 Path-Following Control 9
1.3.2 Coordinated Path-Following Control 11
1.3.3 Consensus and Synchronization in Networks 12
References 14

2 General Framework for Vehicle Cooperation 23
2.1 General Framework 23
2.2 Problem Formulation 24
2.2.1 Cooperative Trajectory Generation 24
2.2.2 Single-Vehicle Path Following 31
2.2.3 Coordination and Communication Constraints 32
2.2.4 Autonomous Vehicles with Inner-Loop Autopilots 35
References 37

Part Two Cooperative Control of Fixed-Wing Air Vehicles 39

3 3D Path-Following Control of Fixed-Wing Air Vehicles 41
3.1 Tracking a Virtual Target on a Path 41
3.2 Path-Following Control Law 47
3.2.1 Nonlinear Control Design at the Vehicle Kinematic Level 47
3.2.2 Stability Analysis with Non-ideal Inner-Loop Performance 49
3.3 Implementation Details 51
3.4 Simulation Example: Shaping the Approach to the Path 52
References 60

4 Time Coordination of Fixed-Wing Air Vehicles **61**
4.1 Coordination States **61**
4.2 Coordination Control Law **64**
4.2.1 Speed Control at the Vehicle Kinematic Level **64**
4.2.2 Convergence Analysis with Non-ideal Inner-Loop Performance **67**
4.3 Combined Path Following and Time Coordination **69**
4.3.1 Stability Analysis at the Kinematic Level **70**
4.3.2 Stability Analysis with Inner-Loop Autopilots **71**
4.4 Implementation Details **73**
4.5 Simulation Examples **74**
4.5.1 Path Following with Simultaneous Arrival **75**
4.5.2 Sequential Auto-Landing **78**
References **86**

5 Meeting Absolute Temporal Specifications **87**
5.1 Strict and Loose Absolute Temporal Constraints **87**
5.2 Coordinating with a Virtual Clock Vehicle **88**
5.2.1 Coordination Control Law **88**
5.2.2 Stability Analysis at the Kinematic Level **91**
5.3 Coordination with Loose Absolute Temporal Constraints **92**
5.4 Illustrative Example: Sequential Auto-Landing with Predefined Arrival Windows **93**
5.4.1 Transition Trajectories and Glide Slope **94**
5.4.2 Mission Execution **95**
References **103**

6 Time Coordination Under Quantization **105**
6.1 Convergence with Quantized Information **105**
6.1.1 Coordination Control Law and Coordination Dynamics **105**
6.1.2 Krasovskii Equilibria **107**
6.1.3 Stability Analysis at the Kinematic Level **108**
6.1.4 Coordination with Fully Quantized Information **109**
6.2 Simulation Example: Sequential Auto-Landing with Quantized Information **111**
References **114**

7 Time Coordination Under Low Connectivity **115**
7.1 Local Estimators and Topology Control **115**
7.1.1 Estimator Dynamics **116**
7.1.2 Coordination Control Law and Link-Weight Dynamics **117**
7.2 Simulation Example: Sequential Auto-Landing Under Severely Limited Communication **119**
References **127**

8 Flight Tests: Cooperative Road Search **129**
8.1 Road Search with Multiple Small Autonomous Air Vehicles **129**
8.1.1 Airborne System Architecture **129**
8.1.2 Flight-Test Results **131**
8.2 Mission Outcomes **137**
References **139**

Part Three Cooperative Control of Multirotor Air Vehicles 141

9 3D Path-Following Control of Multirotor Air Vehicles **143**
9.1 Problem Formulation **143**
9.1.1 6-DoF Model for a Multirotor UAV **144**
9.1.2 Virtual Target and Virtual Time **144**
9.1.3 Path-Following Error **146**
9.2 Path-Following Control Law **149**
9.3 Simulation Example: Following a Virtual Target **154**
References **157**

10 Time Coordination of Multirotor Air Vehicles **159**
10.1 Coordination States and Maps **160**
10.2 Coordination Control Law **161**
10.3 Simulation Results **163**
References **169**

11 Flight Tests of Multirotor UAVs **171**
11.1 System Architecture and Indoor Facility **171**
11.2 Flight-Test Results **172**
11.2.1 Phase on Orbit Coordination **173**
11.2.2 Spatial Coordination Along One Axis **173**
11.2.3 Additional Flight Tests **173**
References **177**

Part Four Final Considerations 179

12 Summary and Concluding Remarks **181**
12.1 Summary **181**
12.2 Open Problems **182**
12.3 Cooperative Control in Future Airspace Scenarios **184**
References **186**

A Mathematical Background **189**
A.1 The *Hat* and *Vee* Maps **189**
A.2 Nonlinear Stability Theory **189**
A.2.1 Lipschitz Functions, Existence and Uniqueness of Solutions **189**

A.2.2 Autonomous Systems 191
A.2.3 The Invariance Principle 192
A.2.4 Nonautonomous Systems 193
A.2.5 Boundedness 195
A.2.6 Input-to-State Stability 196
A.3 Graph Theory 197
A.3.1 Basic Definitions 197
A.3.2 Connectivity 198
A.3.3 Algebraic Graph Theory 198
References 199

B Proofs and Derivations 201
B.1 Proofs and Derivations in Part I 201
B.1.1 The Coordination Projection Matrix 201
B.2 Proofs and Derivations in Part II 202
B.2.1 Time-Derivative of the Coordination States 202
B.2.2 Closed-Loop Coordination Error Dynamics 203
B.2.3 Proof of Lemma 3.1 203
B.2.4 Proof of Lemma 3.2 206
B.2.5 Proof of Lemma 4.1 209
B.2.6 Proof of Lemma 4.2 215
B.2.7 Proof of Theorem 4.1 216
B.2.8 Proof of Theorem 4.2 217
B.2.9 Proof of Lemma 6.1 220
B.2.10 Proof of Proposition 6.1 221
B.2.11 Proof of Theorem 6.1 222
B.2.12 Proof of Lemma 6.2 224
B.3 Proofs and Derivations in Part III 227
B.3.1 Proof of Lemma 9.1 227
B.3.2 Proof of Lemma 9.2 228
B.3.3 Proof of Lemma 9.3 230
B.3.4 Proof of Theorem 10.1 232
B.3.5 Proof of Corollary 10.1 235
References 236

Index 239

List of Figures

Fig. 1.1 Cooperative road search using multiple fixed-wing UAVs. **6**
Fig. 1.2 Cooperative road search using multiple multirotor UAVs. **7**
Fig. 1.3 Sequential auto-landing. **9**

Fig. 2.1 Conceptual control architecture for fixed-wing UAVs. **25**
Fig. 2.2 Conceptual control architecture for multirotor UAVs. **25**
Fig. 2.3 Trajectory generation for two vehicles; spatial deconfliction. **29**
Fig. 2.4 Trajectory generation for two vehicles; temporal deconfliction. **29**
Fig. 2.5 Trajectory generation; arrival margin. **30**
Fig. 2.6 Adaptive augmentation loop for a fixed-wing UAV. **37**

Fig. 3.1 Following a virtual target vehicle; problem geometry. **42**
Fig. 3.2 Shaping the approach to the path. **45**
Fig. 3.3 Path-following kinematic closed-loop system. **47**
Fig. 3.4 Effect of the characteristic distance d on the convergence of the vehicle to the path. **49**
Fig. 3.5 Path-following closed-loop system. **50**
Fig. 3.6 Framed 3D spatial path and main geometric properties. **53**
Fig. 3.7 Path following of a single UAV. **54**
Fig. 3.8 Path-following performance related to Fig. 3.7. **55**
Fig. 3.9 Path following of a single UAV; *fast* convergence to the path ($d = 50$ m). **56**
Fig. 3.10 Path-following performance related to Fig. 3.9. **57**
Fig. 3.11 Path following of a single UAV; *slow* convergence to the path ($d = 250$ m). **58**
Fig. 3.12 Path-following performance related to Fig. 3.11. **59**
Fig. 3.13 Projection of vector $\hat{\mathbf{w}}_1$ onto the tangent vector to the path $\hat{\mathbf{t}}$. **59**

Fig. 4.1 Time-critical cooperative path-following closed-loop system. **70**
Fig. 4.2 Network topologies. **75**
Fig. 4.3 Simultaneous arrival; 3D time-trajectories. **76**
Fig. 4.4 Simultaneous arrival; path deconfliction. **77**
Fig. 4.5 Simultaneous arrival; mission evolution. **78**
Fig. 4.6 Simultaneous arrival; path-following performance. **79**
Fig. 4.7 Simultaneous arrival; time coordination. **80**
Fig. 4.8 Simultaneous arrival; information flow. **80**
Fig. 4.9 Sequential auto-landing; 3D time-trajectories. **81**
Fig. 4.10 Sequential auto-landing; trajectory deconfliction. **82**
Fig. 4.11 Sequential auto-landing; mission evolution. **83**
Fig. 4.12 Sequential auto-landing; path-following performance. **84**
Fig. 4.13 Sequential auto-landing; time coordination. **85**
Fig. 4.14 Sequential auto-landing; information flow. **85**

Fig. 5.1 NASA's IceBridge program; coordination of a DC-8 aircraft with the CryoSat-2 satellite for radar altimeter calibration and validation. **89**

Fig. 5.2 Sequential auto-landing; 3D time-trajectories. **94**
Fig. 5.3 Sequential auto-landing; trajectory deconfliction. **96**
Fig. 5.4 Sequential auto-landing with loose temporal constraints; mission evolution. **98**
Fig. 5.5 Sequential auto-landing with loose temporal constraints; path-following performance. **99**
Fig. 5.6 Sequential auto-landing with loose temporal constraints; time coordination. **100**
Fig. 5.7 Sequential auto-landing with loose temporal constraints; inter-vehicle separation. **100**
Fig. 5.8 Sequential auto-landing with loose temporal constraints; information flow. **101**
Fig. 5.9 Sequential auto-landing with strict temporal constraints; time coordination. **101**
Fig. 5.10 Sequential auto-landing with relative temporal constraints; time coordination. **102**

Fig. 6.1 Sequential auto-landing; partially quantized control law with fine quantization. **111**
Fig. 6.2 Sequential auto-landing; fully quantized control law with fine quantization. **112**
Fig. 6.3 Sequential auto-landing; partially quantized control law with coarse quantization. **113**
Fig. 6.4 Sequential auto-landing; fully quantized control law with coarse quantization. **114**

Fig. 7.1 Extended network with local estimators. **117**
Fig. 7.2 Edge snapping; two-well function and forcing function for link-weight dynamics. **120**
Fig. 7.3 Convergence time of coordination error dynamics with three different algorithms. **121**
Fig. 7.4 Histograms and distribution fits of the normalized convergence time. **122**
Fig. 7.5 Sequential auto-landing; modified coordination control law (hybrid). **123**
Fig. 7.6 Sequential auto-landing: information flow (hybrid). **123**
Fig. 7.7 Sequential auto-landing; link weights and local connectivity (hybrid). **124**
Fig. 7.8 Sequential auto-landing; modified coordination control law (edge snapping). **125**
Fig. 7.9 Sequential auto-landing: information flow (edge snapping). **125**
Fig. 7.10 Sequential auto-landing; link weights and local connectivity (edge snapping). **126**

Fig. 8.1 SIG Rascal UAV with two different onboard cameras. **130**
Fig. 8.2 Network-centric architecture of the airborne platform. **130**
Fig. 8.3 Cooperative road-search; trajectory generation. **132**
Fig. 8.4 Cooperative road-search; time coordination during the transition phase. **133**
Fig. 8.5 Cooperative road-search; cooperative path-following control during the road-search phase. **134**
Fig. 8.6 Cooperative road-search; mission performance. **135**
Fig. 8.7 Cooperative road-search; coordinated vision-based target tracking. **136**
Fig. 8.8 Time-critical cooperation in a road-search mission. **137**
Fig. 8.9 High-resolution image exploitation. **138**

Fig. 9.1 Geometry of the path-following problem. **147**
Fig. 9.2 Shaping the approach to the path. **147**
Fig. 9.3 Following a virtual target; simulation scenario. **150**
Fig. 9.4 Following a virtual target; mission execution. **151**
Fig. 9.5 Following a virtual target; path-following errors and commands. **152**
Fig. 9.6 Following a virtual target; simulation scenario with inner-loop autopilot. **153**
Fig. 9.7 Following a virtual target; mission execution. **154**

Fig. 9.8 Following a virtual target; path-following errors and commands. **155**
Fig. 9.9 Drone drawing a Bézier curve. **156**

Fig. 10.1 Examples of coordinated maneuvers. **160**
Fig. 10.2 Coordinated exchange of positions; simulation results. **164**
Fig. 10.3 Coordinated exchange of positions with ideal communications; time coordination. **165**
Fig. 10.4 Coordinated exchange of positions with range-based communications; estimate of the quality of service (QoS). **166**
Fig. 10.5 Coordinated exchange of positions with range-based communications; time coordination. **167**
Fig. 10.6 Coordinated exchange of positions with range-based communications and inner-loop autopilots; time coordination. **168**
Fig. 10.7 Effect of network connectivity on time-coordination; norm of the coordination error vector. **168**
Fig. 10.8 Effect of path-following correction terms $\bar{\alpha}_i(\cdot)$ on inter-vehicle separation. **169**

Fig. 11.1 Indoor facility at the Naval Postgraduate School. **172**
Fig. 11.2 Parrot's AR.Drone [2]; quadrotor employed in the flight-test experiments. **172**
Fig. 11.3 Phase on orbit coordination; desired and actual vehicle paths in 3D. **174**
Fig. 11.4 Phase on orbit coordination; mission execution. **174**
Fig. 11.5 Phase on orbit coordination; phase shift. **175**
Fig. 11.6 Phase on orbit coordination; coordination error and coordination-state rates. **175**
Fig. 11.7 Spatial coordination along one axis; desired and actual vehicle paths in 3D. **175**
Fig. 11.8 Spatial coordination along one axis; mission execution. **176**
Fig. 11.9 Spatial coordination along one axis; speed profiles. **176**
Fig. 11.10 Spatial coordination along one axis; coordination error and coordination-state rates. **176**

Foreword

We can see on the horizon a new age of autonomous systems in which the opportunities for applications are endless: self-driving cars and trucks, autonomous package and cargo delivery systems, police and military operations, earth exploration, critical infrastructure inspection and maintenance, and medical diagnostics and operating room robotics, to name a few. Autonomous systems with machine reasoning and intelligence will forever change how machines serve humans. We are at the threshold of an era when humans and machines understanding mission context, sharing understanding and situation awareness, and adapting to the needs and capabilities of each other with advanced communication systems, control algorithms, and sensor networks will execute challenging scientific and commercial missions with unprecedented safety, reliability, and performance.

Control theory brings together advanced mathematics, modeling and simulation, and software development, and forms the kernel for the operating systems embedded in autonomous systems. Control of multiple autonomous air vehicles is one of the most challenging control problems due to the nature of flight. Each air vehicle control and mission management system consists of nested control loops which must provide inner-loop stability, command following, and outer-loop trajectory control to accomplish demanding tasks, as well as have the ability to reject/compensate for significant environmental disturbances. The resulting architecture must manage tasks, generate trajectories, enable vehicles to follow the corresponding paths, perform under failures, damage, or when things go wrong, and communicate critical information to all systems that need it.

This book represents a welcome contribution to the literature on the development of autonomous systems. It addresses key topics that are critical to the control of autonomous air vehicles and provides a rigorous mathematical formulation for control systems design and analysis. Clearly, having multiple air vehicles cooperate to execute complex missions requires the ability to follow desired paths and complete tasks at prescribed times, and to be able to do this under limited/constrained communications networks. The authors Isaac Kaminer, António Pascoal, Enric Xargay, Naira Hovakimyan, Venanzio Cichella, and Vladimir Dobrokhodov bring together a wealth of experience and theoretical rigor applied to the time-critical control of autonomous fixed-wing and multirotor air vehicles. The first section of this book gives an overview of a general framework leading to an architecture for trajectory generation, path following, multi-vehicle coordination and communications, and inner-loop autopilot control. The key idea exploited is the decoupling of space and time that allows using the velocities of the vehicles for coordination, while circumventing the performance limitations of trajectory tracking controllers. This decoupling makes it possible to account for heterogeneous vehicle dynamics. The second and third sections use the framework proposed to address time-critical control of autonomous fixed-wing and multirotor air vehicles under very general assumptions on the topology of the underlying communications network. They also provide detailed algorithm development, implementation details, and simulation examples that arise in the course of representative mission scenarios. For the missions described, both relative and absolute temporal constraints are considered. The mathematical machinery adopted to study networked cooperative systems allows for the consideration of challenging issues such as quantization in information sharing and low-connectivity in networks. Flight tests of cooperative road search using fixed-wing UAVs and coordinated flight of multirotors are used to illustrate the theoretical results. By bringing together theory and practice, this book affords researchers and practitioners in the field of cooperative air robotics a solid foundation to develop further theoretical work and pursue the implementation of new methodologies. Both control system academics and aerospace engineers will find the presentation intriguing, illuminating, and inspiring.

Kevin A. Wise
Senior Technical Fellow, Advanced Flight Controls
The Boeing Company

February, 2017

Preface

The advent of powerful embedded systems and communications networks has spawned widespread interest in the problem of coordinated motion control of multiple autonomous vehicles that will be engaged in increasingly challenging mission scenarios. The types of applications considered include, but are not limited to, intelligent surveillance and reconnaissance, environmental monitoring, and data collection by exploiting tools available for networked control of multiple heterogeneous autonomous robots. The latter include air, land, ocean-surface, and underwater vehicles. This book focuses on autonomous air vehicles. However, the tools developed for multiple vehicle coordination are sufficiently general to be extended and applied to the control of other kinds of autonomous platforms. A large subset of the missions proposed for air vehicles requires that they coordinate their motion in relative but not necessarily absolute time. For example, in a rendezvous mission, the times of arrival of the vehicles at their destinations may be required to be the same, but not specified a priori. To meet the requirements imposed by these types of applications, a new control paradigm must be developed that departs considerably from classical control strategies. To this end, this book details a theoretical framework for cooperative autonomous air vehicle control that addresses explicitly new and challenging temporal mission specifications, together with stringent spatial constraints. In particular, the framework integrates algorithms for path following and time-critical coordination that together provide a team of Unmanned Air Vehicles (UAVs) with the ability to meet joint spatial and temporal assignments.

The key contribution of this book is the development of a *cooperative path-following framework* that addresses explicitly time-critical issues. In its essence, the cooperative strategy proposed consists of assigning each vehicle a feasible path that satisfies mission requirements and vehicle dynamic constraints and then having each vehicle follow its assigned path while coordinating its position along that path with the other vehicles in the team. This is accomplished by judiciously decoupling space and time in the formulation of the trajectory-generation, path-following, and time-coordination problems. The result is a systematic framework for integration of various tools and concepts from a broad spectrum of disciplines leading to a streamlined design procedure accessible to a typical control engineer.

The book is organized in four parts with a total of twelve chapters. Two appendices that review the necessary mathematical background and contain proofs of the main theoretical results are also included. The organization of the different parts and chapters is detailed next:

- **Part I** introduces the framework for multiple vehicle cooperation adopted in the book and discusses a set of multi-vehicle mission scenarios that warrant the implementation of the proposed control architectures.

  - **Chapter 1** affords the reader a general description of the problems addressed in the book and how they relate to previous work reported in the literature. The chapter describes the practical motivation for the topics considered and includes two illustrative UAV mission scenarios: cooperative road search and sequential auto-landing.
  - **Chapter 2** formulates the problem of time-critical cooperative path-following control of multiple UAVs in 3D space. The chapter introduces also a set of assumptions and constraints on the supporting communications network as well as on the autopilots mounted onboard the UAVs.

- **Part II** presents the cooperative path-following control framework for fixed-wing UAVs.

  - **Chapter 3** formulates the problem of path-following control for a single UAV, and describes a nonlinear control algorithm that uses angular rates to steer the vehicle along a 3D spatial path for an arbitrary feasible speed assignment along the path.
  - **Chapter 4** presents a strategy for time coordination of multiple UAVs, yielding a framework for cooperative path following. The strategy proposed relies on the adjustment of the speed profile of each vehicle based on coordination information exchanged over the inter-vehicle communications network. In particular, a set of coordination states and coordination maps is proposed that allows for the application of the cooperative path-following framework to the case of path-dependent, desired speed profiles.
  - **Chapter 5** extends the coordination control law presented in Chapter 4 to support time-critical multi-vehicle missions that impose absolute temporal constraints on the trajectories of the vehicles, in addition to relative temporal assignments. The chapter addresses cooperative missions that require the vehicles to strictly observe absolute temporal specifications, as well as missions that only require the fleet to coordinate within a desired temporal window.
  - Motivated by the use of networks with finite-rate communication links, **Chapter 6** analyzes the effect of quantization on the stability and convergence properties of the closed-loop coordination dynamics. The results in this chapter show that, depending on the design of the quantized coordination control law, the closed-loop coordination error dynamics have undesirable "zero-speed" attractors. A modification of the coordination control law presented in Chapter 4 is proposed that retains the origin as the only equilibrium point of the system and prevents the existence of "zero-speed" equilibria.
  - **Chapter 7** proposes a modification of the coordination control law introduced in Chapter 4 with the objective of improving the convergence rate of the closed-loop coordination dynamics in low-connectivity scenarios. The proposed approach, which borrows and expands tools and concepts from the control of complex networks and logic-based communication protocols, leads to an evolving extended network, whose topology depends on the local exchange of information among vehicles.

  - **Chapter 8** presents the results of flight tests for a cooperative road-search mission that show the efficacy of the multi-UAV cooperative framework. The results demonstrate the validity of the proposed theoretical framework in a specific realistic application as well as the feasibility of the onboard implementation of the algorithms developed.

- **Part III** extends the cooperative path-following control framework to multirotor UAVs.

  - **Chapter 9** presents a solution to the problem of path-following control for a single multirotor UAV. The strategy departs from the control law for fixed-wing UAVs described in Chapter 3 in that it is applicable to vehicles that may have a zero velocity vector at some point during the execution of the mission.
  - **Chapter 10** proposes a time-coordination algorithm that, similar to the one discussed in Chapter 4, relies on the exchange of coordination information over the inter-vehicle communications network. However, the different dynamics of multirotor vehicles require a different structure for the coordination control law, which now adjusts the acceleration profile of the vehicles along their corresponding paths, instead of their speed profile.
  - **Chapter 11** presents flight-test results of a team of multirotor UAVs executing time-critical cooperative maneuvers.

- Finally, **Part IV** concludes the book:

  - **Chapter 12** summarizes the current state of development of time-critical cooperative path-following control, identifies some important open problems that deserve further attention, and discusses the application of the proposed framework to present and future multi-vehicle missions.

Acknowledgments

This book is the natural consequence of many years of research by the authors in the general area of networked cooperative motion planning and control of autonomous vehicles. In this endeavor we were brought together by our common passion to bring theoretical results in control and networked systems theory to bear on the development of new practical-oriented methods for time-critical cooperative control of autonomous air vehicles. The writing of the book was in itself a distributed, cooperative process involving human "agents" working at different geographical locations. We hope the final outcome will be looked upon as a comprehensive framework for cooperative UAV control systems design and implementation, capable of inspiring new research and guiding practitioners in various applications of autonomous systems.

The work that we summarize in this book was strongly influenced by and benefited from close interaction with many colleagues, postdoctoral fellows, and graduate students that we were fortunate to meet in our paths. All of them contributed directly or indirectly to creating an extremely rich and motivating collective atmosphere that stimulated visionary thinking, open discussions, and the free interchange of ideas. Among those with whom we shared ideas and concepts, we owe a special word of gratitude to António Aguiar, Reza Ghabcheloo, João Hespanha, and Carlos Silvestre for early joint work on cooperative path-following and networked control under intermittent communication losses. We also owe much to the members, past and present, of the Advanced Controls Research Lab of the University of Illinois for having created and nurtured a pleasant and stimulating working atmosphere. Chengyu Cao deserves special mention for his contribution to the development of path-following control algorithms with $\mathcal{L}_1$ adaptive augmentation. Ronald Choe made significant contributions to our path-following and coordination control laws as he developed a novel framework for distributed cooperative trajectory generation. Javier Puig-Navarro played an important role in advancing the theory of time-critical coordination with absolute temporal constraints. We are also thankful to Dušan Stipanović and Petros Voulgaris for many useful discussions on the topic.

We are extremely grateful to Oleg Yakimenko, who helped open new frontiers in the area of motion planning and control of autonomous systems. The core ideas that he introduced on the use of time and space separation techniques for cooperative motion planning found rich and fertile ground in the vibrant research atmosphere at the Naval Postgraduate School (NPS) and gave impetus to the research described in the present work. We thank Mariano Lizarraga, formerly at NPS and now with MathWorks, for having spent long hours of groundbreaking work on developing the flight critical software that ultimately enabled all of the demonstrated flight-test results. We extend our gratitude to Kevin Jones, also at NPS, for his insight into aircraft flight dynamics, his familiarity with novel aerospace technologies, and his ingenious creativity in building and flying airplanes. Every flight experiment with him as a safety pilot ended long after the aircraft had landed, following stimulating discussions on nearly imperceptible details captured by his attentive eye. Without his dedication to performing experimental flight tests, the level of trust that he conveyed to all of us, and his unfailing support in the implementation of new algorithms, many of the results presented here would have taken much longer to come to fruition.

We are thankful to our collaborators at NASA Langley Research Center, including the team at the Autonomy Incubator. Irene Gregory, Anna Trujillo, and Danette Allen deserve a special word of thanks for their constant attention to our developments, their feedback, and the technology transition opportunities provided by them in various NASA programs. We are also indebted to Kevin Wise from the Boeing Co. for sharing his knowledge and insights on multi-vehicle cooperative motion control, aircraft dynamics, and flight control system design, and for writing the foreword for this book.

A word of deep appreciation is due to our colleagues and close collaborators Pedro Abreu, João Botelho, Bruno Cardeira, Naveen Crasta, Francisco Curado, Pedro Góis, Vahid Hassani, Jorge Ribeiro, Miguel Ribeiro, Manuel Rufino, Luis Sebastiao, and Henrique Silva at the Institute for Systems and Robotics (ISR) for their friendship and for making the atmosphere at the Dynamical Systems and Ocean Robotics Group a truly exciting one. With them we could exploit and test many of the concepts on cooperative vehicle control in the area of marine robotics and establish valuable bridges with the aerial counterpart. We will always cherish the long hours spent together in the laboratory and at sea, engaged in fruitful discussions on cooperative autonomous vehicles and participating in field tests while experiencing collectively the sheer pleasure of watching as the "robotic extensions of our minds" maneuvered gracefully in the real world.

Our colleagues Sanjeev Afzulpurkar, Ehrlich Desa, Elgar Desa, Antony Joseph, R. Madan, Antonio Mascarenhas, and Pramod Maurya at the National Institute of Oceanography, Goa, India provided constant support and encouragement and graciously supported research and field work on autonomous marine vehicles during multiple visits to Goa. Their commitment to seeing beyond what is immediate and stressing the need for a long term vision targeting the scientific and commercial applications of autonomous vehicles has always been a source of inspiration. Their friendship and professionalism are truly appreciated. We thank John Hauser for the enjoyable brain storming sessions on aerial and marine vehicles and for his insight into an endless number of challenging problems in control theory. We also thank Alessandro Saccon for the many illuminating discussions on cooperative motion planning. A special word of thanks goes to former PhD students Joao Almeida, Behzad Bayat, Andreas Hausler, Jorge Soares, and Francesco Vanni for having helped us discover the new territory of cooperative motion estimation and control.

The authors are grateful for the support provided over the years by a number of sponsoring agencies. Not only did their key personnel support enthusiastically various parts of this project but they were also instrumental in suggesting new avenues of research that led ultimately to results with a strong potential for practical applications. We are particularly grateful to the following sponsors:

- NASA Langley Research Center, including the team at the Autonomy Incubator

- Air Force Office of Scientific Research
- Office of Naval Research, Science of Autonomy Program
- US Special Operations Command
- National Science Foundation
- NPS's Consortium for Robotics and Unmanned Systems Education and Research
- European Union's Horizon 2020 Research and Innovation Programme under the Marie Skłodowska-Curie grant agreement No. 642153
- Fundação para a Ciência e a Tecnologia (FCT), through ISR, under the LARSyS FCT (UID/EEA/50009/2013) funding program

The writing of this book made us realize at a very deep level the impact that a number of very special people had on our personal lives and how they ended up shaping our professional careers as well. To them we owe a very special, personal thank you note.

Isaac is profoundly indebted to his parents Olga and Isaac for sacrificing everything when they emigrated to the USA from USSR to make their children's lives better. A special thank you is due to Pramod Khargonekar for having the patience to supervise him during his studies at the University of Michigan. A special word of gratitude goes to Ekaterina for all the support she has provided in the past two years.

António is deeply grateful to his wife Stephanie for her support, patience, and loving presence and to his son Ricardo for his companionship, inquisitive spirit, and thought provoking questions on the usefulness of engineering research. He is indebted to Pramod Khargonekar for having taken him under his guidance in Minneapolis and instilling in him the passion for mathematical systems theory that let ultimately to his involvement in the design of networked systems for air and marine vehicle navigation and control. He is indebted to Michael Athans for his friendship and the passionate and always fruitful discussions they had during the years that he spent at the Instituto Superior Técnico (IST), Univ. Lisbon, Portugal. He would also like to thank João Sentieiro, former Director of the ISR, for his vision, constant encouragement, and guidance.

Enric would like to thank his parents, Enric and Elvira, for their unconditional dedication, love, and support. He greatfully acknowledges the influence of Joaquim Agulló, Ramon Costa-Castelló, Giulio Avanzini, Giorgio Guglieri, Roberto Tempo, and Ricardo Sánchez-Peña, brilliant scholars and researchers that sparked his interest in dynamics and controls.

Naira would like to thank her parents, her sister, and her family for their unbounded love and support. She also acknowledges the influence of Wright Patterson and Eglin Air Force Laboratories for the many stimulating questions, challenging problems, and useful feedback at some instances. She is especially indebted to Siva Banda, David Doman and Johnny Evers. Numerous useful discussions in various AFOSR meetings with Jonathan How, Mark Campbell, Meir Pachter, Magnus Egerstedt, Robert Murphey, Mario Sznaier, Munther Dahleh, Mehran Mesbahi, and many others have influenced very positively her thinking about cooperative control.

Venanzio is thankful to his parents Paolo and Patrizia for their encouragement and unconditional support and to his grandparents Gina, Giovanna, Ferruccio, and Venanzio for their unbounded love that always lives in his heart. His appreciation is extended to his brother Massimo, who has taught him more than he could have imagined a young brother could teach an older one. A very special word of deep appreciation goes to Caterina, for her companionship and loving presence, and for making his life a beautiful journey.

Vladimir would like to thank his parents for their love and support. They have always valued knowledge and education among the best human qualities, and they have made every effort to instill in him the love to life-long learning of anything new and inspiring. He is grateful to his wife Elena for her encouraging support, patience, great sense of humor, and loving presence, and to his daughter Anna for her genuine curiosity in his research. Her interest manifests itself in the form of often naive but thought provoking questions which have always inspired him to become more creative and express his thoughts in a far simpler manner.

António Pascoal
Lisbon, Portugal

Isaac Kaminer, Vladimir Dobrokhodov
Monterey, CA

Enric Xargay
Barcelona, Catalonia

Venanzio Cichella, Naira Hovakimyan
Urbana, IL

May, 2017

Notation and Symbols

$\mathbb{R}$	Field of real numbers.
$\mathbb{Z}$	Set of all integers.
$\mathsf{SO}(3)$	Special orthogonal group of all rotation matrices about the origin of the three-dimensional Euclidean space $\mathbb{R}^3$.
$\mathfrak{so}(3)$	Set of 3×3 skew-symmetric matrices over $\mathbb{R}$.
$\mathbb{I}_n$	Identity matrix of size n.
$\mathbf{0}$	Zero matrix of appropriate dimension.
$\mathbf{1}_n$	Vector of $\mathbb{R}^n$ whose components are all 1.
$\{\mathcal{F}\}$	Reference frame.
$\{\boldsymbol{v}\}_F$	Vector $\boldsymbol{v}$ resolved in frame $\{\mathcal{F}\}$.
$\{\hat{\boldsymbol{e}}\}_F$	Unit vector $\hat{\boldsymbol{e}}$ resolved in frame $\{\mathcal{F}\}$.
$\boldsymbol{\omega}_{F1/F2}$	Angular velocity of frame $\{\mathcal{F}1\}$ with respect to frame $\{\mathcal{F}2\}$.
$\boldsymbol{R}_{F1}^{F2}$	Rotation matrix from frame $\{\mathcal{F}1\}$ to frame $\{\mathcal{F}2\}$.
$\dot{\boldsymbol{v}}]_F$	Time-derivative of vector $\boldsymbol{v}$ computed in frame $\{\mathcal{F}\}$.
$\boldsymbol{v}^\top$	Transpose of vector $\boldsymbol{v}$.
$\lVert\boldsymbol{v}\rVert$	2-norm of vector $\boldsymbol{v}$.
$\lVert\boldsymbol{v}\rVert_\infty$	∞-norm of vector $\boldsymbol{v}$.
$\boldsymbol{M}^\top$	Transpose of matrix $\boldsymbol{M}$.
$\det(\boldsymbol{M})$	Determinant of matrix $\boldsymbol{M}$.
$\mathrm{tr}\,[\boldsymbol{M}]$	Trace of matrix $\boldsymbol{M}$.
$\lambda_{\max}(\boldsymbol{M})$	Maximum eigenvalue of matrix $\boldsymbol{M}$.
$\lambda_{\min}(\boldsymbol{M})$	Minimum eigenvalue of matrix $\boldsymbol{M}$.
$\lVert\boldsymbol{M}\rVert$	Induced 2-norm of matrix $\boldsymbol{M}$.
$\mathrm{card}(\mathcal{S})$	Cardinality of set $\mathcal{S}$.
$\mathrm{K}\,(\cdot)$	Krasovskii operator.
$\mathrm{co}\,\mathcal{S}$	Convex hull of set $\mathcal{S}$.
$\mathrm{sat}\,(\cdot)$	The saturation function.
$(\cdot)^\wedge$	The *hat map*. (See Appendix A.1.)
$(\cdot)^\vee$	The *vee map*. (See Appendix A.1.)
♠	End of definition.
◇	End of corollary, lemma, proposition, or theorem.
□	End of proof.
△	End of remark.
CVBTT	Coordinated Vision-Based Target Tracking.
DoF	Degree of Freedom.
FoV	Field of View.
ISS	Input-to-State Stability.
LoS	Line of Sight.
MANET	Mobile Ad-hoc Network.
NPS	Naval Postgraduate School.
PE	Persistency of Excitation.
QoS	Quality of Service.
UAV	Unmanned Aerial Vehicle.
UTC	Coordinated Universal Time.

General note: Calligraphic, upper-case letters enclosed within curly brackets are used to denote reference frames (e.g. $\{\mathcal{F}\}$). Unless otherwise noted, bold-face, lower-case letters refer to column vectors (e.g. $\boldsymbol{v}$), while bold-face, capital letters refer to matrices (e.g. $\boldsymbol{M}$). In general, the ith component of vector $\boldsymbol{v}$ is denoted by v_i, and the (i, j) entry of matrix $\boldsymbol{M}$ is represented by M_{ij}. The cross product of two vectors $\boldsymbol{v}$ and $\boldsymbol{w}$ is denoted by $\boldsymbol{v} \times \boldsymbol{w}$, while their dot product is denoted by $\boldsymbol{v} \cdot \boldsymbol{w}$.

Part One

Time-Critical Cooperative Control: An Overview

Introduction

1

1.1 General Description

The advent of powerful embedded systems, sensors, and communications networks has drawn widespread interest in the use of unmanned systems to execute missions with limited involvement of human operators. In particular, Unmanned Aerial Vehicles (UAVs) have been playing an increasingly important role in civilian and military applications, thus capturing the interest of academia, industry, and governmental institutions [121]. In simple applications, a single vehicle can be managed by a crew using a ground station provided by the vehicle manufacturer. However, the execution of more challenging missions requires the use of heterogeneous autonomous systems connected by means of a communications network, working in cooperation to achieve common objectives that may be dynamically assigned as the mission unfolds. In the course of these missions, the vehicles must be able to operate safely and execute coordinated tasks in complex, highly uncertain environments while maneuvering in close proximity to each other. In general, the success of multi-vehicle cooperative missions depends on the ability of the vehicle fleet to exchange information in a timely and reliable manner and, therefore, the quality of service (QoS) of the supporting communications network becomes a factor of major importance. In addition, as pointed out in [42] and [47], in many scenarios the flow of information among vehicles may be severely restricted, either for security reasons (stealth requirements) or because of tight bandwidth limitations. As a consequence, no vehicle may be able to communicate with the entire team and, moreover, the amount of information that can be exchanged may be limited. A key enabling element for the effective execution of the envisioned multi-vehicle missions is thus the availability of cooperative motion control strategies that can adapt to changing mission objectives, fleet size, and available autonomy capabilities. These strategies must also be robust against unexpected exogenous events and communication constraints, while ensuring at the same time collision-free maneuvers.

Motivated by these and similar problems, there has been over the past few years increasing interest in the study of multi-agent system networks with application to engineering and science problems. The range of relevant, related topics addressed in the literature includes optimal cooperative control and neural adaptive control of multi-agent systems [68], parallel computing [119], synchronization of oscillators [92,93,101], study of collective behavior and flocking [20,52,83,94,102], multi-system consensus mechanisms [70], multi-vehicle system formations [13,36,37,42,69,84], coordinated motion control [14,59,96,113,122], cooperative path and trajectory planning [22,62, 72,73,87,99,100,120,130], asynchronous protocols [41], dynamic graphs [74,77,86, 115], and graph-related theory [25,43,60]. Especially relevant are the applications of the theory developed in the area of multi-vehicle control: spacecraft formation flying [17,75,95], control of aerial vehicles [18,57,82,103,111,116,121], coordinated

Time-Critical Cooperative Control of Autonomous Air Vehicles, DOI: 10.1016/B978-0-12-809946-9.00002-6

control of land robots [48,97,117], and control of multiple autonomous marine vehicles [10,12,15,23,44,50,51,71,89,107,109,114]. In spite of significant progress in the field, however, much work remains to be done to develop strategies capable of providing the levels of *flexibility*, *performance*, and *safety* required for the envisioned multi-vehicle cooperative missions of the future.

It is against this backdrop of ideas that this book addresses the problem of *steering a fleet of autonomous air vehicles along desired spatial paths while meeting relative and absolute temporal constraints*. In particular, the cooperative missions considered require that each vehicle follow a feasible path, and that the fleet maintain a desired timing plan to ensure that all vehicles execute collision-free maneuvers and arrive at their respective final destinations at the same time, or at different times so as to meet a desired inter-vehicle schedule. Representative examples of such *time-critical missions* are coordinated ground target search and sequential auto-landing. The first refers to the situation where a fleet of UAVs[1] with initial scattered positions must rendezvous at a designated location and fly along a search route in a coordinated manner to maximize the probability of target detection using visual data. In the case of sequential auto-landing, a team of fixed-wing UAVs must break up and arrive at the assigned glide path separated by pre-specified safe-guarding time-intervals, and maintain the required separation as they fly along the glide slope. In medium and high density traffic situations, the UAVs may even be requested to complete the landing within a specific UTC time-slot. In both examples, the missions impose strict *relative* —and possibly *absolute*— temporal constraints, a critical point that needs to be emphasized.

To solve the above problem, a framework for vehicle cooperation is described in this book that brings together concepts and tools from nonlinear control, algebraic graph theory, geometry, topology control, and estimation theory. The framework builds on the approach to multi-vehicle cooperative motion control developed in [65] and on some of the extensions that emerged out of this work, such as [57]. In the setup adopted, the vehicles are assigned nominal trajectories, obtained as solutions to an appropriately formulated cooperative trajectory generation problem. These trajectories meet the spatial and temporal specifications of the mission, and account for the dynamic constraints of the vehicles. At this point, the trajectories are decomposed into spatial paths —without any temporal specifications— and speed profiles along these. Then, each vehicle is requested to converge to and follow its assigned nominal path, while tracking its nominal speed profile. This path-following approach departs from the trajectory-tracking paradigm, in that it does not require a vehicle to be at desired positions along its assigned spatial curve at specified instants of time. Finally, the vehicles negotiate their speeds along the respective paths so as to reach consensus on a set of path-parameterizing variables that are appropriately chosen to capture the collective temporal assignments of the mission. This cooperative path-following strategy allows the vehicles to react to unexpected off-nominal situations, such as adverse environmental disturbances and partial vehicle failures, in response to information exchanged

[1] In this book, we will often refer to autonomous air vehicles simply as UAVs because this acronym is deeply ingrained in the professional lexicon and the popular literature, even though some UAVs may operate under remote control by a human operator.

over the supporting communications network. The proposed solution is thus in striking contrast to a naive, non-cooperative approach in which each vehicle is requested to independently track its assigned trajectory in space and time, simultaneously, regardless of how accurately the other vehicles execute their motion plan. Besides the apparent lack of robustness in terms of inter-vehicle coordination, it has also been shown that, in the presence of unstable zero dynamics, trajectory tracking is subject to fundamental performance limitations that cannot be overcome by any controller structure [8].

The work presented in this book yields a rigorous formulation of the problem of cooperative path-following control and offers solutions to this problem for both fixed-wing and multirotor UAVs. The book further analyzes the degradation in terms of mission performance caused by the presence of dynamic communications networks arising from temporary loss of communication links and switching communication topologies, and proposes distributed coordination algorithms for improved performance in low-connectivity scenarios. Furthermore, motivated by the exchange of information over networks with finite-rate communication links, it also analyzes the effect of quantization on vehicle coordination. In addition, the book addresses the problem of enforcing absolute temporal constraints, such as specifications on the desired final time of a mission. To better motivate the theoretical developments presented in the text, the next section describes two mission scenarios that warrant the use of a team of cooperating UAVs.

1.2 Practical Motivation and Mission Scenarios

1.2.1 Cooperative Road Search

Today's operational scenarios dictate a growing need for up-to-date satellite-like imagery, with enough resolution to detect humans, weapons, and other potential threats. While accurate high-resolution imagery is traditionally provided by satellites and high-end aerial intelligence surveillance and reconnaissance platforms, these assets are not always available to the end-user due to time-of-day, visibility, or mission priority. In such cases, the use of small tactical UAVs outfitted with the ability to capture actionable, high-resolution, geo-referenced imagery and full motion video, represents an economical and expeditious alternative. Moreover, the fact that UAVs can deliver the information to the end-user in seconds or minutes, rather than hours or days, can potentially revolutionize future civilian and military operations.

One of the applications that motivates the use of multiple cooperative UAVs and poses several challenges to systems engineers, both from a theoretical and practical standpoint, is autonomous road search. In what follows we propose two examples of road search mission scenarios, the first one involving fixed-wing UAVs and the second one featuring multirotor UAVs.

Fig. 1.1 illustrates the scenario in which two small tactical fixed-wing UAVs are tasked to execute a coordinated search mission. The two UAVs, equipped with com-

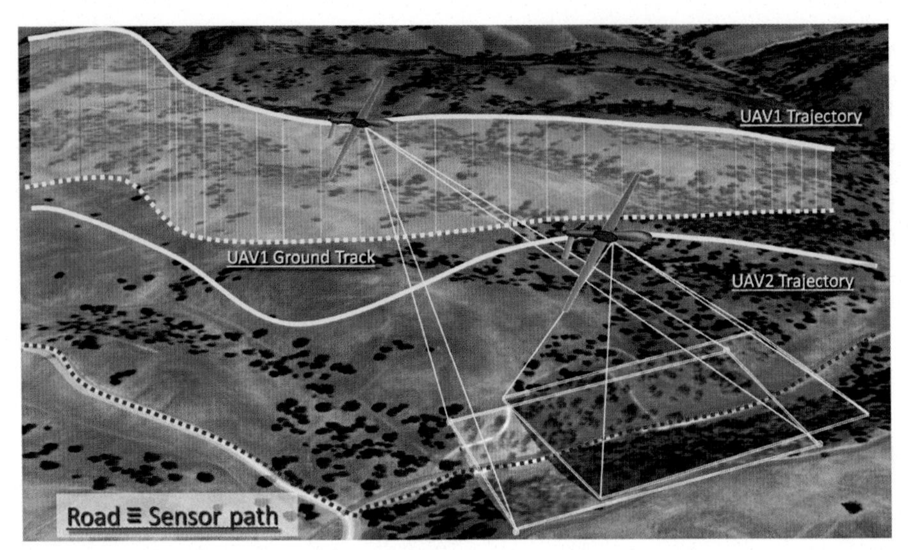

Figure 1.1 Cooperative road search using multiple fixed-wing UAVs. Two small tactical UAVs equipped with complementary vision sensors try to detect and identify an improvised explosive device moving along a road. Cooperative control can ensure a satisfactory overlap of the field-of-view footprints of the vision sensors along the road, thus increasing the probability of target detection.
(Background image © 2011 Google, Landsat/Copernicus.)

plementary vision sensors, try to detect and identify an improvised explosive device moving along a road. The mission is initiated by a minimally trained user who scribbles a path on a digital map, generating a precise continuous ground-track for the airborne sensors to follow. This ground-track is then transmitted over the communications network to the two UAVs. A distributed optimization algorithm generates feasible collision-free flight trajectories that maximize road coverage and account for sensor capabilities —field of view, resolution, and gimbal constraints— as well as inter-vehicle and ground-to-air communication limitations. The team of UAVs is then instructed to start the cooperative road-search mission. During this phase, the information obtained from the sensors mounted onboard the UAVs is shared over the network and retrieved by remote users in near real time. Target detection and identification can thus be done remotely on the ground, based on in-situ imagery data delivered over the network.

Fig. 1.2 shows a road-search mission scenario involving multirotor UAVs. Similarly to the example described above, the mission is triggered by a user who selects an area on an electronic device displaying a digital map. The global coordinates of the selected region are sent to the multirotor UAVs. A distributed optimization algorithm, running onboard the vehicles, computes feasible, collision-free search paths along the road of interest, as well as transition trajectories from the deployment location to the start points of the search paths. At this point, the vehicles start the cooperative search. Coordination along the transition paths ensures that the vehicles do not collide with each other and arrive at the start-points of the road-search paths at the same time, while

(A) 3D representation of the cooperative road search mission. *(Background image © 2017 Google.)*

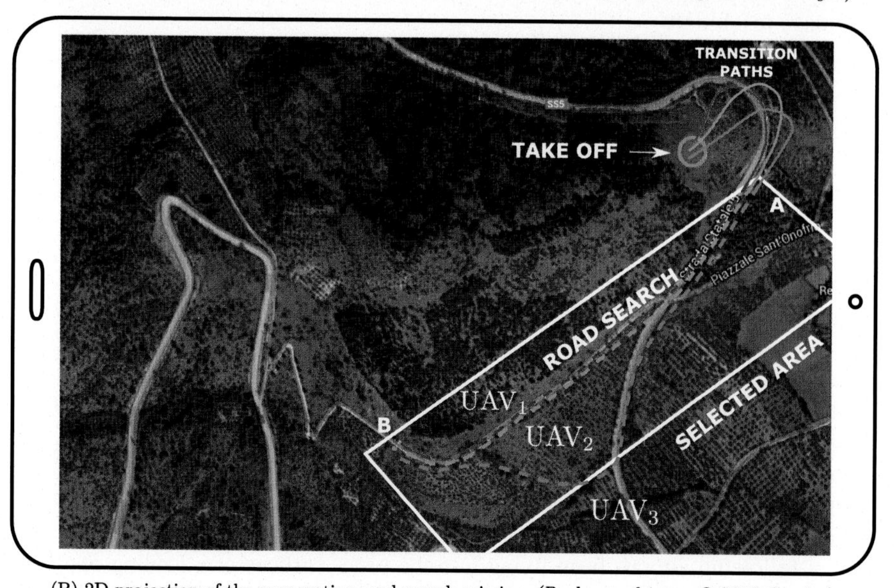

(B) 2D projection of the cooperative road search mission. *(Background image © 2017 Google.)*

Figure 1.2 Cooperative road search using multiple multirotor UAVs. The figures illustrate a scenario in which cooperation among the multirotors is required to accomplish a search mission. The UAVs start from a close-by deployment location (take-off position), follow the transition paths, and arrive at point A. They then proceed along the road search paths, represented by dashed lines, and coordinate with each other as they search for targets of special interest. Temporal coordination along the search paths guarantees non-zero intersection between the fields of view of the cameras mounted onboard the multirotors.

coordination along the search paths guarantees overlapping of the fields of view of the three cameras, as shown pictorially in Fig. 1.2A. In this example, half way through the mission the UAVs detect an uncharted secondary road and one of the vehicles is required to deviate and inspect it; see Fig. 1.2B. Upon inspection of the secondary road, the UAV rejoins the team and together they complete the road search in a coordinated manner.

In the mission scenarios described above, the advantages of using a cooperative group of autonomous vehicles connected by means of a communications network —rather than a single, heavily equipped vehicle— can be immediately identified. In a cooperative scenario, the team can reconfigure the network in response to unplanned events as well as changing mission objectives, and optimize strategies for improved target detection and discrimination. Use of multiple vehicles also improves robustness of mission execution against single-point system failures. Furthermore, in a multi-UAV approach, each vehicle in the team may be required to carry only a reduced number of sensors, making each of the vehicles in the fleet less complex, thus increasing overall system reliability. This cooperative approach requires, however, the implementation of robust cooperative control algorithms to allow the fleet of UAVs to maneuver in a coordinated manner and make use of the complementary capabilities of the onboard sensors. In fact, flying in a coordinated fashion is critical to maximizing the overlap of the fields of view of multiple vision sensors while reliably maintaining a desired image resolution.

1.2.2 Sequential Auto-Landing

In a great number of missions of interest, it is mandatory that a group of air vehicles observe absolute temporal constraints, in addition to satisfying inter-vehicle relative temporal specifications, as determined by some external agent or entity. A compelling example is the case of sequential auto-landing, in which a fleet of vehicles must break up and arrive at an assigned glide path within a desired time window —specified by Air Traffic Management— and separated by pre-specified safe-guarding time-intervals; for safety reasons, the separation between consecutive vehicles must be maintained as they fly along the glide slope. Fig. 1.3 illustrates an auto-landing mission scenario for a fleet of air vehicles.

In the above scenario, a robust coordination strategy is critical to, first, guarantee a safe approach and landing of the fleet of aircraft by enforcing appropriate separation constraints [6,53,78]; and, second, to increase runway throughput by improving time-slot compliance. In fact, the ability to maintain adequate separation is key to preventing a following aircraft from flying into the wake vortices of a leading one, an event that has been identified as one of the contributing factors in aircraft loss of control [19]. Documented incidents and accidents caused by inadequate aircraft separation can be found in [1–3] and the references therein.

More broadly, the adoption of time-coordination control strategies becomes particularly appealing in future airspace operations under both the Next Generation (NextGen) [5] and Single European Sky (SES) [4] air transportation systems. To cope with the rapid growth of air travel and operational diversity, both NextGen

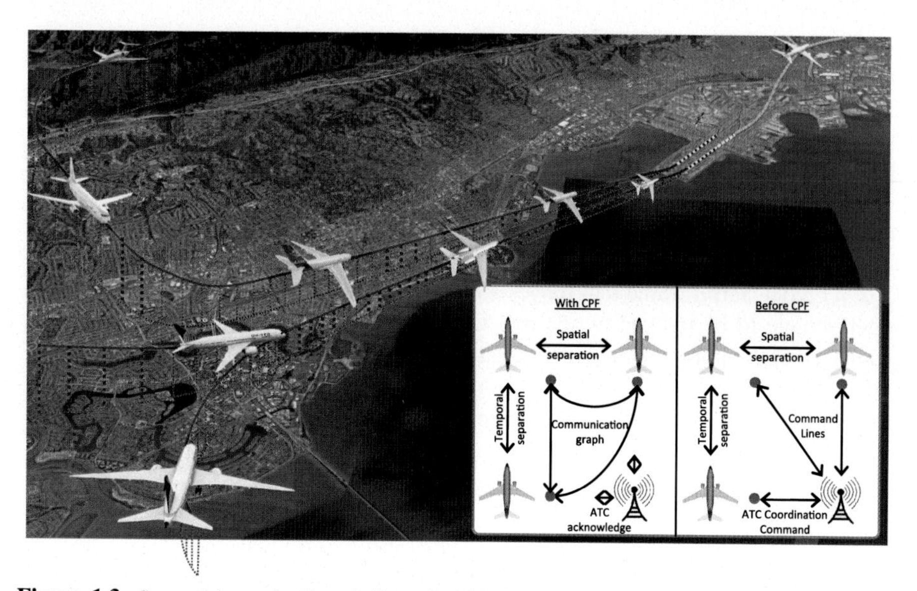

Figure 1.3 Sequential auto-landing. A fleet of aerial vehicles must arrive at the assigned glide path within a desired temporal window and separated by pre-specified safe-guarding time-intervals, which must be maintained as the vehicles fly along the glide slope. Time-critical cooperative control strategies can improve safety and runway throughput in next-generation terminal sequencing and spacing operations. *(Background image © 2016 Google, Landsat/Copernicus.)*

and SES envision a highly integrated airspace, with high-density, all-weather, and self-separation operational concepts. To realize this vision, the two programs adopt 4D trajectory-based operations as well as network-enabled data exchange and motion planning. In this context, it is anticipated that time-critical cooperative control strategies, such as the ones described in this book, will be key enablers in the safe realization of both the NextGen and SES air transportation systems.

1.3 Literature Review

The theoretical framework adopted to solve the general problem of time-critical cooperative control builds upon a number of important concepts and techniques that have been the subject of intensive research. What follows is but a brief review of the literature on the most relevant topics exploited in the present work: Path-Following Control, Coordinated Path-Following Control, and Consensus and Synchronization in Networks. Additional bibliographic references are included throughout the book.

1.3.1 Path-Following Control

The problem of path following can be briefly described as that of making a vehicle converge to and follow a desired spatial path, while tracking a desired speed profile that may be path dependent. The temporal and spatial assignments are therefore separated. Often, it is simply required that the speed of the vehicle be kept constant. Path-following control algorithms are pervasive in many robotic applications and are key to the operation of multiple vehicles undergoing cooperative missions.

There is a wealth of literature on path-following algorithms that defies a short summary. Pioneering work in the area can be found in [76], where an elegant solution to the problem of path-following control was presented for a wheeled robot at the kinematic level. In the setup adopted, the kinematic model of the vehicle is derived with respect to a Frenet-Serret frame that moves along the path and plays the role of a virtual target vehicle to be tracked by the real vehicle. The origin of the Frenet-Serret frame is placed at the point on the path closest to the real vehicle, whenever that point is well defined.

The work in [76] spurred a great deal of research activity, leading to a large number of applications addressing the path-following problem. A popular approach that emerged out of this research effort was to solve a trajectory-tracking problem and then reparameterize the resulting feedback controller using an independent variable other than time. See, for example, the work in [7,11,49] and references therein. The approach proposed in [76] was extended to unmanned air and underwater vehicles in 3D space, with full account of its dynamics, in [54] and [104]. Related results can be found in [39] for autonomous underwater vehicles using a backstepping approach. A common feature of the above papers is to reduce the path-following problem to that of driving the kinematic errors resolved in a Frenet-Serret frame to zero. This approach ensures that path following is essentially done by proper choice of the vehicle's attitude, a strategy that is akin to that used by pilots when they fly airplanes. The same property does not necessarily hold in the case of the strategies that are based on the work reported in [7,11,49].

The setup used in [76] was later reformulated in [110] (and its journal version, [66]), leading to a feedback control law that steers the dynamic model of a wheeled robot along a desired path and overcomes some of the constraints present in [76]. The key to this algorithm is to explicitly control the rate of progression of the virtual target along the path. This approach effectively creates an extra degree of freedom that can be exploited to avoid the singularities that occur when the distance to the path is not well defined —this occurs, for example, when the vehicle is located exactly at the center of curvature of a circular path. Related strategies were exploited in the work in [105] and [108] on output maneuvering and also in the work in [35] and [38]. As will become clear, the path-following algorithms described in this book are an extension of the algorithm presented in [66] to the case of 3D spatial paths. Representatives examples of our previous work on path-following control of autonomous air vehicles include [30,32,55].

Other path-following methods have been presented in the literature that depart from the ideas and concepts of the algorithms described above. In [88], lateral acceleration commands are used to make a UAV converge to and follow planar curved paths. A nonlinear path-following method that generates acceleration commands to steer a holonomic vehicle towards a given 3D path is presented in [46]. Path-following algorithms based on the concept of vector fields can be found in [67] and [80]. The work reported in [90] and [91] presents an elegant approach to path following that uses Lagrange multipliers to derive path-following control laws for mechanical systems subject to both holonomic and nonholonomic constraints. In [129], the authors propose a model predictive control scheme to solve the problem of path following for wheeled robots. Finally, the reader is referred to [118] for a recent overview of the literature on path-following algorithms for fixed-wing UAVs, while recent studies on path-following control for rotary wing UAVs can be found in [24,33,81,98,112].

1.3.2 Coordinated Path-Following Control

The problem of coordinated path following can be briefly described as that of making a group of vehicles converge to and follow a set of desired spatial paths, while meeting pre-specified spatial and temporal constraints. Over the last decade, there has been growing interest in the problem of coordinated path-following control of fleets of autonomous vehicles, mainly for the execution of cooperative marine missions involving multiple autonomous surface and underwater vehicles. Initial work on this topic can be found in [40,65,106,107].

The coordinated path-following control problem was implicit in the early work in [40], where the authors built on and extended the single-vehicle "manoeuvre regulation" approach described in [49], and presented a solution to the problem of coordinated operation of an autonomous surface vehicle and an autonomous underwater vehicle. The strategy adopted, however, required the vehicles to exchange a large amount of information, and could not be easily generalized to larger teams of vehicles. These drawbacks were later overcome in [65], where a leader-follower cooperative approach was proposed that (almost) decouples the temporal and spatial assignments of the mission. The solution adopted is rooted in the results on path-following control of a single vehicle presented in [110], and takes advantage of the fact that, with this path-following algorithm, the speed profile of each vehicle becomes an additional degree of freedom that can be exploited for vehicle coordination. Moreover, in the setup adopted, the vehicles only need to exchange the (scalar) "along-path positions" of their virtual targets, thus reducing drastically the amount of information that must be transmitted among the vehicles when compared to the solution developed in [40]. Interestingly, an approach similar to the one in [65] was proposed at approximately the same time in the work in [106] and [107], where a nonlinear control design method was presented for formation control of a fleet of marine vessels. The approach relies on the maneuvering methodology developed in [108], which is then combined with a centralized guidance system that adjusts the speed profile of each vehicle so as to achieve and maintain the desired formation configuration. The maneuvering strategy in [108] was also exploited in [50], where a passivity framework was used to solve the problem of vehicle coordination and formation maneuvering.

In [47], the authors extended the approach in [65] and addressed the problem of steering a group of vehicles along predefined spatial paths while holding a desired (possibly time-varying) formation pattern. Using results from nonlinear systems and algebraic graph theory, conditions were derived under which the proposed algorithm solves the coordinated path-following control problem in the presence of switching communication topologies and network link latencies. In particular, stability of the closed-loop system was analyzed under two scenarios: first, networks with brief connectivity losses; and second, uniformly jointly connected communication graphs.

The approach in [65] was also extended in [57], where the authors addressed the problem of coordinated control of multiple fixed-wing UAVs. To enforce the temporal constraints of the mission, the coordination algorithm relies on a distributed control law with a proportional-integral structure, which ensures that each vehicle travels along its path at the desired constant speed and also provides disturbance rejection capabilities against steady winds. As will become clear, the approach for vehicle coordination described in Part II of this book is an extension of the algorithm presented in [57] to the case of arbitrary (feasible) desired speed profiles and multiple leaders. The reader is referred to [56,123–125] for intermediate results that ultimately led to the setup adopted in this book. An extension of the work in [65] to the problem of coordinated control of multiple multirotor UAVs can be found in [29] and [31]. As will be shown in Part III of this book, the distinct dynamics of multirotor vehicles require a structure for the coordination control laws that is different from the one used for fixed-wing vehicles. Recently, the setup in [65] was also used to control a team of multirotor UAVs that cooperatively carry a cable-suspended payload [61]; in this work, the authors propose a proportional-derivative control law for team coordination.

Related work can also be found in [9], which proposes a multi-vehicle control architecture aimed at reducing the frequency at which information is exchanged among vehicles by incorporating logic-based communications. To this end, the authors borrowed from and expanded some of the key ideas exposed in [126] and [127], where decentralized controllers for distributed systems were derived by using, for each system, its local state information together with estimates of the states of the systems that it communicates with.

Other relevant cooperative control algorithms have been presented in the literature that address problems akin to that of coordinated path following. In [63] and [64], for example, synchronization techniques were used to develop control laws for ship rendezvous maneuvers. Also, the work in [37] presents a solution to the problem of coordinated path following for multi-agent formation control. In the setup adopted, a reference path is specified for a nonphysical point of the formation, which plays the role of a virtual leader, while a desired formation pattern is defined with respect to this nonphysical point. Control laws are then derived that ensure that the real vehicles converge to the desired reference points of the formation, while the virtual leader follows the reference path.

1.3.3 Consensus and Synchronization in Networks

This book presents algorithms that are key to the solution of a number of technical problems that arise in the course of time-critical cooperative missions, some of which were described earlier in the chapter. At the same time, the book analyzes the stability and convergence properties of these algorithms in the presence of switching network topologies and the exchange of quantized information. The presentation is therefore naturally rooted in theoretical developments and tools from the theories of consensus and synchronization of networked systems. There is an extremely rich body of literature available on both disciplines, and its discussion is well beyond the scope of this section. The reader is referred to [85] for an overview of consensus algorithms and their application to cooperative control of networked multi-agent systems. A thorough review of the major concepts and results in the study of the structure and dynamics of complex networks is presented in [21]. In what follows, for the sake of clarity, we give a brief overview of the work in these disciplines that is most directly and closely related to the developments in this book, namely, proportional-integral consensus protocols and quantized consensus.

Proportional-Integral Consensus Protocols

As mentioned in the previous section, a distributed proportional-integral protocol is used in [57] to enforce the temporal constraints of cooperative missions involving multiple UAVs. The integral term in the consensus algorithm allows the follower vehicles to learn the (constant) reference speed of the leader, and provides disturbance rejection capabilities against steady winds. A generalization of a proportional-integral protocol is proposed in [16], where the authors develop an adaptive algorithm to reconstruct a time-varying reference velocity that is available only to a single leader. The paper uses a passivity framework to show that a network of nonlinear agents with fixed connected topology achieves coordination asymptotically. The work in [26] also uses a (discrete-time) proportional-integral consensus protocol to synchronize networks of clocks with fixed connected topology. In this application, the integral part of the controller is critical to eliminate the different initial clock offsets. A proportional-integral estimation algorithm is also proposed in [45] for dynamic average consensus in sensing networks. In particular, the paper analyzes the stability and convergence properties of the developed proportional-integral estimator, by deriving conditions on both constant and time-varying information flows that ensure stability of the estimator.

Quantized Consensus

The exchange of information over networks with finite-rate communication links motivates the interest in the study of quantized consensus problems. Most of the work on this topic deals with discrete-time dynamics; see, for example, [27,58,79] and references therein. Pioneering work in the area can be found in [58], where the authors analyze the distributed averaging problem on arbitrary connected graphs, and derive bounds on the expected convergence time of the collective dynamics for complete and linear networks. The results in [58] were later extended in [79] to the case of time-varying topologies. Interesting results on quantized consensus can also be found in [27], which proposes a protocol for a network to reach consensus with arbitrarily small precision, at the expense, however, of slow convergence.

The continuous-time quantized averaging problem was studied in detail in [28], where it was proven that Carathéodory solutions might not exist for (continuous-time) quantized consensus problems, implying thus that a weaker concept of solution must be considered. The work in [34] uses a passivity framework to extend the results in [28] to the case of agents with complex dynamics and advanced coordination tasks. Related work can also be found in [128] on the multi-agent rendezvous problem under minimal sensing and actuation.

References

[1] Delta Air Lines, Inc., McDonnell Douglas, DC-9-14, N3305L, Greater Southwest International Airport, Fort Worth, Texas, May 30, 1972, Aircraft Accident Report NTSB-AAR-73-3, National Transportation Safety Board, May 1972.

[2] Safety issues related to wake vortex encounters during visual approach to landing, Special Investigation Report NTSB/SIR-94/01, National Transportation Safety Board, February 1994.

[3] Wake Turbulence Training Aid, § 2, Pilot and Air Traffic Controller Guide to Wake Turbulence; §§ 2.3, Review of Accidents and Incidents, Memorandum DOT/FAA/RD-95/6, Federal Aviation Administration, April 1995.

[4] SESAR Concept of Operations, WP2.2.2/D3 DLT-0612-222-01-00, SESAR Consortium, July 2007. Available at https://www.eurocontrol.int/sites/default/files/field_tabs/content/documents/sesar/20070717-sesar-conops.pdf.

[5] Concept of Operations for the Next Generation Air Transportation System, Joint Planning and Development Office, Version 3.2, September 2010. Available at http://www.dtic.mil/dtic/tr/fulltext/u2/a535795.pdf.

[6] Aeronautical Information Manual: Official Guide to Basic Flight Information and ATC Procedures, Chapter 7, Safety of Flight; § 7-3-9, Air Traffic Wake Turbulence Separations, regulatory information, Federal Aviation Administration, December 2015.

[7] A.P. Aguiar, J.P. Hespanha, Position tracking of underactuated vehicles, in: American Control Conference, Denver, CO, June 2003, pp. 1988–1993.

[8] A.P. Aguiar, J.P. Hespanha, P.V. Kokotović, Performance limitations in reference tracking and path following for nonlinear systems, Automatica 44 (2008) 598–610.

[9] A.P. Aguiar, A.M. Pascoal, Coordinated path-following control for nonlinear systems with logic-based communication, in: IEEE Conference on Decision and Control, New Orleans, LA, December 2007, pp. 1473–1479.

[10] A.P. Aguiar, A.M. Pascoal, Cooperative control of multiple autonomous marine vehicles: theoretical foundations and practical issues, in: G.N. Roberts, R. Sutton (Eds.), Further Advances in Unmanned Marine Vehicles, in: Control Engineering Series, vol. 77, IET, London, UK, 2012, pp. 255–282.

[11] S.A. Al-Hiddabi, N.H. McClamroch, Tracking and maneuver regulation control for nonlinear non-minimum phase systems, IEEE Transactions on Control System Technology 10 (2002) 780–792.

[12] J. Almeida, C. Silvestre, A.M. Pascoal, Cooperative control of multiple surface vessels in the presence of ocean currents and parametric model uncertainty, International Journal of Robust and Nonlinear Control 20 (2010) 1549–1565.

[13] G. Antonelli, F. Arrichiello, F. Caccavale, A. Marino, Decentralized time-varying formation control for multi-robot systems, The International Journal of Robotics Research 33 (2014) 1029–1043.

[14] M. Arcak, Passivity as a design tool for group coordination, IEEE Transactions on Automatic Control 52 (2007) 1380–1390.

[15] F. Arrichiello, S. Chiaverini, T.I. Fossen, Formation control of marine surface vessels using the null-space-based behavioral control, in: K.Y. Pettersen, J.T. Gravdahl, H. Nijmeijer (Eds.), Group Coordination and Cooperative Control, in: Lecture Notes in Control and Information Sciences, vol. 336, Springer-Verlag Berlin Heidelberg, 2006, pp. 1–19.
[16] H. Bai, M. Arcak, J.T. Wen, Adaptive design for reference velocity recovery in motion coordination, Systems & Control Letters 57 (2008) 602–610.
[17] R.W. Beard, J. Lawton, F.Y. Hadaegh, A coordination architecture for spacecraft formation control, IEEE Transactions on Control System Technology 9 (2001) 777–790.
[18] R.W. Beard, T.W. McLain, D.B. Nelson, D. Kingston, D. Johanson, Decentralized cooperative aerial surveillance using fixed-wing miniature UAVs, Proceedings of the IEEE 94 (2006) 1306–1324.
[19] C.M. Belcastro, S.R. Jacobson, Future integrated systems concept for preventing aircraft loss-of-control accidents, in: AIAA Guidance, Navigation, and Control Conference, Toronto, Ontario, Canada, August 2010, AIAA 2010-8142.
[20] V.D. Blondel, J.M. Hendrickx, A. Olshevsky, J.N. Tsitsiklis, Convergence in multiagent coordination, consensus, and flocking, in: IEEE Conference on Decision and Control, Seville, Spain, December 2005, pp. 2996–3000.
[21] S. Boccaletti, V. Latora, Y. Moreno, M. Chavez, D.-U. Hwang, Complex networks: structure and dynamics, Physics Reports 424 (2006) 175–308.
[22] K.P. Bollino, L.R. Lewis, Collision-free multi-UAV optimal path planning and cooperative control for tactical applications, in: AIAA Guidance, Navigation and Control Conference, Honolulu, HI, August 2008, AIAA 2008-7134.
[23] M. Breivik, V.E. Hovstein, T.I. Fossen, Ship formation control: a guided leader-follower approach, in: IFAC World Congress, Seoul, South Korea, July 2008.
[24] D. Cabecinhas, R. Cunha, C. Silvestre, A globally stabilizing path following controller for rotorcraft with wind disturbance rejection, IEEE Transactions on Control Systems Technology 23 (2015) 708–714.
[25] M. Cao, D.A. Spielman, A.S. Morse, A lower bound on convergence of a distributed network consensus algorithm, in: IEEE Conference on Decision and Control, Seville, Spain, December 2005, pp. 2356–2361.
[26] R. Carli, A. Chiuso, L. Schenato, S. Zampieri, A PI consensus controller for networked clocks synchronization, in: IFAC World Congress, Seoul, South Korea, July 2008.
[27] A. Censi, R.M. Murray, Real-valued average consensus over noisy quantized channels, in: American Control Conference, St. Louis, MO, June 2009, pp. 4361–4366.
[28] F. Ceragioli, C. De Persis, P. Frasca, Discontinuities and hysteresis in quantized average consensus, Automatica 47 (2011) 1916–1928.
[29] V. Cichella, R. Choe, S.B. Mehdi, E. Xargay, N. Hovakimyan, V. Dobrokhodov, I. Kaminer, A.M. Pascoal, A.P. Aguiar, Safe coordinated maneuvering of teams of multirotor UAVs, IEEE Control Systems Magazine 36 (2016) 59–82.
[30] V. Cichella, R. Choe, S.B. Mehdi, E. Xargay, N. Hovakimyan, I. Kaminer, V. Dobrokhodov, A 3D path-following approach for a multirotor UAV on $\mathsf{SO}(3)$, IFAC Proceedings Volumes 46 (2013) 13–18.
[31] V. Cichella, I. Kaminer, V. Dobrokhodov, E. Xargay, R. Choe, N. Hovakimyan, A.P. Aguiar, A.M. Pascoal, Cooperative path following of multiple multirotors over time-varying networks, IEEE Transactions on Automation Science and Engineering 12 (2015) 945–957.
[32] V. Cichella, E. Xargay, V. Dobrokhodov, I. Kaminer, A.M. Pascoal, N. Hovakimyan, Geometric 3D path-following control for a fixed-wing UAV on $\mathsf{SO}(3)$, in: AIAA Guidance, Navigation and Control Conference, Portland, OR, August 2011, AIAA 2011-6415.

[33] J.C. Dauer, T. Faulwasser, S. Lorenz, R. Findeisen, Optimization-based feedforward path following for model reference adaptive control of an unmanned helicopter, in: AIAA Guidance, Navigation and Control Conference, Boston, MA, August 2013, AIAA 2013-5002.

[34] C. De Persis, On the passivity approach to quantized coordination problems, in: IEEE Conference on Decision and Control, Orlando, FL, December 2011, pp. 1086–1091.

[35] F. Díaz del Río, G. Jiménez, J.L. Sevillano, C. Amaya, A. Civit Balcells, A new method for tracking memorized paths: application to unicycle robots, in: Mediterranean Conference on Control and Automation, Lisbon, Portugal, July 2002.

[36] W.B. Dunbar, R.M. Murray, Distributed receding horizon control for multi-vehicle formation stabilization, Automatica 42 (2006) 549–558.

[37] M. Egerstedt, X. Hu, Formation constrained multi-agent control, IEEE Transactions on Robotics and Automation 17 (2001) 947–951.

[38] M. Egerstedt, X. Hu, A. Stotsky, Control of mobile platforms using a virtual vehicle approach, IEEE Transactions on Automatic Control 46 (2001) 1777–1782.

[39] P. Encarnação, Nonlinear Path Following Control Systems for Ocean Vehicles, PhD thesis, Instituto Superior Técnico, Lisbon, Portugal, 2002.

[40] P. Encarnação, A.M. Pascoal, Combined trajectory tracking and path following: an application to the coordinated control of autonomous marine craft, in: IEEE Conference on Decision and Control, Orlando, FL, December 2001, pp. 964–969.

[41] L. Fang, P.J. Antsaklis, A. Tzimas, Asynchronous consensus protocols: preliminary results, simulations and open questions, in: IEEE Conference on Decision and Control, Seville, Spain, December 2005, pp. 2194–2199.

[42] J.A. Fax, R.M. Murray, Information flow and cooperative control of vehicle formations, IEEE Transactions on Automatic Control 49 (2004) 1465–1476.

[43] M. Fiedler, Algebraic connectivity of graphs, Czechoslovak Mathematical Journal 23 (1973) 298–305.

[44] T.I. Fossen, Marine Control Systems: Guidance, Navigation and Control of Ships, Rigs and Underwater Vehicles, Marine Cybernetics, Trondheim, Norway, 2002.

[45] R.A. Freeman, P. Yang, K.M. Lynch, Stability and convergence properties of dynamic average consensus estimators, in: IEEE Conference on Decision and Control, San Diego, CA, December 2006, pp. 398–403.

[46] D.J. Gates, Nonlinear path following method, Journal of Guidance, Control and Dynamics 33 (2010) 321–332.

[47] R. Ghabcheloo, A.P. Aguiar, A.M. Pascoal, C. Silvestre, I. Kaminer, J.P. Hespanha, Coordinated path-following in the presence of communication losses and delays, SIAM Journal on Control and Optimization 48 (2009) 234–265.

[48] R. Ghabcheloo, A.M. Pascoal, C. Silvestre, I. Kaminer, Coordinated path following control of multiple wheeled robots using linearization techniques, International Journal of Systems Science 37 (2006) 399–414.

[49] R. Hindman, J. Hauser, Maneuver modified trajectory tracking, in: International Symposium on the Mathematical Theory of Networks and Systems, St. Louis, MO, June 1996.

[50] I.-A.F. Ihle, Coordinated Control of Marine Craft, PhD thesis, Norwegian University of Science and Technology, Trondheim, Norway, September 2006.

[51] I.-A.F. Ihle, J. Jouffroy, T.I. Fossen, Robust formation control of marine craft using Lagrange multipliers, in: K.Y. Pettersen, J.T. Gravdahl, H. Nijmeijer (Eds.), Group Coordination and Cooperative Control, in: Lecture Notes in Control and Information Sciences, vol. 336, Springer-Verlag Berlin Heidelberg, 2006, pp. 113–129.

[52] A. Jadbabaie, J. Lin, A.S. Morse, Coordination of groups of mobile autonomous agents using nearest neighbor rules, IEEE Transactions on Automatic Control 48 (2003) 988–1001.
[53] M. Janic, Toward time-based separation rules for landing aircraft, Transportation Research Record: Journal of the Transportation Research Board 2052 (2008) 79–89.
[54] I. Kaminer, A.M. Pascoal, E. Hallberg, C. Silvestre, Trajectory tracking for autonomous vehicles: an integrated approach to guidance and control, Journal of Guidance, Control and Dynamics 21 (1998) 29–38.
[55] I. Kaminer, A.M. Pascoal, E. Xargay, N. Hovakimyan, C. Cao, V. Dobrokhodov, Path following for unmanned aerial vehicles using $\mathcal{L}_1$ adaptive augmentation of commercial autopilots, Journal of Guidance, Control and Dynamics 33 (2010) 550–564.
[56] I. Kaminer, E. Xargay, V. Cichella, N. Hovakimyan, A.M. Pascoal, A.P. Aguiar, V. Dobrokhodov, R. Ghabcheloo, Time-critical cooperative path following of multiple UAVs: case studies, in: D. Choukroun, Y. Oshman, J. Thienel, M. Idan (Eds.), Advances in Estimation, Navigation, and Spacecraft Control – Itzhack Y. Bar-Itzhack Memorial Symposium, Springer-Verlag Berlin Heidelberg, 2015, pp. 209–233.
[57] I. Kaminer, O.A. Yakimenko, A.M. Pascoal, R. Ghabcheloo, Path generation, path following and coordinated control for time-critical missions of multiple UAVs, in: American Control Conference, Minneapolis, MN, June 2006, pp. 4906–4913.
[58] A. Kashyap, T. Başar, R. Srikant, Quantized consensus, Automatica 43 (2007) 1192–1203.
[59] T. Keviczky, F. Borrelli, K. Fregene, D. Godbole, G.J. Balas, Decentralized receding horizon control and coordination of autonomous vehicle formations, IEEE Transactions on Control System Technology 16 (2008) 19–33.
[60] Y. Kim, M. Mesbahi, On maximizing the second smallest eigenvalue of state-dependent graph Laplacian, IEEE Transactions on Automatic Control 51 (2006) 116–120.
[61] K. Klausen, T.I. Fossen, T.A. Johansen, A.P. Aguiar, Cooperative path-following for multirotor UAVs with a suspended payload, in: IEEE Conference on Control Applications, Sydney, Australia, September 2015, pp. 1354–1360.
[62] Y. Kuwata, J.P. How, Cooperative distributed robust trajectory optimization using receding horizon MILP, IEEE Transactions on Control System Technology 19 (2011) 423–431.
[63] E. Kyrkjebø, K.Y. Pettersen, Ship replenishment using synchronization control, in: IFAC Conference on Manoeuvering and Control of Marine Craft, Girona, Spain, September 2003.
[64] E. Kyrkjebø, M. Wondergem, K.Y. Pettersen, H. Nijmeijer, Experimental results on synchronization control of ship rendezvous operations, in: IFAC Conference on Control Applications in Marine Systems, Ancona, Italy, July 2004, pp. 453–458.
[65] L. Lapierre, D. Soetanto, A.M. Pascoal, Coordinated motion control of marine robots, in: IFAC Conference on Manoeuvering and Control of Marine Craft, Girona, Spain, September 2003.
[66] L. Lapierre, D. Soetanto, A.M. Pascoal, Non-singular path-following control of a unicycle in the presence of parametric modeling uncertainties, International Journal of Robust and Nonlinear Control 16 (2006) 485–503.
[67] D.A. Lawrence, E.W. Frew, W.J. Pisano, Lyapunov vector fields for autonomous unmanned aircraft flight control, Journal of Guidance, Control and Dynamics 31 (2008) 1220–1229.
[68] F.L. Lewis, H. Zhang, K. Hengster-Movric, A. Das, Cooperative Control of Multi-Agent Systems: Optimal and Adaptive Design Approaches, Springer-Verlag London, London, UK, 2014.

[69] F. Liao, R. Teo, J.L. Wang, X. Dong, F. Lin, K. Peng, Distributed formation and reconfiguration control of VTOL UAVs, IEEE Transactions on Control Systems Technology 25 (2017) 270–277.

[70] Z. Lin, B.A. Francis, M. Maggiore, State agreement for continuous-time coupled nonlinear systems, SIAM Journal on Control and Optimization 46 (2007) 288–307.

[71] A. Marino, G. Antonelli, A.P. Aguiar, A.M. Pascoal, A decentralized strategy for multirobot sampling/patrolling: theory and experiments, IEEE Transactions on Control Systems Technology 3 (2014) 313–322.

[72] N. Mathew, S.L. Smith, S.L. Waslander, Optimal path planning in cooperative heterogeneous multi-robot delivery systems, in: Algorithmic Foundations of Robotics XI, Springer, 2015, pp. 407–423.

[73] T.W. McLain, R.W. Beard, Coordination variables, coordination functions, and cooperative timing missions, Journal of Guidance, Control and Dynamics 28 (2005) 150–161.

[74] M. Mesbahi, On state-dependent dynamic graphs and their controllability properties, IEEE Transactions on Automatic Control 50 (2005) 387–392.

[75] M. Mesbahi, F.Y. Hadaegh, Formation flying control of multiple spacecraft via graphs, matrix inequalities, and switching, Journal of Guidance, Control and Dynamics 24 (2001) 369–377.

[76] A. Micaelli, C. Samson, Trajectory Tracking for Unicycle-Type and Two-Steering-Wheels Mobile Robot, Tech. Report 2097 INRIA, Sophia-Antipolis, France, November 1993.

[77] L. Moreau, Stability of multiagent systems with time-dependent communication links, IEEE Transactions on Automatic Control 50 (2005) 169–182.

[78] C. Morris, J. Peters, P. Choroba, Validation of the time based separation concept at London Heathrow airport, in: USA/Europe Air Traffic Management Research and Development Seminar, Chicago, IL, June 2013.

[79] A. Nedić, A. Olshevsky, A. Ozdaglar, J.N. Tsitsiklis, On distributed averaging algorithms and quantization effects, IEEE Transactions on Automatic Control 54 (2009) 2506–2517.

[80] D.R. Nelson, D.B. Barber, T.W. McLain, R.W. Beard, Vector field path following for miniature air vehicles, Transactions on Robotics 23 (2007) 519–529.

[81] P. Niermeyer, V.S. Akkinapalli, M. Pak, F. Holzapfel, B. Lohmann, Geometric path following control for multirotor vehicles using nonlinear model predictive control and 3D spline paths, in: International Conference on Unmanned Aircraft Systems, Arlington, VA, June 2016, pp. 126–134.

[82] K. Nonami, F. Kendoul, S. Suzuki, W. Wang, D. Nakazawa, Autonomous Flying Robots: Unmanned Aerial Vehicles and Micro Aerial Vehicles, Springer Tokyo, Tokyo, Japan, 2010.

[83] R. Olfati Saber, Flocking for multi-agent dynamic systems: algorithms and theory, IEEE Transactions on Automatic Control 51 (2006) 401–420.

[84] R. Olfati Saber, W.B. Dunbar, R.M. Murray, Cooperative control of multi-vehicle systems using cost graphs and optimization, in: American Control Conference, Denver, CO, June 2003, pp. 2217–2222.

[85] R. Olfati Saber, J.A. Fax, R.M. Murray, Consensus and cooperation in networked multi-agent systems, Proceedings of the IEEE 95 (2007) 215–233.

[86] R. Olfati Saber, R.M. Murray, Consensus problems in networks of agents with switching topology and time-delays, IEEE Transactions on Automatic Control 49 (2004) 1520–1533.

[87] J. Ousingsawat, M.E. Campbell, Optimal cooperative reconnaissance using multiple vehicles, Journal of Guidance, Control and Dynamics 30 (2007) 122–132.

[88] S. Park, J. Deyst, J.P. How, Performance and Lyapunov stability of a nonlinear path-following guidance method, Journal of Guidance, Control and Dynamics 30 (2007) 1718–1728.
[89] F.L. Pereira, J.B. de Sousa, Coordinated control of networked vehicles: an autonomous underwater system, Automation and Remote Control 65 (2004) 1037–1045.
[90] E. Peyami, T.I. Fossen, A Lagrangian framework to incorporate positional and velocity constraints to achieve path-following control, in: IEEE Conference on Decision and Control, Orlando, FL, December 2011, pp. 3940–3945.
[91] E. Peyami, T.I. Fossen, Motion control of marine craft using virtual positional and velocity constraints, in: International Conference on Control and Automation, Santiago, Chile, December 2011, pp. 410–416.
[92] Q.-C. Pham, J.-J. Slotine, Stable concurrent synchronization in dynamic system networks, Neural Networks 20 (2007) 62–77.
[93] A. Pikovsky, M. Rosenblum, J. Kurths, Synchronization: A Universal Concept in Nonlinear Sciences, Cambridge University Press, Cambridge, UK, 2001.
[94] Z. Qu, Cooperative Control of Dynamical Systems: Applications to Autonomous Vehicles, Springer-Verlag London, London, UK, 2009.
[95] W. Ren, Formation keeping and attitude alignment for multiple spacecraft through local interactions, Journal of Guidance, Control and Dynamics 30 (2007) 633–638.
[96] W. Ren, Y. Cao, Distributed Coordination of Multi-Agent Networks, Springer-Verlag London, London, UK, 2011.
[97] W. Ren, N. Sorensen, Distributed coordination architecture for multi-robot formation control, Robotics and Autonomous Systems 56 (2008) 324–333.
[98] A. Roza, M. Maggiore, Path following controller for a quadrotor helicopter, in: American Control Conference, Montréal, Canada, June 2012, pp. 4655–4660.
[99] E. Scholte, M.E. Campbell, Robust nonlinear model predictive control with partial state information, IEEE Transactions on Control System Technology 16 (2008) 636–651.
[100] T. Schouwenaars, J.P. How, E. Feron, Decentralized cooperative trajectory planning of multiple aircraft with hard safety guarantees, in: AIAA Guidance, Navigation and Control Conference, Providence, RI, August 2004, AIAA 2004-5141.
[101] R. Sepulchre, D. Paley, N. Leonard, Collective Motion and Oscillator Synchronization, Lecture Notes in Control and Information Sciences, vol. 309, Springer-Verlag, Berlin, 2005, pp. 189–206.
[102] J. Shamma (Ed.), Cooperative Control of Distributed Multi-Agent Systems, John Wiley & Sons, Chichester, UK, 2007.
[103] T. Shima, S. Rasmussen (Eds.), UAV Cooperative Decision and Control: Challenges and Practical Approaches, Advances in Design and Control, Society for Industrial and Applied Mathematics, Philadelphia, PA, 2009.
[104] C. Silvestre, Multi-Objective Optimization Theory with Applications to the Integrated Design of Controllers/Plants for Autonomous Vehicles, PhD thesis, Instituto Superior Técnico, Lisbon, Portugal, June 2000.
[105] R. Skjetne, T.I. Fossen, P.V. Kokotović, Robust output maneuvering for a class of nonlinear systems, Automatica 40 (2004) 373–383.
[106] R. Skjetne, I.-A.F. Ihle, T.I. Fossen, Formation control by synchronizing multiple maneuvering systems, in: IFAC Conference on Manoeuvering and Control of Marine Craft, Girona, Spain, September 2003.
[107] R. Skjetne, S. Moi, T.I. Fossen, Nonlinear formation control of marine craft, in: IEEE Conference on Decision and Control, vol. 2, Las Vegas, NV, December 2002, pp. 1699–1704.

[108] R. Skjetne, A.R. Teel, P.V. Kokotović, Stabilization of sets parameterized by a single variable: application to ship maneuvering, in: Proceedings of the 15th International Symposium on Mathematical Theory of Networks and Systems, Notre Dame, IN, August 2002.
[109] J.M. Soares, A.P. Aguiar, A.M. Pascoal, A. Martinoli, Design and implementation of a range-based formation controller for marine robots, in: M.A. Armada, A. Sanfeliu, M. Ferre (Eds.), ROBOT2013: First Iberian Robotics Conference, in: Advances in Intelligent Systems and Computing, vol. 252, Springer, 2014, pp. 55–67.
[110] D. Soetanto, L. Lapierre, A.M. Pascoal, Adaptive, non-singular path-following control of dynamic wheeled robots, in: International Conference on Advanced Robotics, Coimbra, Portugal, June–July 2003, pp. 1387–1392.
[111] Y.D. Song, Y. Li, X.H. Liao, Orthogonal transformation based robust adaptive close formation control of multi-UAVs, in: American Control Conference, vol. 5, Portland, OR, June 2005, pp. 2983–2988.
[112] S. Spedicato, A. Franchi, G. Notarstefano, From tracking to robust maneuver regulation: an easy-to-design approach for VTOL aerial robots, in: IEEE International Conference on Robotics and Automation, Stockholm, Sweden, May 2016, pp. 2965–2970.
[113] D. Stevenson, M. Wheeler, M.E. Campbell, W.W. Whitacre, R.T. Rysdyk, R. Wise, Cooperative tracking flight test, in: AIAA Guidance, Navigation and Control Conference, Hilton Head, SC, August 2007, AIAA 2007-6756.
[114] D.J. Stilwell, B.E. Bishop, Platoons of underwater vehicles, IEEE Control Systems Magazine 20 (2000) 45–52.
[115] D.J. Stilwell, E.M. Bollt, D.G. Roberson, Sufficient conditions for fast switching synchronization in time-varying network topologies, SIAM Journal of Applied Dynamical Systems 5 (2006) 140–156.
[116] D.M. Stipanović, G. İnalhan, R. Teo, C.J. Tomlin, Decentralized overlapping control of a formation of unmanned aerial vehicles, Automatica 40 (2004) 1285–1296.
[117] A. Stubbs, V. Vladimerou, A.T. Fulford, D. King, J. Strick, G.E. Dullerud, Multivehicle systems control over networks, IEEE Control Systems Magazine 26 (2006) 56–69.
[118] P. Sujit, S. Saripalli, J.B. Sousa, Unmanned aerial vehicle path following: a survey and analysis of algorithms for fixed-wing unmanned aerial vehicles, IEEE Control Systems 34 (2014) 42–59.
[119] J.N. Tsitsiklis, M. Athans, Convergence and asymptotic agreement in distributed decision problems, IEEE Transactions on Automatic Control 29 (1984) 42–50.
[120] A. Tsourdos, B.A. White, M. Shanmugavel, Cooperative Path Planning of Unmanned Aerial Vehicles, John Wiley & Sons, Chichester, UK, 2010.
[121] K.P. Valavanis, G.J. Vachtsevanos (Eds.), Handbook of Unmanned Aerial Vehicles, Springer Dordrecht, Dordrecht, The Netherlands, 2015.
[122] J.H. van Schuppen, T. Villa, Coordination Control of Distributed Systems, Springer, 2015.
[123] E. Xargay, Time-Critical Cooperative Path-Following Control of Multiple Unmanned Aerial Vehicles, PhD thesis, University of Illinois at Urbana-Champaign, Urbana, IL, United States, May 2013.
[124] E. Xargay, V. Dobrokhodov, I. Kaminer, A.M. Pascoal, N. Hovakimyan, C. Cao, Time-critical cooperative control of multiple autonomous vehicles, IEEE Control Systems Magazine 32 (2012) 49–73.
[125] E. Xargay, I. Kaminer, A.M. Pascoal, N. Hovakimyan, V. Dobrokhodov, V. Cichella, A.P. Aguiar, R. Ghabcheloo, Time-critical cooperative path following of multiple unmanned aerial vehicless over time-varying networks, Journal of Guidance, Control and Dynamics 36 (2013) 499–516.

[126] Y. Xu, J.P. Hespanha, Communication logic design and analysis for networked control systems, in: L. Menini, L. Zaccarian, C.T. Abdallah (Eds.), Current Trends in Nonlinear Systems and Control, in: Systems and Control: Foundations & Applications, Birkhäuser Boston, Cambridge, MA, 2006, pp. 495–514.
[127] J.K. Yook, D.M. Tilbury, N.R. Soparkar, Trading computation and bandwidth: reducing communication in distributed control systems using state estimators, IEEE Transactions on Control System Technology 10 (2002) 503–518.
[128] J. Yu, S.M. LaValle, D. Liberzon, Rendezvous without coordinates, IEEE Transactions on Automatic Control 57 (2012) 421–434.
[129] S. Yu, X. Li, H. Chen, F. Allgöwer, Nonlinear model predictive control for path following problems, International Journal of Robust and Nonlinear Control 25 (2015) 1168–1182.
[130] Y. Zhang, J. Yang, S. Chen, J. Chen, Decentralized cooperative trajectory planning for multiple UAVs in dynamic and uncertain environments, in: Intelligent Computing and Information Systems (ICICIS), 2015 IEEE Seventh International Conference on, IEEE, 2015, pp. 377–382.

General Framework for Vehicle Cooperation

2

This chapter presents the framework for vehicle cooperation adopted in this book, and provides a rigorous formulation of the problem of time-critical cooperative path-following control of multiple UAVs in 3D space. The objective is to make a fleet of UAVs converge to and follow a set of desired feasible paths while meeting strict spatial and temporal constraints. The chapter introduces a set of assumptions and constraints on the supporting communications network as well as on the autopilots mounted onboard the UAVs.

2.1 General Framework

The methodology for vehicle cooperation adopted in this book can be traced back to the work reported in [17], [18], and [23], based on the key idea of *decoupling space and time*, and can be summarized in three basic steps. First, given a multi-vehicle cooperative mission, a set of feasible spatial paths together with a set of feasible speed profiles is generated for all the vehicles involved in the mission. This step relies on optimization methods that take explicitly into account initial and final boundary conditions, a general performance criterion to be optimized, simplified vehicle dynamics, safety rules for collision avoidance, and mission-specific constraints. The second step consists of making each vehicle converge to and follow its assigned path, regardless of what the desired speed profile is, as long as the latter is physically feasible. This approach takes advantage of the separation in space and time introduced during trajectory generation, and leaves the speed profile of the vehicle as an additional degree of freedom to be exploited at the time-coordination level. Finally, in the third step, the speed of each vehicle is adjusted about its desired speed profile to enforce the temporal constraints that must be met to coordinate the fleet of vehicles. This last step relies on the underlying communications network as a means to exchange information among vehicles.

A distinctive feature of the framework for cooperative control adopted in this book is that it exhibits a *multiloop control structure* in which an inner-loop controller stabilizes the vehicle dynamics, while a guidance outer-loop controller is designed to control the vehicle kinematics, providing path-following and time-coordination capabilities. To make these ideas more precise, we notice that a typical autonomous vehicle can be modeled as a coupled system of equations describing its kinematics and dynamics. Following standard notation, the kinematics $\mathcal{G}_k$ of the vehicle can be represented as

$$\mathcal{G}_k : \quad \dot{\boldsymbol{x}}_{\boldsymbol{k}}(t) = \boldsymbol{f}(\boldsymbol{x}_{\boldsymbol{k}}(t), \boldsymbol{y}_{\boldsymbol{d}}(t)) , \qquad (2.1)$$

where $\boldsymbol{x}_{\boldsymbol{k}}(t)$ denotes the kinematic state of the vehicle, which usually includes vehi-

Time-Critical Cooperative Control of Autonomous Air Vehicles, DOI: 10.1016/B978-0-12-809946-9.00003-8

cle's position and attitude, $\boldsymbol{y}_d(t)$ represents the vector of variables driving the vehicle kinematics, such as vehicle angular and linear velocities, and $\boldsymbol{f}(\cdot)$ is a known nonlinear function. We assume the vehicle dynamics $\mathcal{G}_d$ can be expressed as

$$\mathcal{G}_d : \quad \begin{cases} \dot{\boldsymbol{x}}_d(t) &= \boldsymbol{g}(\boldsymbol{x}_d(t), \boldsymbol{x}_k(t), \boldsymbol{u}(t), \boldsymbol{d}(t)) \\ \boldsymbol{y}_d(t) &= \boldsymbol{g}_o(\boldsymbol{x}_d(t)) \end{cases}, \tag{2.2}$$

where $\boldsymbol{x}_d(t)$ denotes the dynamic state of the vehicle, $\boldsymbol{u}(t)$ is the control signal that drives the vehicle dynamics, $\boldsymbol{d}(t)$ is an exogenous disturbance, and $\boldsymbol{g}(\cdot)$ and $\boldsymbol{g}_o(\cdot)$ are partially known nonlinear functions. The model above is sufficiently general to capture six-degree-of-freedom (6DoF) dynamics, together with plant uncertainty [8,12, 21]. The cooperative control algorithms presented in this book are primarily derived at the kinematic level for system $\mathcal{G}_k$ in (2.1) and are viewed as guidance outer-loop controllers that provide reference commands to inner-loop controllers. The latter are designed to stabilize the dynamics $\mathcal{G}_d$ in (2.2) and to ensure that the vehicles track the outer-loop commands. This inner/outer loop approach simplifies the design process and affords the designer a systematic approach to seamlessly tailor the algorithms for a wide range of vehicles that come equipped with inner-loop commercial autopilots. The implementation of the various elements of the control architecture still depends, nonetheless, on the particular dynamics of each class of vehicles. Figs. 2.1 and 2.2 show the conceptual architecture of the solutions proposed for cooperative control of fixed-wing and multirotor UAVs, respectively. As explained later, in the case of fixed-wing UAVs the path-following and coordination control laws run in parallel, sending their respective commands directly to the inputs of the inner-loop augmentation system of the vehicle. On the other hand, the dynamics of multirotor UAVs require that the two algorithms be run in series; in this case, the time-coordination control law generates a speed command that is fed to the path-following algorithm, which generates attitude and thrust commands to be sent to the inner-loop controller of the multirotor UAV. In the figures, each vehicle communicates with a number of neighboring vehicles via a communications network.

The remainder of this chapter formulates the general problem of cooperative vehicle control that is the main topic of the book.

2.2 Problem Formulation

2.2.1 Cooperative Trajectory Generation

Given a set of n vehicles tasked to execute a desired cooperative mission, the problem of *cooperative trajectory generation* can be defined as follows:

Definition 2.1 (Cooperative Trajectory-Generation Problem)**.** Compute a set of n 3D time-trajectories $\boldsymbol{p}_{d,i} : [0, t_d^{\mathrm{f}}] \to \mathbb{R}^3$, conveniently parameterized by a single time-variable $t_d \in [0, t_d^{\mathrm{f}}]$, that minimize a given cost function, satisfy desired boundary

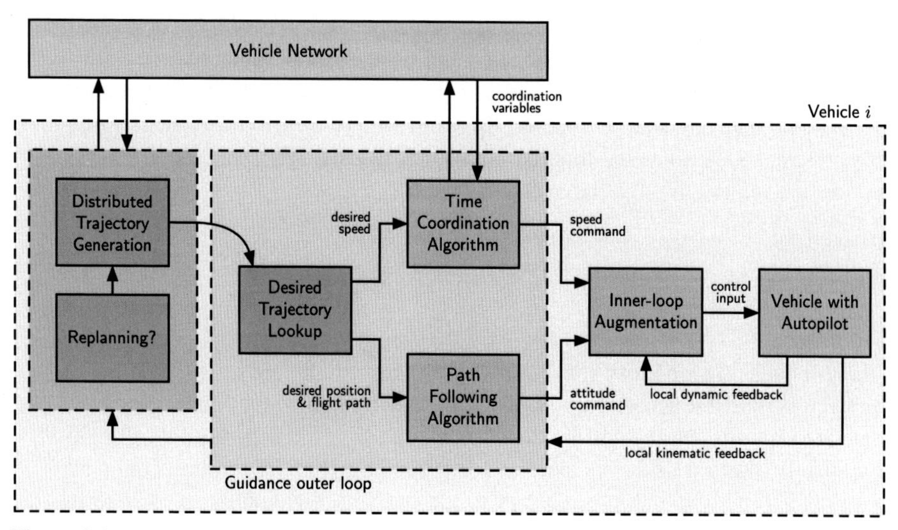

Figure 2.1 Conceptual architecture of the cooperative control framework adopted for fixed-wing UAVs.

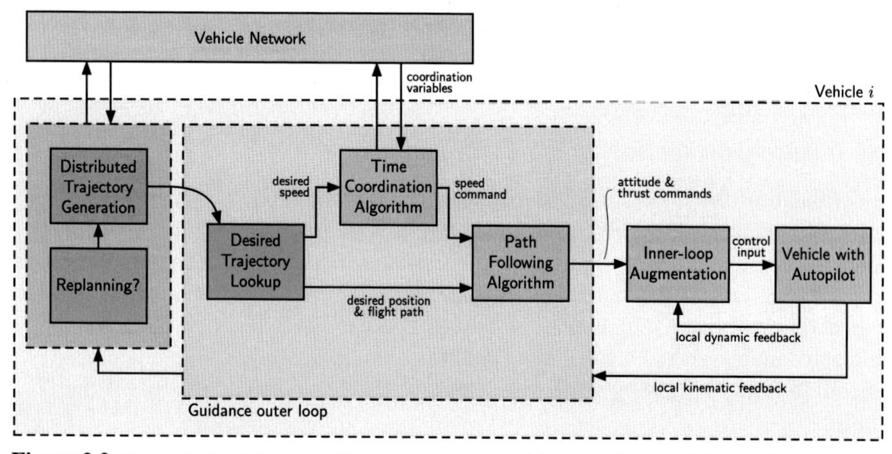

Figure 2.2 Conceptual architecture of the cooperative control framework adopted for multirotor UAVs.

conditions, do not violate the dynamic constraints of the vehicles, ensure that the vehicles maintain a predefined spatial clearance, and satisfy pre-specified mission-specific constraints. ♠

In this formulation, the variable t_d represents a *desired mission time*, with t_d^{f} being the *desired mission duration*. For a given t_d, $\boldsymbol{p}_{d,i}(t_d)$ defines the desired position of the ith vehicle t_d seconds after the initiation of the cooperative mission. The variable t_d is used during the trajectory-generation phase and is distinct from the actual mission time that evolves as the mission unfolds. Recall that many of the missions considered in this book impose only relative temporal constraints on their execution, and therefore it is not strictly required that the vehicles track their respective trajectories. In fact, in such missions, a time drift —uniform to all vehicles— is acceptable. As will become clear later, it is sometimes convenient to consider a (re)parameterization of these trajectories in terms of their *arc length*, denoted by $\tau_{\ell,i}$, $i = 1, \ldots, n$, to obtain desired *spatial paths* $\boldsymbol{p}_{d,i} : [0, \ell_{fi}] \to \mathbb{R}^3$, thus dropping the temporal specifications of the mission explicitly.[1] Here, ℓ_{fi} denotes the total length of the ith path. Arc length and desired mission time are related through the expression

$$\tau_{\ell,i} = \int_0^{t_d} v_{d,i}(\sigma)\, d\sigma ,$$

where $v_{d,i} : [0, t_d^{\mathrm{f}}] \to \mathbb{R}$ represents the desired speed profile, parameterized by t_d, along the ith path. Throughout the book, and with a slight abuse of notation, a (time-)trajectory will be denoted by $\boldsymbol{p}_{d,i}(t_d)$, while the corresponding spatial path (parameterized by arc length) will be denoted by $\boldsymbol{p}_{d,i}(\tau_{\ell,i})$. This notation will help us distinguish between trajectories and the corresponding paths. The same comment applies to other functions that are relevant to the problem formulation presented in this chapter. Next, we formulate the problem of cooperative trajectory generation to compute feasible trajectories for multiple autonomous vehicles that satisfy collision-avoidance constraints. The interested reader is referred to [1,3,5,6,11,13,15,16,20, 22] and the references therein for different approaches to the generation of feasible, collision-free trajectories subject to collective temporal specifications.

Feasible Trajectory Generation for a Single Vehicle

Before formulating the cooperative trajectory-generation problem for multiple vehicles, we first address the problem of generating a feasible trajectory for a single vehicle. In the context of this book, we define a *feasible trajectory* as one that satisfies desired geometric constraints (such as curvature and flight-path-angle bounds), and can be tracked by a given vehicle without having it exceed pre-specified bounds on the vehicle dynamic state and control input (such as speed limits, angular-rate bounds, or acceleration limits). Therefore, the trajectory $\boldsymbol{p}_{d,i}(t_d)$ is said to be feasible if the conditions

$$\begin{aligned} \boldsymbol{c}_g\big(\boldsymbol{p}_{d,i}(\tau_{\ell,i}), \boldsymbol{p}'_{d,i}(\tau_{\ell,i}), \boldsymbol{p}''_{d,i}(\tau_{\ell,i}) \ldots\big) &\geq 0 \\ \boldsymbol{c}_d\big(\boldsymbol{x}_{d,i}(t_d), \boldsymbol{u}_{d,i}(t_d)\big) &\geq 0 \end{aligned} \tag{2.3}$$

are satisfied for all $\tau_{\ell,i} \in [0, \ell_{fi}]$ and all $t_d \in [0, t_d^{\mathrm{f}}]$. In the above expression, $\boldsymbol{p}_{d,i}^{(k)}(\tau_{\ell,i})$ denotes the kth partial derivative of path $\boldsymbol{p}_{d,i}(\tau_{\ell,i})$ with respect to $\tau_{\ell,i}$, $\boldsymbol{x}_{d,i}(t_d)$ and $\boldsymbol{u}_{d,i}(t_d)$ represent the desired time-histories of the dynamic state and

[1] As discussed in [9], parameterization of a curve by its arc length does not, in general, admit a closed-form expression in terms of elementary functions. The result for the case of real rational curves is proven in [10]. However, in the context of this book, this parameterization is advantageous for the theoretical derivations of the path-following control algorithm for fixed-wing air vehicles described later in Chapter 3. The chapter will also discuss details about the practical implementation of the resulting path-following control law.

control signal of the vehicle, as introduced in (2.2), and $\boldsymbol{c_g}(\cdot)$ and $\boldsymbol{c_d}(\cdot)$ are nonlinear functions of their arguments. For the case of fixed-wing UAVs, standard geometric and dynamic feasibility constraints include the following:

$$\begin{aligned} \gamma_{d\,\min} \;\leq\; \gamma_{d,i}(\tau_{\ell,i}) \;\leq\; \gamma_{d\,\max}\,, \\ 0 \;<\; v_{d\,\min} \;\leq\; v_{d,i}(t_d) \;\leq\; v_{d\,\max}\,, \\ |\dot{\psi}_{d,i}(t_d)| \;\leq\; \dot{\psi}_{d\,\max}\,, \qquad |\dot{\gamma}_{d,i}(t_d)| \;\leq\; \dot{\gamma}_{d\,\max}\,, \end{aligned} \tag{2.4}$$

where $\gamma_{d,i}$ and $\dot{\gamma}_{d,i}$ denote, respectively, the desired flight path angle and its rate of change, and $\dot{\psi}_{d,i}$ represents the desired vehicle's turn rate. The bounds in (2.4) may be vehicle dependent. For clarity of exposition, and in order not to clutter the presentation, this dependency will not be stated explicitly.

A feasible trajectory for the ith vehicle can thus be obtained by solving, for example, the optimization problem

$$\min_{\Xi_i} J(\cdot)$$

subject to initial and final boundary conditions as well as the feasibility conditions in (2.3). In the problem above, $J(\cdot)$ is an appropriate cost function and Ξ_i represents the vector of optimization parameters that characterize the trajectory $\boldsymbol{p}_{d,i}(t_d)$. In the work in [6], for example, the vector Ξ_i includes parameters related to the control points of a spatial Pythagorean Hodograph Bézier curve and a Bézier timing law along it. In [15], the vector Ξ_i includes the length of the domain of the variable that parameterizes a polynomial path, its initial parametric jerk, and also the coefficients of a polynomial timing law along the path. Finally, we note that the cost function $J(\cdot)$ may include terms related to mission-specific goals, while additional constraints can also be added to account for vehicle-to-ground communication limitations, sensory capabilities, obstacle collision avoidance, and no-fly zones.

Feasible Collision-Free Trajectory Generation for Multiple Vehicles

We now formulate the problem of cooperative trajectory generation for multiple vehicles. In particular, the time-critical missions described in this book require that each vehicle follow a collision-free trajectory, and that the fleet maintain a desired cooperative timing plan in order to ensure that all vehicles arrive at their respective final destinations at the same time, or at different times so as to meet a desired inter-vehicle schedule. Without loss of generality, in this section we consider the problem of simultaneous arrival. For these missions, the generation of collision-free trajectories can be addressed using two complementary approaches. The first one, referred to as *spatial deconfliction*, ensures that no feasible paths intersect. Alternatively, the second approach —*temporal deconfliction*— implies that no two vehicles are at the same place at the same time. The first approach may be particularly useful in military applications, where jamming prevents vehicles from communicating with each other, and is preferable to the current practice of separating vehicles by altitude. On the other hand, the second approach relies heavily on inter-vehicle communications to properly coordinate the vehicle motions and is thus a function of the QoS of the underlying communications network. Formally, these two strategies lead to two alternative constraints. For spatial deconfliction, the trajectories for the n vehicles must satisfy the constraint

$$\min_{\substack{j,k=1,\dots,n \\ j\neq k}} \|\boldsymbol{p_{d,j}}(\tau_{\ell,j}) - \boldsymbol{p_{d,k}}(\tau_{\ell,k})\|^2 \geq E^2 , \quad \text{for all } (\tau_{\ell,j}, \tau_{\ell,k}) \in [0, \ell_{fj}] \times [0, \ell_{fk}] ,$$

whereas, for temporal deconfliction, the trajectories need to obey the restrictions

$$\min_{\substack{j,k=1,\dots,n \\ j\neq k}} \|\boldsymbol{p_{d,j}}(t_d) - \boldsymbol{p_{d,k}}(t_d)\|^2 \ \geq \ E^2 , \quad \text{for all } t_d \in [0, t_d^{\mathrm{f}}] ,$$

where $E > 0$ is the desired distance for spatial clearance between any two distinct vehicles. Figs. 2.3 and 2.4 illustrate the types of trajectories obtained using these two approaches in the case of two vehicles. The trajectories in these two figures are generated using the same start and end points, the same mission duration, and also the same geometric and dynamic constraints. However, the resulting trajectories differ significantly. In Fig. 2.3, the distance between any two points on the paths is greater than the desired spatial clearance. In contrast, in Fig. 2.4, the two paths intersect at one point, and the desired clearance for vehicle separation is achieved through the speed profiles of the vehicles.

In addition to trajectory deconfliction, the simultaneous time-of-arrival requirement adds an additional constraint on the trajectory-generation problem. Let $\delta w_i := [t^{\mathrm{f}}_{d\min,i}, \ t^{\mathrm{f}}_{d\max,i}]$ be the *arrival-time window* for the ith vehicle, where $t^{\mathrm{f}}_{d\min,i}$ and $t^{\mathrm{f}}_{d\max,i}$ represent the minimal and maximal possible durations of the mission for the ith vehicle, defined as $t^{\mathrm{f}}_{d\min,i} := \ell_{fi}/v_{d\,\max}$ and $t^{\mathrm{f}}_{d\max,i} := \ell_{fi}/v_{d\,\min}$. Then, the simultaneous arrival problem has a solution if and only if the intersection of the arrival-time windows is nonempty, that is, $\delta w_i \cap \delta w_j \neq 0$ for all $i, j \in \{1, \dots, n\}, i \neq j$. In particular, if we define the *arrival margin* δw_{am} as

$$\delta w_{\mathrm{am}} := \min_{i=1,\dots,n} t^{\mathrm{f}}_{d\max,i} - \max_{i=1,\dots,n} t^{\mathrm{f}}_{d\min,i} ,$$

then nonemptiness of the intersection of arrival-time windows is implied by enforcing a positive arrival margin; see Fig. 2.5. Moreover, enlarging the arrival margin adds robustness to mission execution at the coordination level.

Letting T^{f}_d be a predefined upper bound on the final time for the mission to be completed, and defining a cost function $J(\cdot)$ to be minimized, the cooperative *trajectory-generation problem* can be cast in the form of two alternative optimization problems. The first optimization problem addresses spatial deconfliction and is formulated as follows:

$$\min_{\boldsymbol{\Xi_1} \times \cdots \times \boldsymbol{\Xi_n}} J(\cdot) \tag{2.5}$$

subject to initial and final boundary conditions and the feasibility conditions (2.3) for

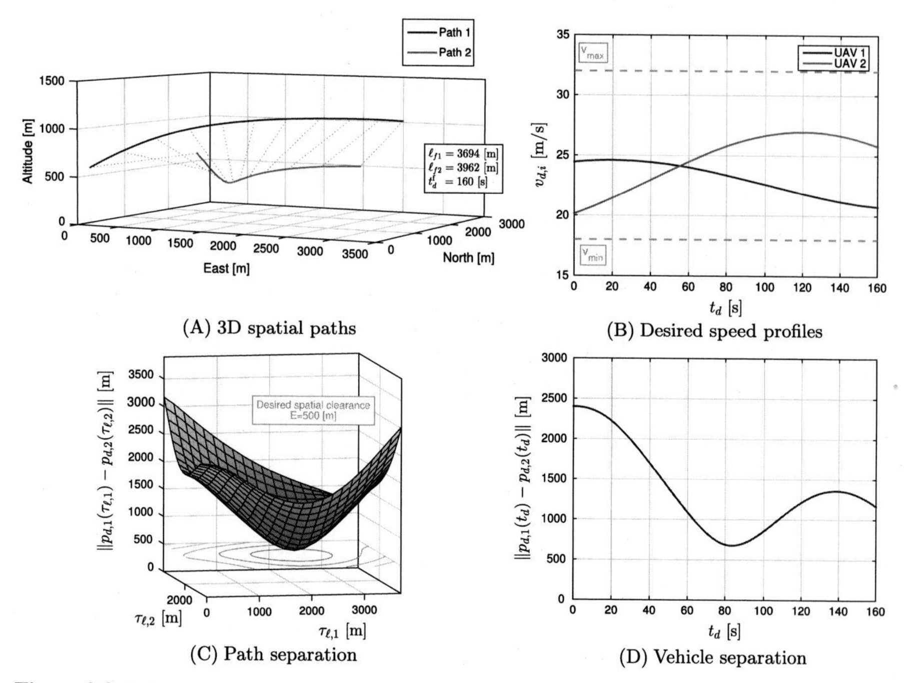

Figure 2.3 Trajectory generation for two vehicles; spatial deconfliction.

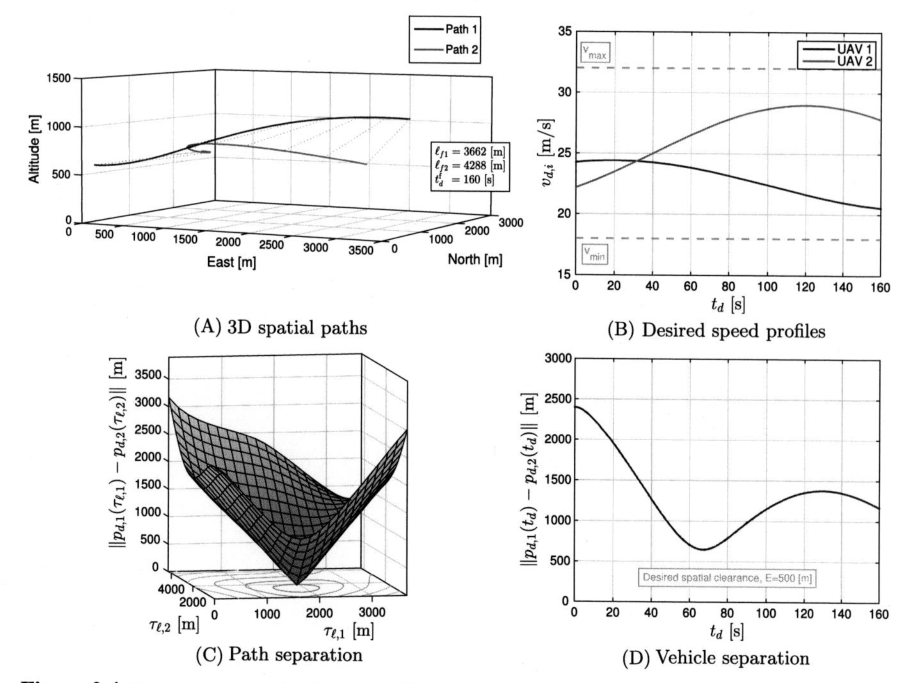

Figure 2.4 Trajectory generation for two vehicles; temporal deconfliction.

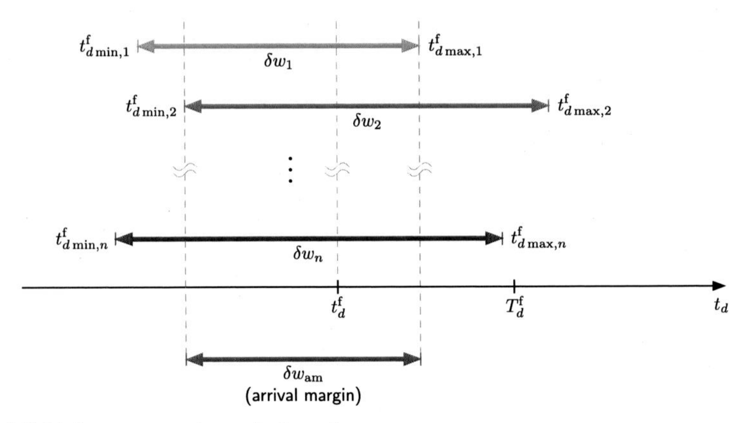

Figure 2.5 Trajectory generation; arrival margin.

all vehicles $i \in \{1, \ldots, n\}$, as well as the constraints

$$\min_{\substack{j,k=1,\ldots,n \\ j\neq k}} \|\boldsymbol{p}_{d,j}(\tau_{\ell,j}) - \boldsymbol{p}_{d,k}(\tau_{\ell,k})\|^2 \geq E^2 , \quad \text{for all } (\tau_{\ell,j}, \tau_{\ell,k}) \in [0, \ell_{fj}] \times [0, \ell_{fk}] ,$$

$$\ell_{fi} = \int_0^{t_d^{\mathrm{f}}} v_{d,i}(t_d)\, \mathrm{d}t_d , \quad \text{for all } i \in \{1, \ldots, n\} ,$$

$$\delta w_{\mathrm{am}} \geq \delta W_{\mathrm{am}} ,$$

$$t_d^{\mathrm{f}} \leq T_d^{\mathrm{f}} ,$$

where $\boldsymbol{\Xi}_i$ is the set of optimization parameters for the ith trajectory, $E > 0$ is the minimal allowable separation distance between paths, and $\delta W_{\mathrm{am}} > 0$ represents a requirement in the arrival margin. The second optimization problem accounts for temporal deconfliction and is posed as:

$$\min_{\boldsymbol{\Xi}_1 \times \cdots \times \boldsymbol{\Xi}_n} J(\cdot) \tag{2.6}$$

subject to initial and final boundary conditions and the feasibility conditions (2.3) for all vehicles $i \in \{1, \ldots, n\}$, as well as the constraints

$$\min_{\substack{j,k=1,\ldots,n \\ j\neq k}} \|\boldsymbol{p}_{d,j}(t_d) - \boldsymbol{p}_{d,k}(t_d)\|^2 \geq E^2 , \quad \text{for all } t_d \in [0, t_d^{\mathrm{f}}] ,$$

$$\ell_{fi} = \int_0^{t_d^{\mathrm{f}}} v_{d,i}(t_d)\, \mathrm{d}t_d , \quad \text{for all } i \in \{1, \ldots, n\} ,$$

$$\delta w_{\mathrm{am}} \geq \delta W_{\mathrm{am}} ,$$

$$t_d^{\mathrm{f}} \leq T_d^{\mathrm{f}} ,$$

where $\boldsymbol{\Xi}_i$ is again the set of optimization parameters for the ith trajectory, while $E > 0$ is now the minimal allowable separation distance between vehicles. In the cooperative trajectory-generation problems above, the cost function $J(\cdot)$ includes terms related to mission-specific goals and cooperative performance criteria, while additional constraints can also be added to account for inter-vehicle and vehicle-to-ground communication limitations, sensory capabilities, task allocation under resource constraints, and collision avoidance with obstacles.

The outcome of the optimization problems (2.5) and (2.6) is a set of n feasible spatial paths $\boldsymbol{p}_{d,i}(\tau_{\ell,i})$ and corresponding desired speed profiles $v_{d,i}(t_d)$ such that, if each agent follows its assigned path and speed profile, then the time-critical mission is executed optimally. However, the presence of disturbances, modeling uncertainty, and failures in the communications network require the synthesis of robust feedback laws to ensure that the mission can be accomplished with a high degree of confidence. In the remainder of this chapter, we present a general framework to synthesize path-following and coordination control laws that can address the performance of the overall time-critical mission in the presence of system uncertainty and a faulty time-varying communications network.

2.2.2 Single-Vehicle Path Following

The path-following control strategies described in this book use the vehicle's attitude control effectors to follow a *virtual target vehicle* that runs along the path and whose rate of progression is an additional variable that can be manipulated at will. To this end, following the formulation developed in [18], it is convenient to introduce a frame attached to this virtual target and define a generalized error vector between this moving coordinate system and a frame attached to the actual vehicle. With this setup, the path-following control problem is reduced to that of driving the generalized error vector to zero, while following an arbitrary feasible speed profile. This approach leaves the vehicle's linear speed as an extra degree of freedom to be used at the coordination level.

The generalized error vector, denoted here by $\boldsymbol{x}_{pf,i}(t)$, must take into account errors in position and direction of the velocity vector, defined as

$$\begin{aligned} \boldsymbol{e}_{p,i}(t) &:= \boldsymbol{h}_1(\boldsymbol{p}_{d,i}(\theta_i(t)), \boldsymbol{p}_i(t)), \\ \boldsymbol{e}_{R,i}(t) &:= \boldsymbol{h}_2(\boldsymbol{v}_i(t), \hat{\mathbf{t}}_i(\theta_i(t))), \end{aligned}$$

where $\boldsymbol{h}_1(\cdot)$ and $\boldsymbol{h}_2(\cdot)$ are known nonlinear functions, $\theta_i(t)$ is a free (length- or time-) variable that defines the position of the ith virtual target vehicle along the ith path and, therefore, $\boldsymbol{p}_{d,i}(\theta_i(t))$ denotes the position of this target vehicle in an inertial frame at time t, $\boldsymbol{p}_i(t)$ and $\boldsymbol{v}_i(t)$ denote the position and velocity of the center of mass of the actual vehicle in the same inertial frame, and $\hat{\mathbf{t}}_i(\theta_i(t))$ is a unit vector that defines the tangent direction to the path at the point determined by $\theta_i(t)$. Once an appropriate generalized error vector has been defined, the kinematics of the vehicle in (2.1) can be used to derive the path-following (kinematic) error dynamics:

$$\mathcal{G}_{pf,i}: \quad \dot{\boldsymbol{x}}_{pf,i}(t) = \boldsymbol{h}(\boldsymbol{x}_{pf,i}(t), \theta_i(t), \dot{\theta}_i(t), \boldsymbol{\nu}_i(t)), \tag{2.7}$$

where $\boldsymbol{\nu}_i(t)$ represents the vector of variables driving the path-following error dynamics and generally comprises a subset of the dynamic states of the ith vehicle $\boldsymbol{x}_{d,i}(t)$, while $\boldsymbol{h}(\cdot)$ is a known nonlinear function. In the kinematic model above, the vector $\boldsymbol{v}_i(t)$ plays the role of a control input, while the rate of progression of variable $\theta_i(t)$ becomes an additional degree of freedom. In general, the path-following error dynamics are also a function of the kinematic state of the vehicle $\boldsymbol{x}_{k,i}(t)$. For clarity of exposition, this dependency is not stated explicitly. Additionally, depending on the equations of motion of the vehicle, it may be convenient to extend the path-following generalized error vector to include (part of) the vehicle's dynamic state $\boldsymbol{x}_{d,i}(t)$. In this case, the path-following error dynamics will also depend on the exogenous disturbance $\boldsymbol{d}_i(t)$, while the control input $\boldsymbol{v}_i(t)$ will also include the vehicle's control input $\boldsymbol{u}_i(t)$. As will become clear later in the book, this extended formulation is convenient when dealing with multirotor air vehicles.

Using the above formulation, and given a spatially defined feasible path $\boldsymbol{p}_{d,i}(\cdot)$, the problem of *path following* for a single vehicle can now be defined accordingly:

Definition 2.2 (Path-Following Problem)**.** Let $\boldsymbol{p}_{d,i}(\cdot)$ be a path, conveniently parameterized by a scalar variable, and let $\boldsymbol{p}_{d,i}(\theta_i(t))$ denote the position of a virtual target vehicle along this path. For a vehicle with equations of motion given by (2.1)-(2.2), design feedback control laws for the path-following control input $\boldsymbol{v}_{c,i}(t)$ and the rate of progression $\dot{\theta}_i(t)$ of the virtual target vehicle along the path that guarantee that all closed-loop signals —with the exception of $\theta_i(t)$— are bounded, and that there exist a class $\mathcal{KL}$ function $\beta_{pf}(\cdot)$ and class $\mathcal{K}$ functions $\alpha^{v}_{pf}(\cdot)$ and $\alpha^{d}_{pf}(\cdot)$ such that the path-following generalized error vector $\boldsymbol{x}_{pf,i}(t)$ satisfies the condition

$$\|\boldsymbol{x}_{pf,i}(t)\| \leq \beta_{pf}\left(\|\boldsymbol{x}_{pf,i}(0)\|, t\right) + \alpha^{v}_{pf}\left(\sup_{0\leq\tau\leq t} \|\boldsymbol{v}_{c,i}(\tau) - \boldsymbol{v}_i(\tau)\|\right) + \alpha^{d}_{pf}\left(\sup_{0\leq\tau\leq t} \|\boldsymbol{d}_i(\tau)\|\right),$$

regardless of the (feasible) temporal assignments of the mission. ♠

Stated in simple terms, the problem above amounts to designing feedback laws so that a vehicle converges to and remains inside a tube centered at the desired path assigned to this vehicle, for an arbitrary speed profile (subject to feasibility constraints). The "size" of this tube will depend on the ability of the vehicle (and its autopilot) to track the control commands $\boldsymbol{v}_{c,i}(t)$ and also on the severity of the disturbance $\boldsymbol{d}_i(t)$.

2.2.3 Coordination and Communication Constraints

To enforce the temporal constraints that must be met to coordinate the entire fleet of vehicles, following the approach presented in [15], the speed profile of each vehicle is adjusted based on coordination information exchanged among the vehicles over a supporting communications network. To generate this speed correction, the coordination problem is formulated as a consensus problem, in which the objective of the fleet of vehicles is to reach agreement on some distributed variables of interest. An appropriate coordination variable needs thus to be defined (for each vehicle) that captures the objective of the cooperative mission, which in the context of this book translates to satisfying temporal constraints.

The coordination variable for the ith vehicle, denoted here by $\xi_i(t)$, will thus be a time-variable, defined through a coordination map $\eta_i(\cdot)$ as

$$\xi_i(t) := \eta_i(\theta_i(t)), \qquad \xi_i(t) \in [0, t_d^{\mathrm{f}}].$$

The coordination variables thus defined must ensure that, for any two vehicles i and j, if $\xi_i(t) = \xi_j(t)$, the target vehicles corresponding to these two vehicles have the desired relative position along their paths at time t and, therefore, satisfy the relative temporal constraints as specified by the trajectory-generation algorithm. Additionally, the definition of the coordination maps must be such that, if the ith coordination variable evolves at rate 1 at a given time t, that is $\dot{\xi}_i(t) = 1$, then the corresponding virtual target is traveling at the desired speed $v_{d,i}(\xi_i(t))$.

To reach agreement on these coordination states and ensure that the desired temporal assignments of the mission are met, coordination information must be exchanged among the vehicles over the supporting communications network. In what follows, we use tools and facts from algebraic graph theory to model the information exchange over the network as well as the constraints imposed by the communication topology. Key concepts on algebraic graph theory are presented in Appendix A.3. The reader is referred to [4] for a more thorough discussion of this topic.

It is assumed throughout the book that the ith vehicle can only exchange information with a neighboring set of vehicles, denoted here by $\mathcal{N}_i(t)$. It is also assumed that communications between two vehicles are bidirectional and, for simplicity, that information is transmitted continuously with no delays. Moreover, since the flow of information among vehicles may be severely restricted, either for security reasons or because of tight bandwidth limitations, we impose the constraint that each vehicle only exchanges its coordination state $\xi_i(t)$ with its neighbors. Finally, we assume that the connectivity of the communication graph $\Gamma(t)$ that captures the underlying bidirectional communications network topology of the fleet at time t satisfies the persistency of excitation (PE)-like condition [2]

$$\frac{1}{n}\frac{1}{T}\int_t^{t+T} \boldsymbol{Q}\,\boldsymbol{L}(\tau)\,\boldsymbol{Q}^\top \mathrm{d}\tau \geq \mu\,\mathbb{I}_{n-1}, \qquad \text{for all } t \geq 0, \tag{2.8}$$

where $\boldsymbol{L}(t) \in \mathbb{R}^{n\times n}$ is the Laplacian of the graph $\Gamma(t)$ and $\boldsymbol{Q}$ is an $(n-1)\times n$ matrix such that $\boldsymbol{Q}\mathbf{1}_n = \mathbf{0}$ and $\boldsymbol{Q}\boldsymbol{Q}^\top = \mathbb{I}_{n-1}$, with $\mathbf{1}_n$ being the vector in $\mathbb{R}^n$ whose components are all 1. The parameters $T > 0$ and $\mu \in (0, 1]$ characterize the QoS of the communications network, which in the context of this book represents a measure of the level of connectivity of the communication graph. We note that the PE-like condition (2.8) requires the communication graph $\Gamma(t)$ to be connected only in an integral sense, not pointwise in time. In fact, the graph may be disconnected during some interval of time or may even fail to be connected at all times. A similar type of condition can be found, for example, in [19].

Using the formulation above, we next define the problem of time-critical cooperative path following for a fleet of n vehicles.

Definition 2.3 (Time-Critical Cooperative Path-Following Problem). Let $\boldsymbol{p}_{d,i}(\cdot)$, $i = 1, \ldots, n$, be a set of desired 3D paths, each conveniently parameterized by a scalar variable, and let $\xi_i(t)$, $i = 1, \ldots, n$, denote a set of coordination variables that capture the temporal assignments of the cooperative mission. Given a fleet of n vehicles with equations of motion (2.1)-(2.2) and supported by an inter-vehicle communications network, design feedback control laws for the control input $\boldsymbol{v}_{c,i}(t)$ and the rate of progression $\dot{\theta}_i(t)$ of each virtual target vehicle such that:

1. All closed-loop signals —with the exceptions of $\theta_i(t)$ and $\xi_i(t)$— are bounded.
2. There exist a class $\mathcal{KL}$ function $\beta_{pf}(\cdot)$ and class $\mathcal{K}$ functions $\alpha^{\nu}_{pf}(\cdot)$ and $\alpha^{d}_{pf}(\cdot)$ such that, for every vehicle, the corresponding path-following generalized error vector $\boldsymbol{x}_{pf,i}(t)$ satisfies the condition

$$\|\boldsymbol{x}_{pf,i}(t)\| \leq \beta_{pf}\left(\|\boldsymbol{x}_{pf,i}(0)\|, t\right) + \alpha^{\nu}_{pf}\left(\sup_{0\leq\tau\leq t} \|\boldsymbol{v}_{c,i}(\tau) - \boldsymbol{v}_i(\tau)\|\right) + \alpha^{d}_{pf}\left(\sup_{0\leq\tau\leq t} \|\boldsymbol{d}_i(\tau)\|\right).$$

3. There exist class $\mathcal{KL}$ functions $\beta^{\xi}_{cd,1}(\cdot)$, $\beta^{\xi}_{cd,2}(\cdot)$, $\beta^{\dot{\xi}}_{cd,1}(\cdot)$ and $\beta^{\dot{\xi}}_{cd,2}(\cdot)$, and class $\mathcal{K}$ functions $\alpha^{\xi\nu}_{cd}(\cdot)$, $\alpha^{\xi d}_{cd}(\cdot)$, $\alpha^{\dot{\xi}\nu}_{cd}(\cdot)$, and $\alpha^{\dot{\xi}d}_{cd}(\cdot)$ such that, for every pair of vehicles i and j, $i, j = 1, \ldots, n$, the coordination errors $(\xi_i(t) - \xi_j(t))$ and $(\dot{\xi}_i(t) - 1)$ satisfy the conditions

$$|\xi_i(t) - \xi_j(t)| \leq \beta^{\xi}_{cd,1}\left(\|\boldsymbol{x}_{cd}(0)\|, t\right) + \beta^{\xi}_{cd,2}\left(\max_k \|\boldsymbol{x}_{pf,k}(0)\|, t\right) + \alpha^{\xi\nu}_{cd}\left(\sup_{0\leq\tau\leq t} \max_k \|\boldsymbol{v}_{c,k}(\tau) - \boldsymbol{v}_k(\tau)\|\right) + \alpha^{\xi d}_{cd}\left(\sup_{0\leq\tau\leq t} \max_k \|\boldsymbol{d}_k(\tau)\|\right),$$

$$|\dot{\xi}_i(t) - 1| \leq \beta^{\dot{\xi}}_{cd,1}\left(\|\boldsymbol{x}_{cd}(0)\|, t\right) + \beta^{\dot{\xi}}_{cd,2}\left(\max_k \|\boldsymbol{x}_{pf,k}(0)\|, t\right) + \alpha^{\dot{\xi}\nu}_{cd}\left(\sup_{0\leq\tau\leq t} \max_k \|\boldsymbol{v}_{c,k}(\tau) - \boldsymbol{v}_k(\tau)\|\right) + \alpha^{\dot{\xi}d}_{cd}\left(\sup_{0\leq\tau\leq t} \max_k \|\boldsymbol{d}_k(\tau)\|\right),$$

where $\boldsymbol{x}_{cd}(0)$ is a vector that characterizes the initial coordination error of the entire fleet of vehicles. ♠

In this case, in addition to solving the path-following problem for each vehicle in the fleet, the design of the feedback control laws must ensure that, after a transient response due to initial path-following and coordination errors, the vehicles reach agreement in a practical sense. In particular, the final disagreement on the coordination variables will depend on the ability of the vehicles (and their autopilots) to track the control commands $\boldsymbol{v}_{c,k}(t)$ and also on the severity of the disturbances $\boldsymbol{d}_k(t)$.

2.2.4 Autonomous Vehicles with Inner-Loop Autopilots

At this point, it is important to stress that this book addresses the design of control algorithms for path following and time coordination of a fleet of vehicles executing time-critical cooperative missions. The design of inner-loop onboard autopilots that are capable of tracking reference commands generated by outer-loop controllers and providing uniform performance across the operational envelope is, however, beyond the scope of the work presented here. This section presents a set of assumptions on the inner closed-loop performance of the vehicles with their autopilots, which will be useful to analyze the convergence properties of the path-following and coordination control laws developed later.

We assume that each vehicle involved in a cooperative mission is equipped with an autopilot designed to stabilize the vehicle and to provide input/output tracking capabilities of the path-following control signals $\boldsymbol{v}_{c,i}(t)$. In particular, we express the closed-loop dynamics of the vehicle with its autopilot as follows:

$$\mathcal{G}_{D,i} : \quad \begin{cases} \dot{\boldsymbol{x}}_{D,i}(t) &= \hat{\boldsymbol{g}}_i(\boldsymbol{x}_{D,i}(t), \boldsymbol{x}_{k,i}(t), \boldsymbol{v}_{c,i}(t), \boldsymbol{d}_i(t)) \\ \boldsymbol{v}_i(t) &= \hat{\boldsymbol{g}}_{o,i}(\boldsymbol{x}_{D,i}(t)) \end{cases} ,$$

where $\boldsymbol{x}_{D,i}(t)$ denotes the dynamic state of the augmented vehicle and $\hat{\boldsymbol{g}}_i(\cdot)$ and $\hat{\boldsymbol{g}}_{o,i}(\cdot)$ are partially known nonlinear functions. In addition, we make the assumption that the onboard autopilots ensure that each vehicle is capable of tracking $\boldsymbol{v}_{c,i}(t)$ with guaranteed uniform performance bound γ_v, that is,

$$\|\boldsymbol{v}_{c,i}(t) - \boldsymbol{v}_i(t)\| \le \gamma_v , \qquad \text{for all } t \ge 0 ,$$

provided these commands satisfy suitable feasibility conditions of the form

$$\boldsymbol{c}_{\boldsymbol{v}}\left(\boldsymbol{v}_{c,i}(t)\right) \ge 0 ,$$

where $\boldsymbol{c}_{\boldsymbol{v}}(\cdot)$ is, often, a linear function of its arguments.

For the case of fixed-wing UAVs, we will assume that the vehicles are equipped with autopilots that provide angular-rate as well as speed tracking capabilities. More specifically, we will make the assumption that the onboard autopilots ensure that each vehicle can track bounded pitch-rate and yaw-rate commands, denoted here by $q_c(t)$ and $r_c(t)$, with guaranteed performance bounds γ_q and γ_r. Stated mathematically,

$$\begin{aligned} |q_{c,i}(t) - q_i(t)| &\le \gamma_q , \qquad && \text{for all } t \ge 0 , \\ |r_{c,i}(t) - r_i(t)| &\le \gamma_r , \qquad && \text{for all } t \ge 0 . \end{aligned} \tag{2.9}$$

We will also assume that if the speed command $v_c(t)$ satisfies the bounds

$$v_{\min} \le v_{c,i}(\tau) \le v_{\max} , \qquad \text{for all } \tau \in [0, t] , \tag{2.10}$$

then the autopilots ensure that each vehicle tracks its corresponding speed command with guaranteed performance bound γ_v, that is,

$$|v_{c,i}(\tau) - v_i(\tau)| \le \gamma_v , \qquad \text{for all } \tau \in [0, t] . \tag{2.11}$$

The bounds γ_q, γ_r, and γ_v characterize the level of tracking performance that an inner-loop autopilot can provide. It is important to note that, in this setup, it is the autopilot that determines the bank angle required to track the angular-rate commands $q_c(t)$ and $r_c(t)$. Therefore, it is justified to assume that the roll dynamics of the fixed-wing UAV (roll rate and bank angle) are bounded for bounded angular-rate commands corresponding to the set of feasible paths considered.

Similarly, for the case of multirotor UAVs, we will assume that the vehicles are equipped with autopilots that provide roll-rate, pitch-rate, and yaw-rate tracking capabilities defined by

$$\begin{aligned} |p_{c,i}(t) - p_i(t)| &\leq \gamma_p\,, && \text{for all } t \geq 0\,, \\ |q_{c,i}(t) - q_i(t)| &\leq \gamma_q\,, && \text{for all } t \geq 0\,, \\ |r_{c,i}(t) - r_i(t)| &\leq \gamma_r\,, && \text{for all } t \geq 0\,. \end{aligned} \tag{2.12}$$

It will also be assumed that the overall thrust $T_i(t)$ generated by the propellers mounted on the ith vehicle tracks the thrust command $T_{c,i}(t)$ with the following uniform performance bound:

$$|T_{c,i}(t) - T_i(t)| \leq \gamma_T\,, \qquad \text{for all } t \geq 0\,. \tag{2.13}$$

Remark 2.1. The performance bounds above will be used later in the book to set constraints on the inner-loop tracking performance requirements that guarantee stability of the complete cooperative control architecture. As will become clear from the algorithms for path following and time coordination proposed, a proper choice of the initial boundary conditions for the trajectory-generation problem may be required to ensure that these bounds can be satisfied at all times. A more relaxed —and realistic— assumption would be *ultimate boundedness* of the inner-loop tracking errors; under this assumption, the results presented in this book would still hold with a few modifications, especially affecting the initial transient phase. For simplicity, however, we assume that the performance bounds introduced in this section hold *uniformly* in time. From a practical perspective, these performance bounds —as well as the constraints on them derived in the following sections— should be seen as guidelines/specifications for the design of the inner-loop autopilots. △

Remark 2.2. For many missions of interest, typical off-the-shelf autopilots are capable of providing uniform performance across the operational envelope of small autonomous vehicles while operating in nominal conditions. However, these commercial autopilots may fail to provide adequate performance in the event of actuator failures, partial vehicle damage, or in the presence of adverse environmental disturbances. Under these unfavorable circumstances, adaptive augmentation loops are seen as an appealing technology that can improve vehicle performance. In [7] and [14], for example, an $\mathcal{L}_1$ adaptive control architecture for fixed-wing UAV autopilot augmentation is presented that retains the properties of the onboard commercial autopilot and adjusts the autopilot commands only when the tracking performance degrades. Fig. 2.6 shows the inner-loop control architecture considered in [7] and [14], with the

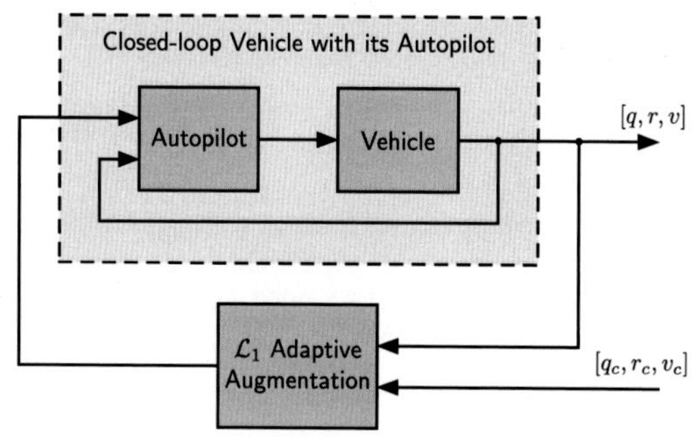

Figure 2.6 Inner-loop control structure with an $\mathcal{L}_1$ adaptive augmentation loop for a fixed-wing UAV [7,14].

adaptive augmentation loop wrapped around the autopilot. In this setup, the adaptive controller uses angular-rate and speed measurements to modify the commands generated by the outer-loop algorithms, which are then sent to the autopilot as references to be tracked. This structure for autopilot augmentation does not require any modifications to the autopilot itself, and at the same time it does not use internal states of the autopilot for control design purposes. △

References

[1] K.P. Bollino, L.R. Lewis, Collision-free multi-UAV optimal path planning and cooperative control for tactical applications, in: AIAA Guidance, Navigation and Control Conference, Honolulu, HI, August 2008, AIAA 2008-7134.

[2] M. Arcak, Passivity as a design tool for group coordination, IEEE Transactions on Automatic Control 52 (2007) 1380–1390.

[3] F. Augugliaro, A. Schoellig, R. D'Andrea, Generation of collision-free trajectories for a quadrocopter fleet: a sequential convex programming approach, in: IEEE/RSJ International Conference on Intelligent Robots and Systems, Vilamoura, Portugal, October 2012, pp. 1917–1922.

[4] N. Biggs, Algebraic Graph Theory, Cambridge University Press, New York, NY, 1993.

[5] Y. Chen, M. Cutler, J.P. How, Decoupled multiagent path planning via incremental sequential convex programming, in: IEEE International Conference on Robotics and Automation, Seattle, WA, May 2015, pp. 5954–5961.

[6] R. Choe, Distributed Cooperative Trajectory Generation for Multiple Autonomous Vehicles Using Pythagorean Hodograph Bézier Curves, PhD thesis, University of Illinois at Urbana-Champaign, Urbana, IL, United States, April 2017.

[7] V. Dobrokhodov, I. Kaminer, I. Kitsios, E. Xargay, N. Hovakimyan, C. Cao, I.M. Gregory, L. Valavani, Experimental validation of $\mathcal{L}_1$ adaptive control: the Rohrs counterexample in flight, Journal of Guidance, Control and Dynamics 34 (2011) 1311–1328.

[8] B. Etkin, Dynamics of Atmospheric Flight, Dover Publications, Mineola, NY, 2005.

[9] R.T. Farouki, Pythagorean-Hodograph Curves: Algebra and Geometry Inseparable, Springer-Verlag Berlin Heidelberg, 2008.

[10] R.T. Farouki, T. Sakkalis, Real rational curves are not "unit speed", Computer Aided Geometric Design 8 (1991) 151–157.
[11] A.J. Häusler, A. Saccon, A.P. Aguiar, J. Hauser, A.M. Pascoal, Energy-optimal motion planning for multiple robotic vehicles with collision avoidance, IEEE Transactions on Control System Technology 24 (2016) 867–883.
[12] G.M. Hoffmann, H. Huang, S.L. Waslander, C.J. Tomlin, Quadrotor helicopter flight dynamics and control: theory and experiment, in: AIAA Guidance, Navigation and Control Conference, Hilton Head, SC, August 2007, AIAA 2007-6461.
[13] M.A. Hurni, P. Sekhavat, M. Karpenko, I.M. Ross, A pseudospectral optimal motion planner for autonomous unmanned vehicles, in: American Control Conference, Baltimore, MD, June–July 2010, pp. 1591–1598.
[14] I. Kaminer, A.M. Pascoal, E. Xargay, N. Hovakimyan, C. Cao, V. Dobrokhodov, Path following for unmanned aerial vehicles using $\mathcal{L}_1$ adaptive augmentation of commercial autopilots, Journal of Guidance, Control and Dynamics 33 (2010) 550–564.
[15] I. Kaminer, O.A. Yakimenko, A.M. Pascoal, R. Ghabcheloo, Path generation, path following and coordinated control for time-critical missions of multiple UAVs, in: American Control Conference, Minneapolis, MN, June 2006, pp. 4906–4913.
[16] Y. Kuwata, J.P. How, Cooperative distributed robust trajectory optimization using receding horizon MILP, IEEE Transactions on Control System Technology 19 (2011) 423–431.
[17] L. Lapierre, D. Soetanto, A.M. Pascoal, Coordinated motion control of marine robots, in: IFAC Conference on Manoeuvering and Control of Marine Craft, Girona, Spain, September 2003.
[18] L. Lapierre, D. Soetanto, A.M. Pascoal, Non-singular path-following control of a unicycle in the presence of parametric modeling uncertainties, International Journal of Robust and Nonlinear Control 16 (2006) 485–503.
[19] Z. Lin, B.A. Francis, M. Maggiore, State agreement for continuous-time coupled nonlinear systems, SIAM Journal on Control and Optimization 46 (2007) 288–307.
[20] J. Ousingsawat, M.E. Campbell, Optimal cooperative reconnaissance using multiple vehicles, Journal of Guidance, Control and Dynamics 30 (2007) 122–132.
[21] G.D. Padfield, Helicopter Flight Dynamics, Blackwell Publishing, Oxford, UK, 2007.
[22] M. Shanmugavel, A. Tsourdos, R. Żbikowski, B.A. White, C.-A. Rabbath, N. Léchevin, A solution to simultaneous arrival of multiple UAVs using Pythagorean Hodograph curves, in: American Control Conference, Minneapolis, MN, June 2006, pp. 2813–2818.
[23] R. Skjetne, I.-A.F. Ihle, T.I. Fossen, Formation control by synchronizing multiple maneuvering systems, in: IFAC Conference on Manoeuvering and Control of Marine Craft, Girona, Spain, September 2003.

Part Two

Cooperative Control of Fixed-Wing Air Vehicles

3D Path-Following Control of Fixed-Wing Air Vehicles

3

This chapter describes an outer-loop path-following nonlinear control algorithm for fixed-wing UAVs. The algorithm uses angular-rate commands to steer a vehicle along a 3D spatial path for an arbitrary feasible speed assignment along the path. Controller design follows the formulation described in Chapter 2, and involves the derivation of a control law on $\mathsf{SO}(3)$ that avoids the geometric singularities and complexities that appear when dealing with local parameterizations of the vehicle's attitude. First, we address only the kinematic equations of the vehicle by taking pitch rate and yaw rate as virtual outer-loop control inputs. In particular, we show that there exist stabilizing functions for pitch and yaw rates that yield local exponential stability of the origin of the kinematic error dynamics with a prescribed domain of attraction. Then, we perform a stability analysis for the case of non-ideal inner-loop tracking and show that the path-following errors are locally uniformly ultimately bounded with the same domain of attraction. The results lead to an efficient methodology to design path-following controllers for autonomous fixed-wing vehicles with due account for the vehicle kinematics and the characteristics of their inner-loop autopilots. For notational simplicity, in this chapter we drop the subscript i used to denote a particular vehicle.

The control law described next is rooted in the results on planar path-following control for unicycle robots developed in [7], and builds upon our previous results on 3D path following for fixed-wing UAVs in [5] and [6]. In particular, the work in [6] used a Frenet-Serret coordinate system to frame the path and relied on Euler angles to represent the vehicle's attitude. In [5], the Frenet-Serret frame was replaced by a parallel transport frame to prevent singularities when the path has a vanishing second derivative. The path-following algorithm in this chapter was first reported in [2], while the proofs of the main theoretical results were first published in [10]. An overview of different approaches used for the derivation of path-following control algorithms can be found in Section 1.3.1 of Chapter 1.

3.1 Tracking a Virtual Target on a Path

The solution to the path-following problem proposed in this chapter extends the algorithm presented in [7] to the case of 3D spatial paths, and relies on the insight that a fixed-wing UAV can follow a given path using only its attitude, thus leaving its linear speed as an extra degree of freedom to be used at the coordination level. The key idea of the algorithm is to use the vehicle's attitude actuators to track a *virtual target vehicle* running along the path. To this end, following the approach described in Chapter 2, we introduce a frame attached to the virtual target and define a generalized error vector between the corresponding moving coordinate system and a frame attached to the actual vehicle. In this section, we characterize the dynamics of the generalized kine-

Time-Critical Cooperative Control of Autonomous Air Vehicles, DOI: 10.1016/B978-0-12-809946-9.00005-1

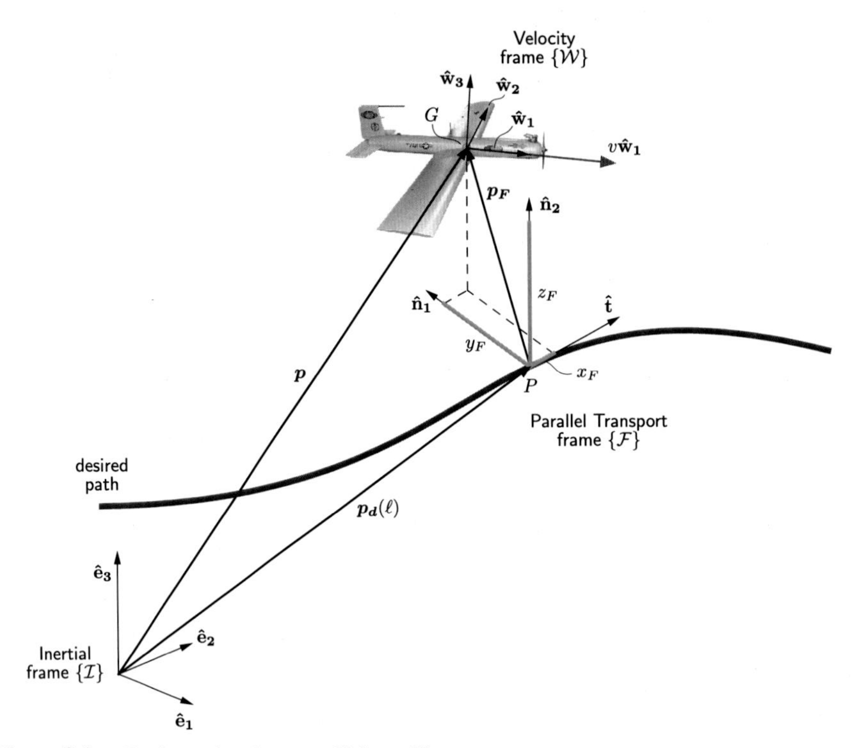

Figure 3.1 Following a virtual target vehicle; problem geometry.

matic error between one of the n vehicles involved in a cooperative mission and the corresponding virtual target.

Fig. 3.1 captures the geometry of the problem at hand. The symbol $\{\mathcal{I}\}$ denotes an inertial reference frame with unit vectors $\{\hat{\mathbf{e}}_1, \hat{\mathbf{e}}_2, \hat{\mathbf{e}}_3\}$ and $\boldsymbol{p}_d(\tau_\ell)$ is the desired path assigned to one of the vehicles, which —for theoretical purposes— we consider parameterized by its arc length τ_ℓ. Vector $\boldsymbol{p}(t)$ denotes the position of the center of mass G of the vehicle in the inertial frame $\{\mathcal{I}\}$. Further, we let P be an arbitrary point on the desired path that plays the role of virtual target, and denote by $\boldsymbol{p}_d(\ell)$ its position in the inertial frame. Here $\ell \in [0, \ell_f]$, with ℓ_f being the total length of the path, is a free length-variable that defines the position of the virtual target vehicle along the path.[1] In this approach, the total rate of progression of the virtual target along the path, $\frac{\mathrm{d}\ell(t)}{\mathrm{d}t}$, is an additional design parameter. This strategy is in contrast with the approach used in the path-following algorithm introduced in [9], where P was defined as the point on the path that is closest to the vehicle. Endowing point P with an extra degree of freedom is the key to the path-following algorithm presented in [7] and its extension to the 3D case described in this chapter.

[1] In this setup, the length-variable $\ell(t)$ plays the role of the free variable $\theta(t)$ introduced in Section 2.2.2, Chapter 2.

For our purposes, it is convenient to define a *parallel transport frame* $\{\mathcal{F}\}$ attached to point P on the path and characterized by the orthonormal vectors $\{\hat{\mathbf{t}}(\ell), \hat{\mathbf{n}}_1(\ell), \hat{\mathbf{n}}_2(\ell)\}$, which satisfy the frame equations [1,4]

$$\begin{bmatrix} \frac{\mathrm{d}\hat{\mathbf{t}}}{\mathrm{d}\ell}(\ell) \\ \frac{\mathrm{d}\hat{\mathbf{n}}_1}{\mathrm{d}\ell}(\ell) \\ \frac{\mathrm{d}\hat{\mathbf{n}}_2}{\mathrm{d}\ell}(\ell) \end{bmatrix} = \begin{bmatrix} 0 & k_1(\ell) & k_2(\ell) \\ -k_1(\ell) & 0 & 0 \\ -k_2(\ell) & 0 & 0 \end{bmatrix} \begin{bmatrix} \hat{\mathbf{t}}(\ell) \\ \hat{\mathbf{n}}_1(\ell) \\ \hat{\mathbf{n}}_2(\ell) \end{bmatrix},$$

where $k_1(\ell)$ and $k_2(\ell)$ define the *normal development* of the path, and are related to the polar coordinates of curvature $\kappa(\ell)$ and torsion $\tau(\ell)$ as

$$\kappa(\ell) = \left(k_1^2(\ell) + k_2^2(\ell)\right)^{\frac{1}{2}}, \qquad \tau(\ell) = -\frac{\mathrm{d}}{\mathrm{d}\ell}\left(\tan^{-1}\left(\frac{k_2(\ell)}{k_1(\ell)}\right)\right).$$

Vectors $\hat{\mathbf{t}}(\ell)$, $\hat{\mathbf{n}}_1(\ell)$, and $\hat{\mathbf{n}}_2(\ell)$ define an orthonormal basis for $\{\mathcal{F}\}$, in which the unit vector $\hat{\mathbf{t}}(\ell)$ defines the tangent direction to the path at the point determined by ℓ, while $\hat{\mathbf{n}}_1(\ell)$ and $\hat{\mathbf{n}}_2(\ell)$ define the normal plane perpendicular to $\hat{\mathbf{t}}(\ell)$. We note that, unlike the Frenet-Serret frame, parallel transport frames are well defined even when the path has a vanishing second derivative and the normal components do not exhibit sudden "flips" on passing through inflection points [4]. This orthonormal basis can be used to construct the rotation matrix $\boldsymbol{R}_F^I(\ell) = [\{\hat{\mathbf{t}}(\ell)\}_I, \{\hat{\mathbf{n}}_1(\ell)\}_I, \{\hat{\mathbf{n}}_2(\ell)\}_I]$ from $\{\mathcal{F}\}$ to $\{\mathcal{I}\}$. Furthermore, the angular velocity of $\{\mathcal{F}\}$ with respect to $\{\mathcal{I}\}$, resolved in $\{\mathcal{F}\}$, can be easily expressed in terms of the parameters $k_1(\ell)$ and $k_2(\ell)$ as

$$\{\boldsymbol{\omega}_{F/I}\}_F = \begin{bmatrix} 0, & -k_2(\ell)\,\dot{\ell}, & k_1(\ell)\,\dot{\ell} \end{bmatrix}^\top.$$

The position of the vehicle's center of mass G in the parallel transport frame $\{\mathcal{F}\}$ is denoted by $\boldsymbol{p}_F(t)$, and $x_F(t)$, $y_F(t)$, and $z_F(t)$ are the components of this vector with respect to the basis $\{\hat{\mathbf{t}}, \hat{\mathbf{n}}_1, \hat{\mathbf{n}}_2\}$, that is,

$$\{\boldsymbol{p}_F\}_F = \begin{bmatrix} x_F, & y_F, & z_F \end{bmatrix}^\top.$$

Finally, we let $\{\mathcal{W}\}$ denote a vehicle-carried velocity frame $\{\hat{\mathbf{w}}_1, \hat{\mathbf{w}}_2, \hat{\mathbf{w}}_3\}$ with its origin at the vehicle's center of mass G and the x axis aligned with the velocity vector of the vehicle. The z axis is chosen to lie in the vertical plane of symmetry of the vehicle, and the y axis is determined by completing the right-hand system. In this chapter, $q(t)$ and $r(t)$ are the y-axis and z-axis components, respectively, of the vehicle's rotational velocity resolved in the $\{\mathcal{W}\}$ frame. With a slight abuse of notation, $q(t)$ and $r(t)$ will be referred here to as *pitch rate* and *yaw rate*, respectively, in the $\{\mathcal{W}\}$ frame.

With the above notation, we next characterize the path-following kinematic error dynamics of the vehicle with respect to the virtual target. We start by deriving the position-error dynamics. To this end, we note that

$$\boldsymbol{p} = \boldsymbol{p}_d(\ell) + \boldsymbol{p}_F,$$

from which it follows that

$$\dot{\boldsymbol{p}}]_I = \dot{\ell}\,\hat{\mathbf{t}} + \boldsymbol{\omega}_{F/I} \times \boldsymbol{p}_F + \dot{\boldsymbol{p}}_F]_F\,,$$

where $\cdot\,]_I$ and $\cdot\,]_F$ are used to indicate that the derivatives are computed in the inertial and parallel transport frames, respectively. Because

$$\dot{\boldsymbol{p}}]_I = v\,\hat{\mathbf{w}}_1\,,$$

where $v(t)$ denotes the magnitude of the vehicle's ground velocity vector, the path-following kinematic position-error dynamics of the vehicle with respect to the virtual target can be written as

$$\dot{\boldsymbol{p}}_F]_F = -\,\dot{\ell}\,\hat{\mathbf{t}} - \boldsymbol{\omega}_{F/I} \times \boldsymbol{p}_F + v\,\hat{\mathbf{w}}_1\,. \tag{3.1}$$

With respect to the basis $\{\hat{\mathbf{t}}, \hat{\mathbf{n}}_1, \hat{\mathbf{n}}_2\}$, the above equation takes the following form:

$$\begin{bmatrix} \dot{x}_F \\ \dot{y}_F \\ \dot{z}_F \end{bmatrix} = -\begin{bmatrix} \dot{\ell} \\ 0 \\ 0 \end{bmatrix} - \left(\begin{bmatrix} 0 \\ -k_2(\ell)\,\dot{\ell} \\ k_1(\ell)\,\dot{\ell} \end{bmatrix} \times \begin{bmatrix} x_F \\ y_F \\ z_F \end{bmatrix} \right) + \boldsymbol{R}_W^F \begin{bmatrix} v \\ 0 \\ 0 \end{bmatrix}.$$

To derive the attitude-error dynamics of the vehicle with respect to its virtual target, we first introduce an auxiliary frame $\{\mathcal{D}\}$ that, as in [9], will be used to shape the approach attitude to the path as a function of the *cross-track error* components y_F and z_F. Frame $\{\mathcal{D}\}$ has its origin at the center of mass of the vehicle and is characterized by vectors $\hat{\mathbf{b}}_{1\mathbf{D}}(t)$, $\hat{\mathbf{b}}_{2\mathbf{D}}(t)$, and $\hat{\mathbf{b}}_{3\mathbf{D}}(t)$, which are defined as

$$\hat{\mathbf{b}}_{1\mathbf{D}} := \frac{d\,\hat{\mathbf{t}} - y_F\,\hat{\mathbf{n}}_1 - z_F\,\hat{\mathbf{n}}_2}{\left(d^2 + y_F^2 + z_F^2\right)^{\frac{1}{2}}}\,, \qquad \hat{\mathbf{b}}_{2\mathbf{D}} := \frac{y_F\,\hat{\mathbf{t}} + d\,\hat{\mathbf{n}}_1}{\left(d^2 + y_F^2\right)^{\frac{1}{2}}}\,, \qquad \hat{\mathbf{b}}_{3\mathbf{D}} := \hat{\mathbf{b}}_{1\mathbf{D}} \times \hat{\mathbf{b}}_{2\mathbf{D}}\,, \tag{3.2}$$

where $d > 0$ is a constant *characteristic distance* that plays the role of a design parameter, as will become clear later. In particular, the basis vector $\hat{\mathbf{b}}_{1\mathbf{D}}(t)$ defines the desired direction of the vehicle's velocity vector. As illustrated in Fig. 3.2, when the vehicle is far from the desired path vector $\hat{\mathbf{b}}_{1\mathbf{D}}(t)$ becomes (nearly) perpendicular to $\hat{\mathbf{t}}(\ell)$. As the vehicle comes closer to the path and the cross-track error becomes smaller, then $\hat{\mathbf{b}}_{1\mathbf{D}}(t)$ tends to $\hat{\mathbf{t}}(\ell)$. Vectors $\hat{\mathbf{b}}_{1\mathbf{D}}(t)$, $\hat{\mathbf{b}}_{2\mathbf{D}}(t)$, and $\hat{\mathbf{b}}_{3\mathbf{D}}(t)$ can be used to compute the rotation matrix $\boldsymbol{R}_D^F(t)$, which is given by

$$\boldsymbol{R}_D^F = \begin{bmatrix} \dfrac{d}{(d^2+y_F^2+z_F^2)^{\frac{1}{2}}} & \dfrac{y_F}{(d^2+y_F^2)^{\frac{1}{2}}} & \dfrac{z_F d}{(d^2+y_F^2+z_F^2)^{\frac{1}{2}}(d^2+y_F^2)^{\frac{1}{2}}} \\ \dfrac{-y_F}{(d^2+y_F^2+z_F^2)^{\frac{1}{2}}} & \dfrac{d}{(d^2+y_F^2)^{\frac{1}{2}}} & \dfrac{-y_F z_F}{(d^2+y_F^2+z_F^2)^{\frac{1}{2}}(d^2+y_F^2)^{\frac{1}{2}}} \\ \dfrac{-z_F}{(d^2+y_F^2+z_F^2)^{\frac{1}{2}}} & 0 & \dfrac{(d^2+y_F^2)^{\frac{1}{2}}}{(d^2+y_F^2+z_F^2)^{\frac{1}{2}}} \end{bmatrix}.$$

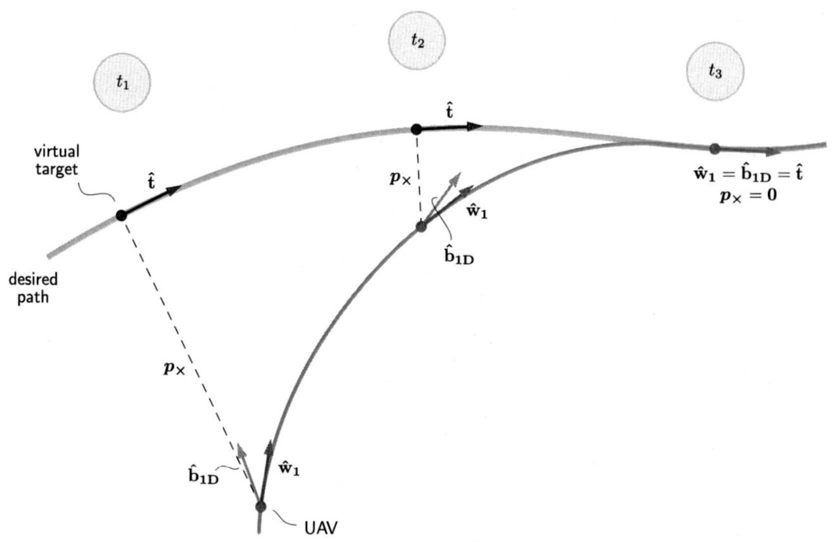

Figure 3.2 Shaping the approach to the path as a function of the cross-track error $\boldsymbol{p}_\times$. (For simplicity, the plot above assumes that the along-path error $x_F(t)$ is zero.)

Next, let $\tilde{\boldsymbol{R}}$ be the rotation matrix from $\{\mathcal{W}\}$ to $\{\mathcal{D}\}$, that is,

$$\tilde{\boldsymbol{R}} := \boldsymbol{R}_W^D = \boldsymbol{R}_F^D \, \boldsymbol{R}_W^F = (\boldsymbol{R}_D^F)^\top \, \boldsymbol{R}_W^F ,$$

and note that the velocity vector of the vehicle is aligned with the basis vector $\hat{\mathbf{b}}_{\mathbf{1D}}(t)$ if, and only if, the (1, 1) entry of $\tilde{\boldsymbol{R}}$, denoted here by $\tilde{R}_{11}$, is equal to one. Then, using notation similar to that in [8], we define the real-valued error function on $\mathsf{SO}(3)$

$$\Psi(\tilde{\boldsymbol{R}}) := \frac{1}{2}\mathrm{tr}\left[\left(\mathbb{I}_3 - \boldsymbol{\Pi}_R^\top \boldsymbol{\Pi}_R\right)\left(\mathbb{I}_3 - \tilde{\boldsymbol{R}}\right)\right], \tag{3.3}$$

where $\boldsymbol{\Pi}_R := \left[\begin{smallmatrix} 0 & 1 & 0 \\ 0 & 0 & 1 \end{smallmatrix}\right]$. The function $\Psi(\tilde{\boldsymbol{R}})$ in (3.3) can be expressed in terms of the entries of $\tilde{\boldsymbol{R}}$ as

$$\Psi(\tilde{\boldsymbol{R}}) = \frac{1}{2}\left(1 - \tilde{R}_{11}\right)$$

and, therefore, $\Psi(\tilde{\boldsymbol{R}})$ is a positive-definite function on a neighborhood of $\tilde{R}_{11} = 1$. Letting $(\cdot)^\wedge : \mathbb{R}^3 \to \mathfrak{so}(3)$ denote the *hat map* (see Appendix A, Section A.1), the attitude kinematics equation

$$\dot{\tilde{\boldsymbol{R}}} = \dot{\boldsymbol{R}}_W^D = \boldsymbol{R}_W^D \left(\{\boldsymbol{\omega}_{W/D}\}_W\right)^\wedge = \tilde{\boldsymbol{R}}\left(\{\boldsymbol{\omega}_{W/D}\}_W\right)^\wedge$$

can now be used to derive the time-derivative of $\Psi(\tilde{\boldsymbol{R}})$, given by

$$\dot{\Psi}(\tilde{\boldsymbol{R}}) = -\frac{1}{2}\mathrm{tr}\left[\left(\mathbb{I}_3 - \boldsymbol{\Pi}_R^\top \boldsymbol{\Pi}_R\right)\dot{\tilde{\boldsymbol{R}}}\right] = -\frac{1}{2}\mathrm{tr}\left[\left(\mathbb{I}_3 - \boldsymbol{\Pi}_R^\top \boldsymbol{\Pi}_R\right)\tilde{\boldsymbol{R}}\left(\{\boldsymbol{\omega}_{W/D}\}_W\right)^\wedge\right].$$

Property (A.1) of the hat map leads to

$$\dot{\Psi}(\tilde{\boldsymbol{R}}) = \frac{1}{2}\left(\left(\left(\mathbb{I}_3 - \boldsymbol{\Pi}_R^\top \boldsymbol{\Pi}_R\right)\tilde{\boldsymbol{R}} - \tilde{\boldsymbol{R}}^\top\left(\mathbb{I}_3 - \boldsymbol{\Pi}_R^\top \boldsymbol{\Pi}_R\right)\right)^\vee\right)^\top \{\boldsymbol{\omega}_{W/D}\}_W\,,$$

where $(\cdot)^\vee : \mathfrak{so}(3) \to \mathbb{R}^3$ denotes the *vee map*, which is defined as the inverse of the hat map. Moreover, since the first component of $\left((\mathbb{I}_3 - \boldsymbol{\Pi}_R^\top \boldsymbol{\Pi}_R)\tilde{\boldsymbol{R}} - \tilde{\boldsymbol{R}}^\top(\mathbb{I}_3 - \boldsymbol{\Pi}_R^\top \boldsymbol{\Pi}_R)\right)^\vee$ is equal to zero, we can also write

$$\begin{aligned}
\dot{\Psi}(\tilde{\boldsymbol{R}}) &= \frac{1}{2}\left(\left(\left(\mathbb{I}_3 - \boldsymbol{\Pi}_R^\top \boldsymbol{\Pi}_R\right)\tilde{\boldsymbol{R}} - \tilde{\boldsymbol{R}}^\top\left(\mathbb{I}_3 - \boldsymbol{\Pi}_R^\top \boldsymbol{\Pi}_R\right)\right)^\vee\right)^\top \boldsymbol{\Pi}_R^\top \boldsymbol{\Pi}_R \{\boldsymbol{\omega}_{W/D}\}_W \\
&= \left(\frac{1}{2}\boldsymbol{\Pi}_R\left(\left(\mathbb{I}_3 - \boldsymbol{\Pi}_R^\top \boldsymbol{\Pi}_R\right)\tilde{\boldsymbol{R}} - \tilde{\boldsymbol{R}}^\top\left(\mathbb{I}_3 - \boldsymbol{\Pi}_R^\top \boldsymbol{\Pi}_R\right)\right)^\vee\right)^\top \boldsymbol{\Pi}_R \{\boldsymbol{\omega}_{W/D}\}_W\,.
\end{aligned} \tag{3.4}$$

We now define the attitude error $\boldsymbol{e}_{\tilde{R}}(t)$ as

$$\boldsymbol{e}_{\tilde{R}} := \frac{1}{2}\boldsymbol{\Pi}_R\left(\left(\mathbb{I}_3 - \boldsymbol{\Pi}_R^\top \boldsymbol{\Pi}_R\right)\tilde{\boldsymbol{R}} - \tilde{\boldsymbol{R}}^\top\left(\mathbb{I}_3 - \boldsymbol{\Pi}_R^\top \boldsymbol{\Pi}_R\right)\right)^\vee,$$

which allows us to rewrite (3.4) in the more compact form

$$\dot{\Psi}(\tilde{\boldsymbol{R}}) = \boldsymbol{e}_{\tilde{R}} \cdot \left(\boldsymbol{\Pi}_R \{\boldsymbol{\omega}_{W/D}\}_W\right)\,.$$

We note that the attitude error $\boldsymbol{e}_{\tilde{R}}(t)$ can also be expressed in terms of the entries of $\tilde{\boldsymbol{R}}(t)$ as

$$\boldsymbol{e}_{\tilde{R}} = \frac{1}{2}\left[\ \tilde{R}_{13}\,, \quad -\tilde{R}_{12}\ \right]^\top$$

and, therefore, within the region where $\Psi(\tilde{\boldsymbol{R}}) < 1$, we have that if $\|\boldsymbol{e}_{\tilde{R}}\| = 0$, then $\Psi(\tilde{\boldsymbol{R}}) = 0$. Finally, noting that $\{\boldsymbol{\omega}_{W/D}\}_W$ can be expressed as

$$\begin{aligned}
\{\boldsymbol{\omega}_{W/D}\}_W &= \{\boldsymbol{\omega}_{W/I}\}_W + \{\boldsymbol{\omega}_{I/F}\}_W + \{\boldsymbol{\omega}_{F/D}\}_W \\
&= \{\boldsymbol{\omega}_{W/I}\}_W - \boldsymbol{R}_F^W \{\boldsymbol{\omega}_{F/I}\}_F - \boldsymbol{R}_D^W \{\boldsymbol{\omega}_{D/F}\}_D \\
&= \{\boldsymbol{\omega}_{W/I}\}_W - \tilde{\boldsymbol{R}}^\top\left(\boldsymbol{R}_F^D \{\boldsymbol{\omega}_{F/I}\}_F + \{\boldsymbol{\omega}_{D/F}\}_D\right),
\end{aligned}$$

we obtain

$$\dot{\Psi}(\tilde{\boldsymbol{R}}) = \boldsymbol{e}_{\tilde{R}} \cdot \left(\boldsymbol{\Pi}_R\left(\{\boldsymbol{\omega}_{W/I}\}_W - \tilde{\boldsymbol{R}}^\top\left(\boldsymbol{R}_F^D \{\boldsymbol{\omega}_{F/I}\}_F + \{\boldsymbol{\omega}_{D/F}\}_D\right)\right)\right),$$

or equivalently, by noting that $\boldsymbol{\Pi}_R \{\boldsymbol{\omega}_{W/I}\}_W = \left[\begin{smallmatrix} q \\ r \end{smallmatrix}\right]$,

$$\dot{\Psi}(\tilde{\boldsymbol{R}}) = \boldsymbol{e}_{\tilde{R}} \cdot \left(\begin{bmatrix} q \\ r \end{bmatrix} - \boldsymbol{\Pi}_R \tilde{\boldsymbol{R}}^\top\left(\boldsymbol{R}_F^D \{\boldsymbol{\omega}_{F/I}\}_F + \{\boldsymbol{\omega}_{D/F}\}_D\right)\right). \tag{3.5}$$

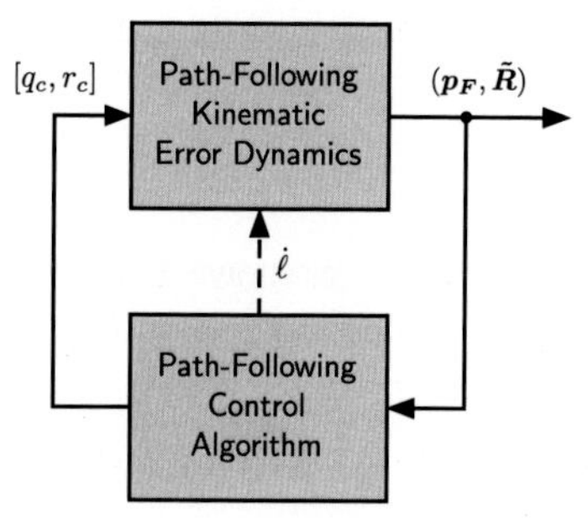

Figure 3.3 Path-following closed-loop system for a single vehicle solved at the kinematic level.

This equation describes the path-following kinematic attitude-error dynamics of frame $\{\mathcal{W}\}$ with respect to frame $\{\mathcal{D}\}$. The overall path-following kinematic error dynamics can now be obtained by combining (3.1) and (3.5), yielding

$$\begin{aligned} \dot{\boldsymbol{p}}_F]_F &= -\,\dot{\ell}\,\hat{\mathbf{t}} \,-\, \boldsymbol{\omega}_{F/I} \times \boldsymbol{p}_F \,+\, v\,\hat{\mathbf{w}}_1\,, \\ \dot{\Psi}(\tilde{\boldsymbol{R}}) &= \boldsymbol{e}_{\tilde{\boldsymbol{R}}} \cdot \left(\begin{bmatrix} q \\ r \end{bmatrix} - \boldsymbol{\Pi}_R \tilde{\boldsymbol{R}}^\top \left(\boldsymbol{R}_F^D \{\boldsymbol{\omega}_{F/I}\}_F + \{\boldsymbol{\omega}_{D/F}\}_D \right) \right) . \end{aligned} \tag{3.6}$$

In the kinematic error model (3.6), $q(t)$ and $r(t)$ play the role of control inputs, while the rate of progression $\dot{\ell}(t)$ of point P along the path becomes an additional variable that can be manipulated at will. At this point, we formally define the path-following generalized error vector $\boldsymbol{x}_{pf}(t)$ as

$$\boldsymbol{x}_{pf} := [\ \boldsymbol{p}_F^\top,\ \ \boldsymbol{e}_{\tilde{\boldsymbol{R}}}^\top\ ]^\top .$$

Notice that, within the region where $\Psi(\tilde{\boldsymbol{R}}) < 1$, if $\boldsymbol{x}_{pf} = \boldsymbol{0}$, then both the path-following position error and the path-following attitude error are equal to zero, that is, $\boldsymbol{p}_F = \boldsymbol{0}$ and $\Psi(\tilde{\boldsymbol{R}}) = 0$.

Using the above formulation, we now design feedback laws that solve the path-following control problem, as defined earlier in Chapter 2.

3.2 Path-Following Control Law

3.2.1 *Nonlinear Control Design at the Vehicle Kinematic Level*

At the kinematic level, the path-following problem can be solved by determining feedback control laws for $q(t)$, $r(t)$, and $\dot{\ell}(t)$ that ensure that the origin of the kinematic error dynamics (3.6) is exponentially stable with a given domain of attraction. Fig. 3.3 presents the kinematic closed-loop system considered in this section.

To solve the path-following problem, we first let the rate of progression of point P along the path be governed by

$$\dot{\ell} = \left(v\,\hat{\mathbf{w}}_1 + k_\ell \boldsymbol{p}_F\right) \cdot \hat{\mathbf{t}}\,, \tag{3.7}$$

where k_ℓ is a positive constant gain. Then, as shown later, the pitch- and yaw-rate commands, denoted respectively by $q_c(t)$ and $r_c(t)$, given by

$$\begin{bmatrix} q_c \\ r_c \end{bmatrix} := \boldsymbol{\Pi}_R \tilde{\boldsymbol{R}}^\top \left(\boldsymbol{R}_F^D \{\boldsymbol{\omega}_{F/I}\}_F + \{\boldsymbol{\omega}_{D/F}\}_D\right) \; - \; 2k_{\tilde{R}} \boldsymbol{e}_{\tilde{\boldsymbol{R}}}\,, \tag{3.8}$$

where $k_{\tilde{R}}$ is also a positive constant gain, drive the path-following generalized error vector $\boldsymbol{x}_{pf}(t)$ to zero with a guaranteed rate of convergence. A formal statement of this result is given in the lemma below.

Lemma 3.1. *Assume that the vehicle speed $v(t)$ verifies the bounds*

$$0 < v_{\min} \leq v(t) \leq v_{\max}\,, \qquad \textit{for all } t \geq 0\,. \tag{3.9}$$

If, for given positive constants $c < \frac{1}{\sqrt{2}}$ and c_1, the path-following control parameters k_ℓ, $k_{\tilde{R}}$, and d are chosen so that

$$k_{\tilde{R}}\, \tilde{k}_\ell > \frac{v_{\max}^2}{c_1^2 (1 - 2c^2)^2}\,, \tag{3.10}$$

where $\tilde{k}_\ell$ is defined as

$$\tilde{k}_\ell := \min\left\{k_\ell,\ \frac{v_{\min}}{(d^2 + c^2 c_1^2)^{\frac{1}{2}}}\right\}, \tag{3.11}$$

then the rate commands (3.8)*, together with the law* (3.7) *for the rate of progression of the virtual target along the path, ensure that the origin of the kinematic error equations* (3.6) *is exponentially stable with guaranteed rate of convergence*

$$\bar{\lambda}_{pf} := \frac{\tilde{k}_\ell + k_{\tilde{R}}(1 - c^2)}{2} - \frac{1}{2}\left(\left(\tilde{k}_\ell - k_{\tilde{R}}(1 - c^2)\right)^2 + \frac{4(1 - c^2)}{c_1^2(1 - 2c^2)^2}\, v_{\max}^2\right)^{\frac{1}{2}} \tag{3.12}$$

and domain of attraction

$$\Omega_{pf} := \left\{(\boldsymbol{p}_F, \tilde{\boldsymbol{R}}) \in \mathbb{R}^3 \times \mathsf{SO(3)} \mid \Psi(\tilde{\boldsymbol{R}}) + \frac{1}{c_1^2}\|\boldsymbol{p}_F\|^2 \leq c^2 < \frac{1}{2}\right\}. \tag{3.13}$$

◇

Proof. The proof of this result, which uses valuable insight from [8], is given in Appendix B.2.3. □

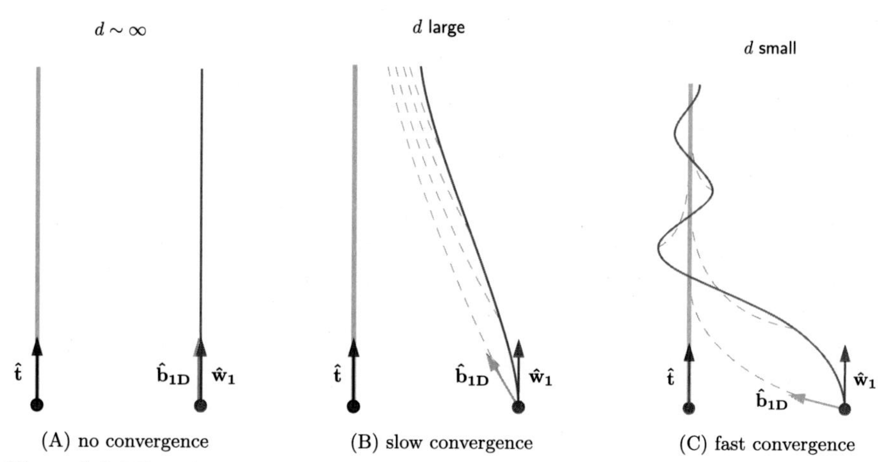

(A) no convergence (B) slow convergence (C) fast convergence

Figure 3.4 Effect of the characteristic distance d on the convergence of the vehicle to the path. Large values of d result in slow convergence to the path, whereas small values of d lead to fast path captures, with UAV trajectories that run nearly perpendicular to the path. Overly small values of d may yield oscillatory path-following behavior.

Remark 3.1. The choice of the characteristic distance d in the definition of the auxiliary frame $\{\mathcal{D}\}$ (see Eq. (3.2)) can be used to adjust the rate of convergence of the path-following error vector to zero. This is consistent with the fact that a large parameter d reduces the penalty for cross-track position errors, which results in a small rate of convergence of the vehicle to the path. Fig. 3.4 illustrates this point. When $d \sim \infty$, the vehicle never converges to the path, since $\boldsymbol{\omega}_{D/F} = \mathbf{0}$. For large values of d, the term $\boldsymbol{\omega}_{D/F}$ introduces only small corrections to the "feedforward" term $\boldsymbol{\omega}_{F/I}$, and therefore the convergence of the vehicle to the desired path is slow. On the other hand, small values of d allow for higher rates of convergence (subject to the design of the gains k_ℓ and $k_{\tilde{R}}$), which however might result in oscillatory path-following behavior. △

3.2.2 Stability Analysis with Non-ideal Inner-Loop Performance

The stabilizing control laws (3.7) and (3.8) lead to local exponential stability of the origin of the path-following kinematic error dynamics (3.6) with a prescribed domain of attraction. In general, this result does not hold when the dynamics of the vehicle are included in the problem formulation; see Fig. 3.5. In this section, we perform a stability analysis of the path-following closed-loop system for the case of non-ideal inner-loop tracking. In particular, we assume that the onboard autopilot ensures that the vehicle is able to track bounded pitch-rate and yaw-rate commands with guaranteed performance bounds and show that the path-following errors $\boldsymbol{p}_F(t)$ and $\boldsymbol{e}_{\tilde{R}}(t)$ are locally uniformly ultimately bounded with the same domain of attraction Ω_{pf}. The next lemma states this result formally.

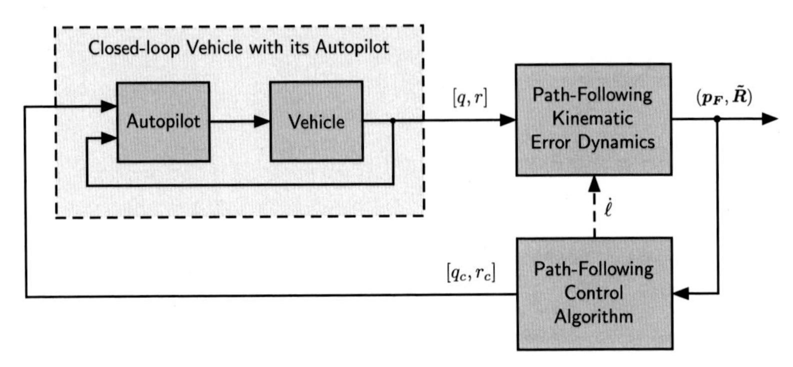

Figure 3.5 Path-following closed-loop system for a single vehicle.

Lemma 3.2. *Assume that the vehicle speed $v(t)$ verifies the bounds in* (3.9). *For given positive constants $c < \frac{1}{\sqrt{2}}$ and c_1, choose the path-following control parameters k_ℓ, $k_{\tilde{R}}$, and d according to the design constraint* (3.10). *Further, let $\lambda_{pf} := \bar{\lambda}_{pf}(1-\delta_\lambda)$, where $\bar{\lambda}_{pf}$ was defined in* (3.12) *and δ_λ is a positive constant verifying $0 < \delta_\lambda < 1$. If the performance bounds γ_q and γ_r in* (2.9) *satisfy the inequality*

$$\gamma_\omega := \left(\gamma_q^2 + \gamma_r^2\right)^{\frac{1}{2}} < \frac{2c}{(1-c^2)^{\frac{1}{2}}} \bar{\lambda}_{pf} \delta_\lambda , \tag{3.14}$$

then, for any initial state $(\boldsymbol{p}_F(0), \tilde{\boldsymbol{R}}(0)) \in \Omega_{pf}$, the rate commands (3.8), *together with the law* (3.7) *for the rate of progression of the virtual target along the path, ensure that there is a time $T_b \geq 0$ such that the path-following errors $\boldsymbol{p}_F(t)$ and $\boldsymbol{e}_{\tilde{R}}(t)$ satisfy*

$$\|\boldsymbol{e}_{\tilde{R}}(t)\|^2 + \frac{1}{c_1^2}\|\boldsymbol{p}_F(t)\|^2 \leq \left(\frac{1}{1-c^2}\|\boldsymbol{e}_{\tilde{R}}(0)\|^2 + \frac{1}{c_1^2}\|\boldsymbol{p}_F(0)\|^2\right) \mathrm{e}^{-2\lambda_{pf} t} , \quad \textit{for all } 0 \leq t < T_b , \tag{3.15a}$$

$$\|\boldsymbol{e}_{\tilde{R}}(t)\|^2 + \frac{1}{c_1^2}\|\boldsymbol{p}_F(t)\|^2 \leq \frac{(1-c^2)\gamma_\omega^2}{4\bar{\lambda}_{pf}^2 \delta_\lambda^2} , \quad \textit{for all } t \geq T_b . \tag{3.15b}$$

◇

Proof. The proof of this result is given in Appendix B.2.4. □

Remark 3.2. Inequalities (3.15) show that the path-following errors $\boldsymbol{p}_F(t)$ and $\boldsymbol{e}_{\tilde{R}}(t)$ are uniformly bounded for all $t \geq 0$ and uniformly ultimately bounded with ultimate bounds

$$\|\boldsymbol{e}_{\tilde{R}}(t)\| \leq \frac{(1-c^2)^{\frac{1}{2}}}{2\bar{\lambda}_{pf}\delta_\lambda}\gamma_\omega , \qquad \|\boldsymbol{p}_F(t)\| \leq \frac{c_1(1-c^2)^{\frac{1}{2}}}{2\bar{\lambda}_{pf}\delta_\lambda}\gamma_\omega , \qquad \text{for all } t \geq T_b .$$

These ultimate bounds are proportional to the inner-loop tracking performance bound γ_ω and, in the limiting ideal case of perfect inner-loop tracking, one recovers the exponential stability result derived in Lemma 3.1. △

Remark 3.3. An implicit assumption in the results and derivations above is that the presence of wind and gusts does not result in the UAV flying at zero or “negative” groundspeed. This assumption also holds throughout the remainder of this part of the book for all of the UAVs involved in the cooperative mission. In the case of strong winds that would violate this assumption, trajectory replanning with due account for wind conditions will be required. △

3.3 Implementation Details

In this section we discuss briefly some details about the practical implementation of the path-following control law proposed in the chapter:

- The derivations above assume that the desired path is parameterized by its arc length. However, this type of parameterization does not, in general, yield a closed-form expression in terms of elementary functions [3]. This implies that the actual implementation of the path-following control law —as described in the chapter— requires the use of a numerical approach to trace the path $\boldsymbol{p}_d(\ell)$ and determine its relevant features, such as the tangent vector $\hat{\mathbf{t}}(\ell)$ and the normal development $(k_1(\ell), k_2(\ell))$.
  If the path to be followed admits a regular analytical representation in terms of a known parameterizing variable, then it is possible to define an alternative formulation of the path-following control problem that does not rely on a numerical procedure to trace the path. This can be achieved by simply defining the position of the virtual target vehicle along the path in terms of this new parameterizing variable. In this alternative setup, and letting τ denote the new free variable that defines the position of the virtual target, the control law (3.7) must be replaced by the expression
  $$\dot{\tau} = \frac{\left[v\,\hat{\mathbf{w}}_1 + k_\ell(\boldsymbol{p} - \boldsymbol{p}_d(\tau))\right] \cdot \hat{\mathbf{t}}(\tau)}{\|\boldsymbol{p}'_d(\tau)\|},$$
  where $\boldsymbol{p}'_d(\tau)$ denotes here the partial derivative of the path with respect to the new parameterizing variable, evaluated at τ.
- The strategy for path-following control requires the definition of a parallel transport frame $\{\mathcal{F}\}$ attached to the virtual target on the path. This frame need not be computed online as the mission unfolds; instead, the path can be framed before the actual execution of the mission. Path framing can, in fact, be implemented as a post-processing routine of trajectory generation. Algorithms for computation of parallel transport frames on spatial 3D curves are discussed in [4]. Similarly, the normal development of the path, required to determine $\{\boldsymbol{\omega}_{F/I}\}_F$, can also be pre-computed as part of the trajectory-generation algorithm.

- The implementation of the control law (3.7) and the rate commands (3.8) also requires the estimation of the inertial velocity vector of the UAV, which is necessary to define frame $\{\mathcal{W}\}$ and compute the rotation matrix $\tilde{\boldsymbol{R}}(t)$ as well as the attitude error $\boldsymbol{e}_{\tilde{R}}(t)$. The inertial velocity vector can be estimated from inertial sensors and GPS measurements, generally available on UAVs.
- We also note that the control law (3.8) produces angular-rate commands defined in the $\{\mathcal{W}\}$ frame. However, typical commercial off-the-self autopilots accept rate commands defined in the body-fixed frame. These two frames can differ significantly, especially for the case of small UAVs operating in high winds, which implies that the rate commands (3.8) need to be transformed to the body-fixed frame before they are sent to the onboard autopilot.
- Finally, since the path-following control law (3.7)-(3.8) is only guaranteed to work locally (see Lemmas 3.1 and 3.2), a secondary guidance loop must be implemented that complements the angular-rate command (3.8) and, as soon as external disturbances push the vehicle outside the domain of attraction Ω_{pf} defined in (3.13), drives the vehicle's position and velocity vector back inside this domain.

3.4 Simulation Example: Shaping the Approach to the Path

This section presents simulation results that illustrate the performance of the path-following algorithm proposed in this chapter. The simulations are based on the following kinematic model of the UAV:

$$\begin{aligned} \dot{\boldsymbol{p}}]_I &= v\,\hat{\mathbf{w}}_1\,, \\ \dot{\boldsymbol{R}}_W^I &= \boldsymbol{R}_W^I\left(\{\boldsymbol{\omega}_{W/I}\}_W\right)^\wedge\,, \end{aligned} \tag{3.16}$$

along with a simplified, decoupled linear model describing the roll, pitch, yaw, and speed dynamics of the closed-loop UAV with its autopilot. In particular, the linear model used here corresponds to an identified second-order model of the SIG Rascal 110 research aircraft operated by the Naval Postgraduate School; see Chapter 8.

Fig. 3.6 presents the desired path with the parallel transport frame. The path consists of a left turn followed by a right turn, climbing steadily from 200 m to 400 m. The total length of the path is 2998.5 m, and its beginning is indicated with a circle. The figure also shows the normal development of the path (parameters k_1 and k_2) as well as the flight path angle along the path. For speeds within the operating range of the small UAVs considered here (ranging from 18 m/s to 32 m/s), the angular rates required to follow the path are well within feasible bounds.

Simulation results for this scenario are presented next. The path-following controller gains are selected as follows:

$$k_\ell = 0.20\,[1/\mathrm{s}]\,, \qquad k_{\tilde{R}} = 0.50\,[1/\mathrm{s}]\,, \qquad d = 125\,[\mathrm{m}]\,,$$

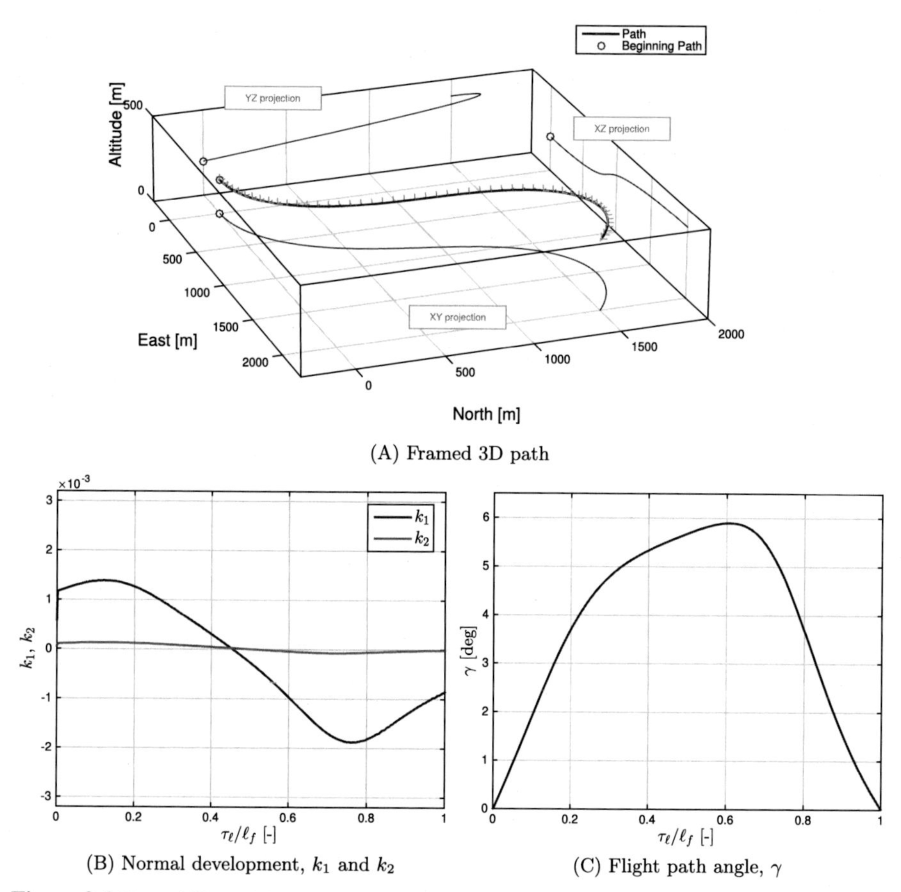

Figure 3.6 Framed 3D spatial path and main geometric properties.

while the speed command is set to $v_c = 20$ m/s. The angular-rate commands are saturated to ± 0.3 rad/s. Fig. 3.7 illustrates the evolution of the UAV as well as the virtual target moving along the path. This figure also includes the $\{\mathcal{W}\}$ frame attached to the UAV as well as the $\{\mathcal{F}\}$ frame attached to the virtual target. Both the initial position and attitude of the UAV present an initial offset with respect to the beginning of the desired framed path. As can be seen in the figure, the path-following control algorithm eliminates this initial offset and steers the UAV along the path. Details about the performance of the path-following algorithm are shown in Fig. 3.8; the path-following attitude and position errors, $\Psi(\tilde{\boldsymbol{R}}(t))$ and $\boldsymbol{p}_F(t)$, converge to a neighborhood of zero within 10 s and 40 s, respectively. The figure also presents the angular-rate commands, $q_c(t)$ and $r_c(t)$, the actual angular rates, $q(t)$ and $r(t)$, as well as the rate of progression, $\dot{\ell}(t)$, of the virtual target along the path.

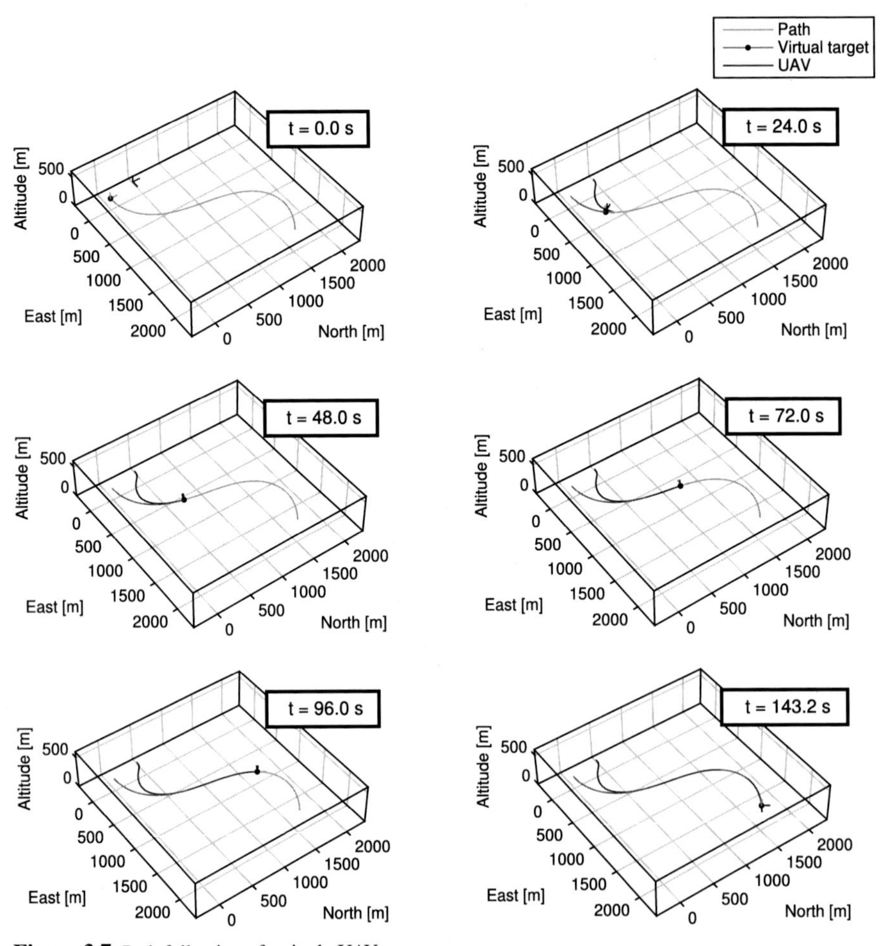

Figure 3.7 Path following of a single UAV.

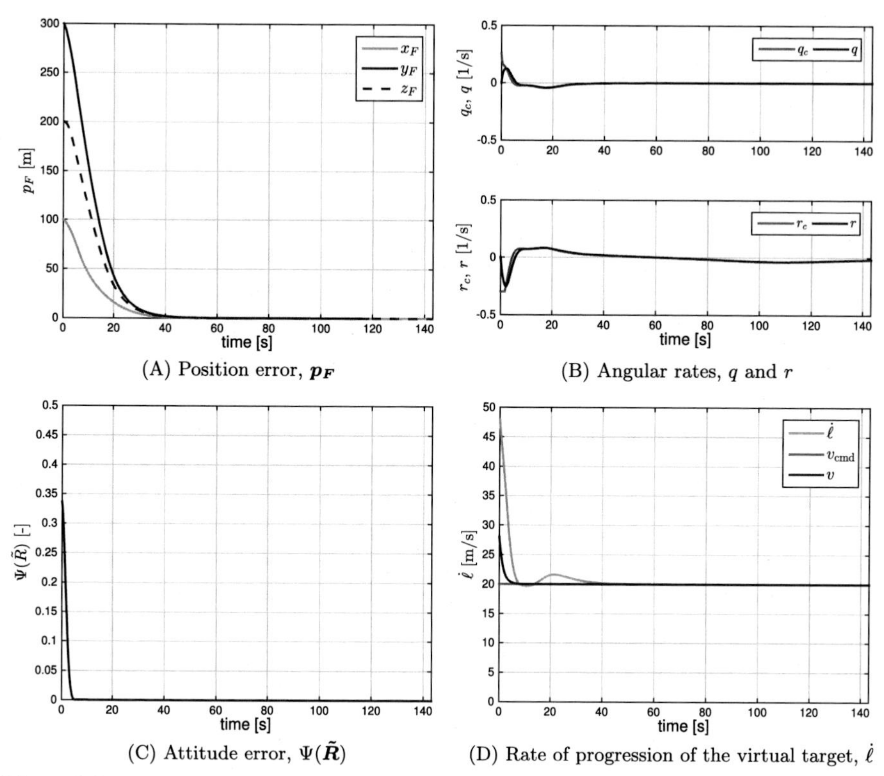

(A) Position error, $\boldsymbol{p_F}$

(B) Angular rates, q and r

(C) Attitude error, $\Psi(\tilde{\boldsymbol{R}})$

(D) Rate of progression of the virtual target, $\dot{\ell}$

Figure 3.8 Path-following performance related to Fig. 3.7.

To illustrate the effect of the characteristic distance d on the convergence of the UAV to the path, Figs. 3.9-3.10 and 3.11-3.12 show simulation results for the same scenario as Figs. 3.7-3.8, but now with $d = 50$ m and $d = 250$ m, respectively. As expected, a smaller characteristic distance leads to faster convergence to the path; in fact, as can be seen in Fig. 3.10, the cross-track error components $y_F(t)$ and $z_F(t)$ converge now to a neighborhood of zero in about 30 s, which also requires more aggressive angular-rate commands in order to align the UAV's velocity vector with the tangent vector to the path. Instead, a larger characteristic distance yields slower convergence to the path; Fig. 3.12 shows that the path-following cross-track error converges to a neighborhood of zero in about 60 s, with a much gentler angular-rate response. The ability of the characteristic distance d to shape the approach to the path can also be illustrated through the time-history of the projection of vector $\hat{\mathbf{w}}_1(t)$ onto the tangent vector to the path $\hat{\mathbf{t}}(t)$; see Fig. 3.13. Clearly, as the characteristic distance decreases, path capture becomes more aggressive, yielding UAV trajectories that run nearly perpendicular to the desired path.

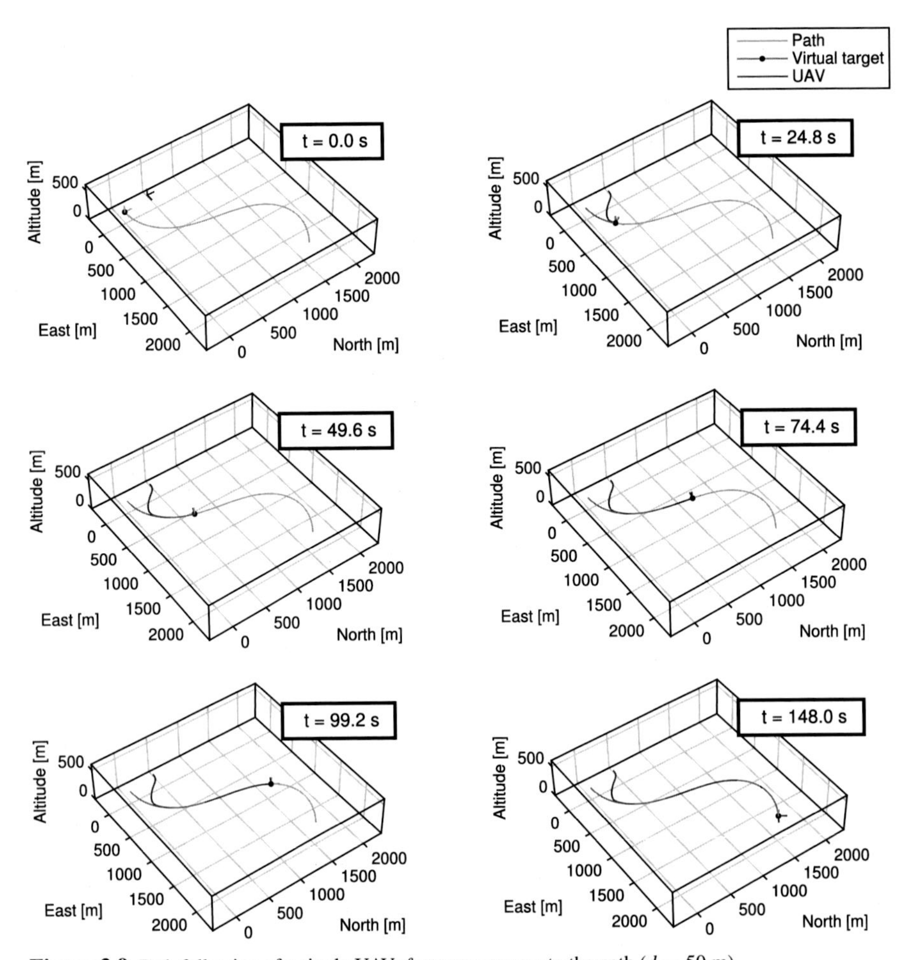

Figure 3.9 Path following of a single UAV; *fast* convergence to the path ($d = 50$ m).

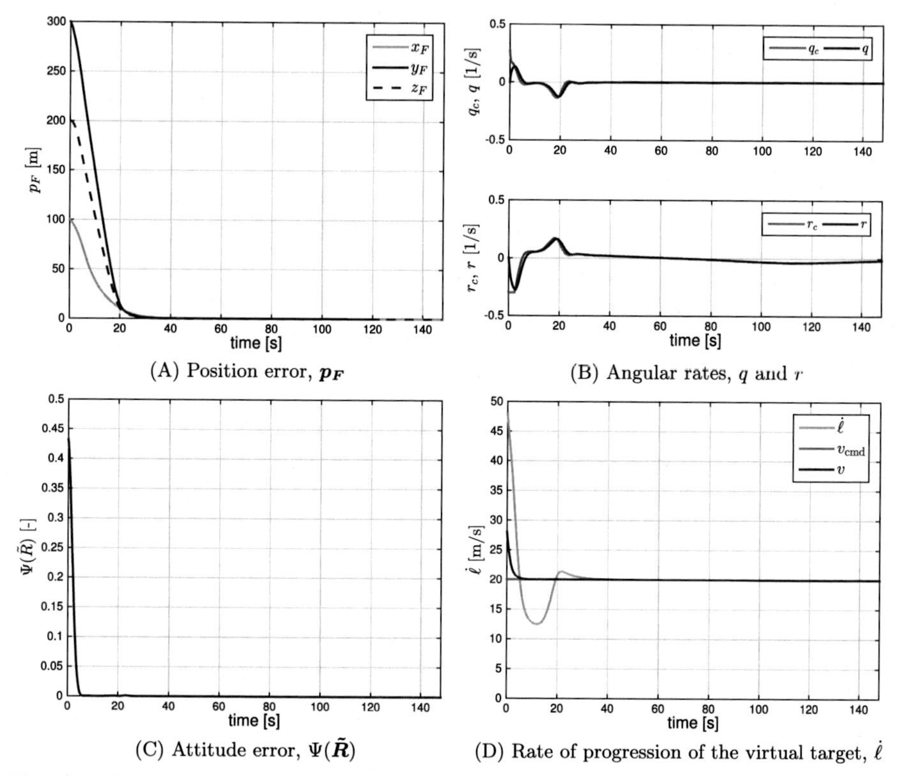

Figure 3.10 Path-following performance related to Fig. 3.9.

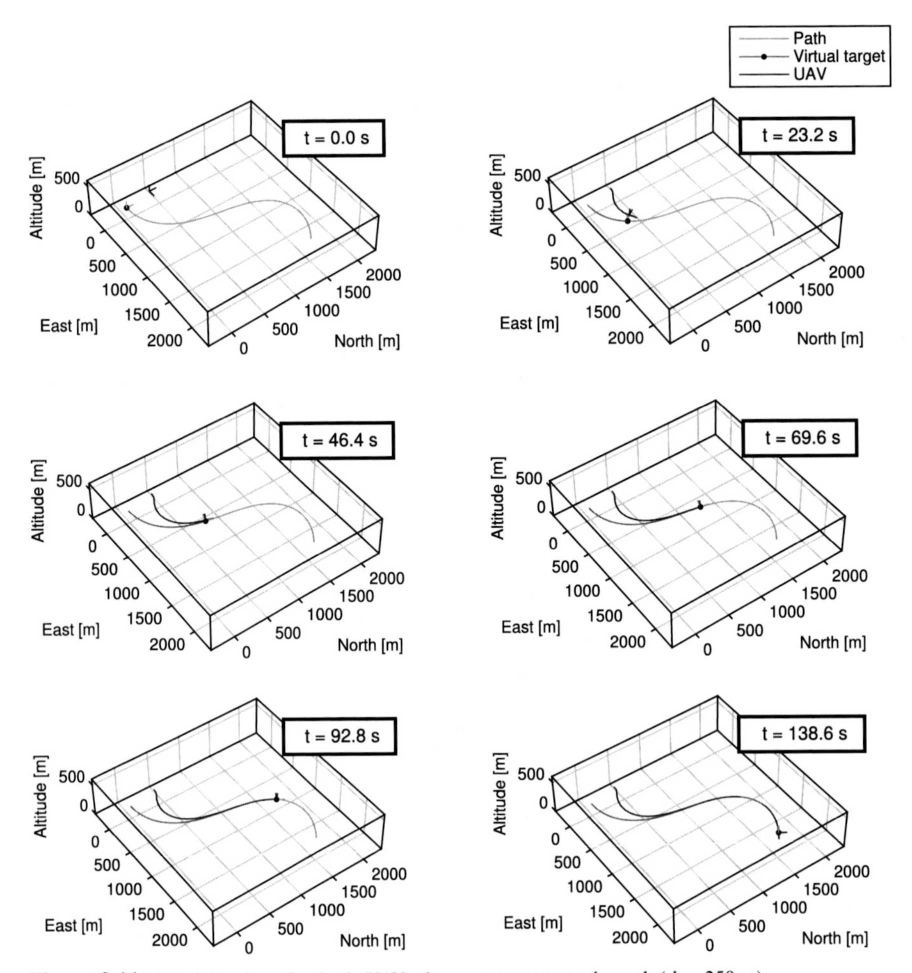

Figure 3.11 Path following of a single UAV; *slow* convergence to the path ($d = 250$ m).

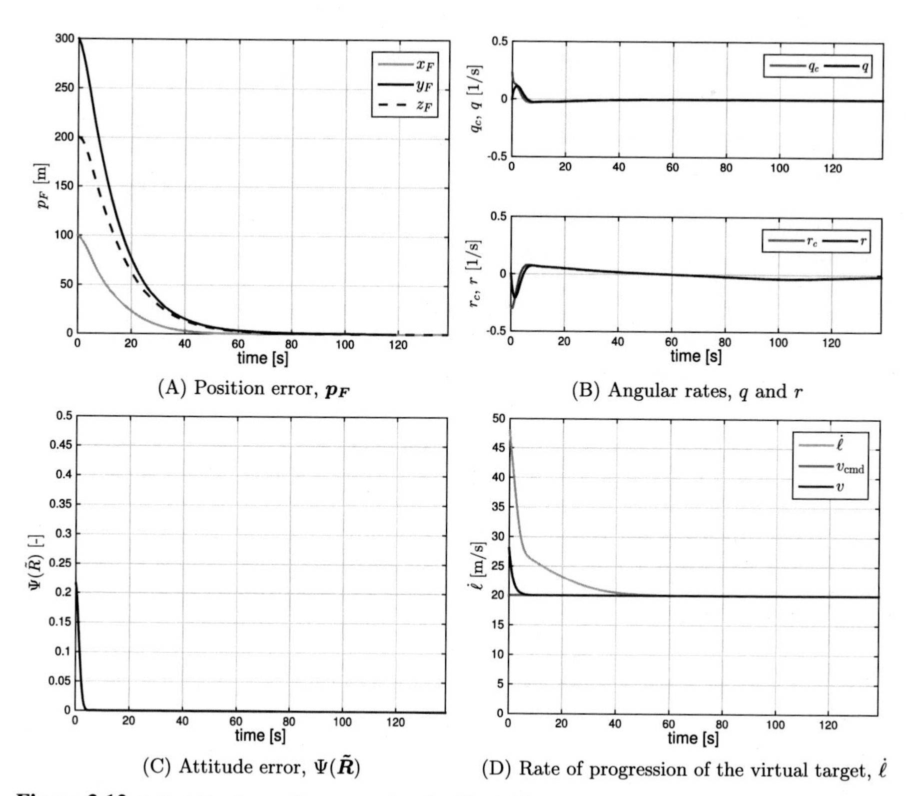

Figure 3.12 Path-following performance related to Fig. 3.11.

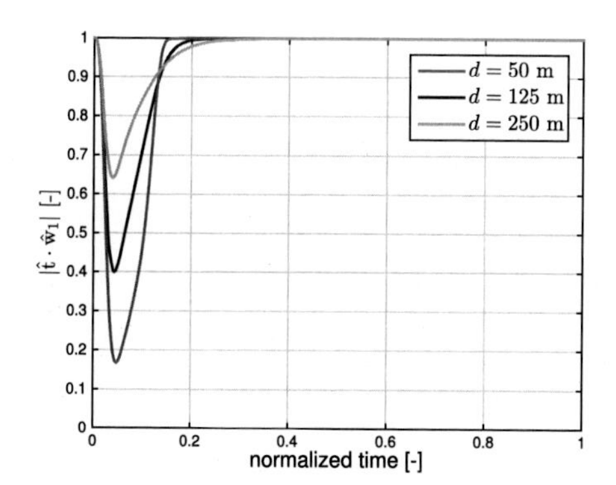

Figure 3.13 Projection of vector $\hat{\mathbf{w}}_1$ onto the tangent vector to the path $\hat{\mathbf{t}}$.

References

[1] R.L. Bishop, There is more than one way to frame a curve, The American Mathematical Monthly 82 (1975) 246–251.
[2] V. Cichella, E. Xargay, V. Dobrokhodov, I. Kaminer, A.M. Pascoal, N. Hovakimyan, Geometric 3D path-following control for a fixed-wing UAV on $\mathsf{SO}(3)$, in: AIAA Guidance, Navigation and Control Conference, Portland, OR, August 2011, AIAA 2011-6415.
[3] R.T. Farouki, Pythagorean-Hodograph Curves: Algebra and Geometry Inseparable, Springer-Verlag Berlin Heidelberg, 2008.
[4] A.J. Hanson, H. Ma, Parallel Transport Approach to Curve Framing, Tech. Rep. TR425, Computer Science Department, Indiana University, 1995.
[5] I. Kaminer, A.M. Pascoal, E. Xargay, N. Hovakimyan, C. Cao, V. Dobrokhodov, Path following for unmanned aerial vehicles using $\mathcal{L}_1$ adaptive augmentation of commercial autopilots, Journal of Guidance, Control and Dynamics 33 (2010) 550–564.
[6] I. Kaminer, O.A. Yakimenko, A.M. Pascoal, R. Ghabcheloo, Path generation, path following and coordinated control for time-critical missions of multiple UAVs, in: American Control Conference, Minneapolis, MN, June 2006, pp. 4906–4913.
[7] L. Lapierre, D. Soetanto, A.M. Pascoal, Non-singular path-following control of a unicycle in the presence of parametric modeling uncertainties, International Journal of Robust and Nonlinear Control 16 (2006) 485–503.
[8] T. Lee, M. Leok, N.H. McClamroch, Control of complex maneuvers for a quadrotor UAV using geometric methods on $\mathsf{SE}(3)$. Available online arXiv:1003.2005v4, 2011.
[9] A. Micaelli, C. Samson, Trajectory Tracking for Unicycle-Type and Two-Steering-Wheels Mobile Robot, Tech. Rep. 2097, INRIA, Sophia-Antipolis, France, November 1993.
[10] E. Xargay, I. Kaminer, A. Pascoal, N. Hovakimyan, V. Dobrokhodov, V. Cichella, A.P. Aguiar, R. Ghabcheloo, Time-critical cooperative path following of multiple unmanned aerial vehicles over time-varying networks, Journal of Guidance, Control and Dynamics 36 (2013) 499–516.

Time Coordination of Fixed-Wing Air Vehicles

The previous chapter offered a solution to the path-following problem for a single fixed-wing UAV and an arbitrary feasible speed profile by using a control strategy in which the vehicle's attitude control effectors are used to make the vehicle follow a virtual target running along the path. We now address the problem of time-critical cooperative control of multiple vehicles for missions with strict relative temporal constraints. To this end, following the approach in [3], the speed of the vehicles is adjusted based on coordination information exchanged among them over a supporting communications network. In particular, the outer-loop coordination control law derived in this chapter is intended to provide a correction to the desired speed profile $v_{d,i}(\cdot)$ obtained in the trajectory-generation step described in Chapter 2, and to generate a total speed command $v_{c,i}(t)$. This speed command is then to be tracked by the ith vehicle to achieve coordination in time.

As mentioned above, the results in this chapter are an extension of our work in [3], which in its turn was rooted in the results presented in [4]. In particular, the work in [3] proposed a proportional-integral coordination control law that is suitable for trajectories with constant speed profiles, and derived stability results for networks with topologies connected pointwise in time. A preliminary stability analysis for the case of time-varying communication topologies with integral connectivity was presented in [1]. Related work can be found in [2], where stability results were derived for a proportional coordination control law in the presence of switching communication topologies and network link latencies; in particular, this work considered the case of networks with brief connectivity losses and also networks that are uniformly connected in mean. The control law in [3] was later generalized in [8] to the case of path-dependent speed profiles. In [8], we also provided a more elegant stability analysis for the case of switching communication topologies with integral connectivity. The coordination control law proposed in this chapter modifies the control law in [8] by allowing the use of multiple fleet leaders, but the ensuing stability analysis follows closely the methodology described in that reference. The results in this chapter were previously reported in [5]. An overview of complementary work on time-critical coordination and cooperative path-following control is provided in Section 1.3.2 of Chapter 1.

4.1 Coordination States

As described in Chapter 1, in the framework for cooperative control adopted in this book, the vehicles execute time-coordinated path-following maneuvers. Compared to "open-loop" approaches, such as having each vehicle independently track its assigned trajectory, the proposed cooperative solution leads to improved robustness against off-

Time-Critical Cooperative Control of Autonomous Air Vehicles, DOI: 10.1016/B978-0-12-809946-9.00006-3

nominal situations. Furthermore, because in this chapter we deal with missions that impose only relative temporal constraints, it is not required that the fleet of vehicles observe the absolute temporal assignments of the generated trajectories. In this context, as indicated earlier in Chapter 2, the coordination problem is cast in the form of a consensus problem. In line with this approach, we now take advantage of the formulation of the path-following control problem discussed in the previous chapter to define appropriate coordination variables and coordination maps for vehicle cooperation.

We start by defining $\ell'_{d,i}(t_d)$ as the desired normalized curvilinear abscissa of the ith vehicle along its corresponding path at the desired mission time t_d, given by

$$\ell'_{d,i}(t_d) := \frac{1}{\ell_{fi}} \int_0^{t_d} v_{d,i}(\tau)\, \mathrm{d}\tau \,, \qquad i = 1, \ldots, n \,,$$

with ℓ_{fi} and $v_{d,i}(\cdot)$ being, respectively, the length of the path and the desired speed profile corresponding to the ith vehicle. The trajectory-generation algorithm ensures that the desired speed profiles $v_{d,i}(\cdot)$ satisfy feasibility conditions, which implies that the following bounds hold for all vehicles:

$$\begin{aligned} 0 < v_{\min} \leq v_{d\,\min} \leq v_{d,i}(t_d) \leq v_{d\,\max} \leq v_{\max} \,, \\ \text{for all } t_d \in [0, t_d^{\mathrm{f}}] \,, \quad \text{and all } i \in \{1, \ldots, n\} \,, \end{aligned} \tag{4.1}$$

where $v_{\min}$ and $v_{\max}$ denote, respectively, the minimum and maximum operating speeds of the vehicle, while $v_{d\,\min}$ and $v_{d\,\max}$ denote lower and upper bounds on the desired speed profiles of the vehicles; see Section 2.2.1. From the definition of $\ell'_{d,i}(t_d)$ and the bounds in (4.1), it follows that $\ell'_{d,i}(t_d)$ is a strictly increasing continuous function of t_d mapping $[0, t_d^{\mathrm{f}}]$ onto $[0, 1]$, and satisfying $\ell'_{d,i}(0) = 0$ and $\ell'_{d,i}(t_d^{\mathrm{f}}) = 1$. We also define $\eta_i : [0, 1] \to [0, t_d^{\mathrm{f}}]$ to be the inverse function of $\ell'_{d,i}(t_d)$. Clearly, $\eta_i(\cdot)$ is also a strictly increasing continuous function of its argument. Then, letting $\ell'_i(t)$ be the normalized curvilinear abscissa at time t of the ith virtual target vehicle running along its path, defined as

$$\ell'_i := \frac{\ell_i}{\ell_{fi}} \,, \qquad i = 1, \ldots, n \,,$$

we define the time-variables

$$\xi_i := \eta_i(\ell'_i) \,, \qquad i = 1, \ldots, n \,. \tag{4.2}$$

From this definition, it follows that $\xi_i(t) \in [0, t_d^{\mathrm{f}}]$, and therefore this variable characterizes the status of the mission for the ith vehicle at time t in terms of the desired mission time t_d.

We note that, for any two vehicles i and j, if $\xi_i(t) = \xi_j(t) = t'_d$ at a given time t, then $\ell'_i(t) = \ell'_{d,i}(t'_d)$ and $\ell'_j(t) = \ell'_{d,j}(t'_d)$, which implies that at time t the target vehicles corresponding to vehicles i and j have the desired relative position along their paths at the desired mission time t'_d. Clearly, if $\xi_i(t) = \xi_j(t)$ for all $t \geq 0$, then the ith and jth virtual target vehicles maintain desired relative position along their paths at all times and, therefore, these two target vehicles satisfy the relative temporal constraints as provided by the trajectory-generation algorithm. Also, in the case of temporal deconfliction (see Section 2.2.1), if $\xi_i(t) = \xi_j(t)$ for all $t \geq 0$, then the solution to the trajectory-generation problem ensures that the virtual targets i and j are not at the same place at the same time during the entire duration of the mission. Moreover, it can be shown that if the ith coordination state evolves at rate 1 at a given time t, that is $\dot{\xi}_i(t) = 1$, then the ith virtual target travels at the desired speed, $\dot{\ell}_i(t) = v_{d,i}(\xi_i(t))$ (see Appendix B.2.1). Hence, these variables comply with the requirements detailed in Section 2.2.3, Chapter 2, and are therefore suitable to be used as coordination states.

Remark 4.1. If the desired speed profiles $v_{d,i}(t_d)$ obtained from the trajectory-generation algorithm are constant along the corresponding paths, that is,

$$v_{d,i}(t_d) = v_{d,i}\,, \qquad \text{for all } t_d \in [0, t_d^{\mathrm{f}}]\,, \quad \text{and all } i \in \{1, \ldots, n\}\,,$$

then the following equalities hold:

$$\ell'_{d,i}(t_d) = \ell'_{d,j}(t_d)\,, \qquad \text{for all } t_d \in [0, t_d^{\mathrm{f}}]\,, \text{ and all } i, j \in \{1, \ldots, n\},$$

$$\frac{v_{d,i}}{\ell_{fi}} = \frac{\mathrm{d}\ell'_{d,i}(t_d)}{\mathrm{d}t_d} = \frac{\mathrm{d}\ell'_{d,j}(t_d)}{\mathrm{d}t_d} = \frac{v_{d,j}}{\ell_{fj}}\,, \quad \text{for all } t_d \in [0, t_d^{\mathrm{f}}]\,, \text{ and all } i, j \in \{1, \ldots, n\}.$$

This implies that, in the case of constant desired speed profiles, the normalized curvilinear abscissas $\ell'_i(t)$ can be equivalently used as coordination states, while $(\ell'_i(t) - \ell'_j(t))$ and $(\dot{\ell}'_i(t) - \frac{v_{d,i}}{\ell_{fi}})$ can be used to characterize the coordination errors. This is, in fact, the setup for vehicle coordination adopted in [3]. △

Remark 4.2. In mission scenarios in which the various vehicles are required to follow their respective paths separated by pre-specified time-intervals, the coordination states in (4.2) need to be defined differently so as to capture the desired inter-vehicle schedule. This would be the case, for example, in a sequential auto-landing scenario, in which several UAVs are required to follow a common glide slope with a given speed profile and separated by pre-specified safe-guarding time-intervals. Later in this chapter, we will illustrate through an example how the coordination states can be defined for this type of time-critical missions. △

Remark 4.3. The formulation of the time-critical cooperative path-following problem described above assumes only relative temporal constraints in the execution of a given mission. Absolute temporal constraints, such as specifications on the desired final time of the mission, are not considered. As we will show in Chapter 5, such constraints can be easily incorporated in the problem formulation, and enforced by judiciously modifying the coordination control laws presented later in this chapter. △

4.2 Coordination Control Law

4.2.1 Speed Control at the Vehicle Kinematic Level

To solve the coordination problem, we first note that the evolution of the ith coordination state is given by (see Appendix B.2.1)

$$\dot{\xi}_i = \frac{\dot{\ell}_i}{v_{d,i}(\xi_i)}\,.$$

Next, we recall from the solution to the path-following problem in Chapter 3 that the evolution of the ith virtual target vehicle along the path is given by

$$\dot{\ell}_i = \left(v_i\,\hat{\mathbf{w}}_{1,i} + k_\ell\,\boldsymbol{p}_{F,i}\right)\cdot\hat{\mathbf{t}}_i\,,$$

where for simplicity we keep k_ℓ without indexing. The dynamics of the ith coordination state can thus be rewritten as

$$\dot{\xi}_i = \frac{\left(v_i\,\hat{\mathbf{w}}_{1,i} + k_\ell\,\boldsymbol{p}_{F,i}\right)\cdot\hat{\mathbf{t}}_i}{v_{d,i}(\xi_i)}\,.$$

At this point, it is important to note that if the path-following control law can guarantee that, for every vehicle, the quantity $(\hat{\mathbf{w}}_{1,i}\cdot\hat{\mathbf{t}}_i)$ is positive and bounded away from zero for all $t \geq 0$, that is,

$$\hat{\mathbf{w}}_{1,i}\cdot\hat{\mathbf{t}}_i \geq c_2 > 0\,, \qquad \text{for all } t \geq 0\,, \quad \text{and all } i \in \{1,\ldots,n\}\,, \tag{4.3}$$

where c_2 is any constant satisfying $0 < c_2 \leq 1$, then we can use dynamic inversion and define the speed command for the ith vehicle as

$$v_{c,i} := \frac{u_{\text{coord},i}\,v_{d,i}(\xi_i) - k_\ell\,\boldsymbol{p}_{F,i}\cdot\hat{\mathbf{t}}_i}{\hat{\mathbf{w}}_{1,i}\cdot\hat{\mathbf{t}}_i}\,, \tag{4.4}$$

where $u_{\text{coord},i}(t)$ is a coordination control law to be defined later. At the kinematic level, this speed command leads to the following dynamics for the ith coordination state:

$$\dot{\xi}_i = u_{\text{coord},i}\,. \tag{4.5}$$

In the remainder of this section we assume that the bound in (4.3) holds for every vehicle and derive a coordination control law $u_{\text{coord},i}(t)$ that achieves coordination for the entire fleet of vehicles. This assumption will be verified later in Section 4.3, where we prove stability of the combined time-critical cooperative path-following closed-loop system and derive an expression for the constant c_2.

Recall now that each vehicle is only allowed to exchange its coordination parameter $\xi_i(t)$ with its neighbors $\mathcal{N}_i(t)$, which are defined by the (possibly time-varying) inter-vehicle communication topology; see Section 2.2.3. To satisfy this constraint, we propose the distributed coordination law

$$u_{\text{coord},i}(t) = -k_P \sum_{j \in \mathcal{N}_i(t)} (\xi_i(t) - \xi_j(t)) + 1\,, \qquad i = 1, \ldots, n_\ell\,,$$

$$\left.\begin{aligned} u_{\text{coord},i}(t) &= -k_P \sum_{j \in \mathcal{N}_i(t)} (\xi_i(t) - \xi_j(t)) + \chi_{I,i}(t) \\ \dot{\chi}_{I,i}(t) &= -k_I \sum_{j \in \mathcal{N}_i(t)} (\xi_i(t) - \xi_j(t))\,, \qquad \chi_{I,i}(0) = 1 \end{aligned}\right., \qquad i = n_\ell + 1, \ldots, n\,, \tag{4.6}$$

where vehicles 1 through n_ℓ, $n_\ell \leq n$, are elected as fleet leaders (they can be *virtual leaders*) and k_P and k_I are positive coordination control gains. Note that the coordination control law for the follower vehicles has a proportional-integral structure, which provides disturbance rejection capabilities at the coordination level [7]. We also note that the leaders adjust their dynamics according to information exchanged with their neighboring vehicles, rather than running as isolated agents. Finally, we notice that the presence of multiple leaders can improve the robustness of the cooperative control architecture to a single-point vehicle failure.

The coordination law (4.6) can be rewritten in compact form as

$$\begin{aligned} \boldsymbol{u}_{\textbf{coord}} &= -k_P \boldsymbol{L}(t)\,\boldsymbol{\xi} + \begin{bmatrix} \mathbf{1}_{n_\ell} \\ \boldsymbol{\chi}_I \end{bmatrix}, \\ \dot{\boldsymbol{\chi}}_I &= -k_I\, \boldsymbol{C}^\top \boldsymbol{L}(t)\,\boldsymbol{\xi}\,, \qquad \boldsymbol{\chi}_I(0) = \mathbf{1}_{n-n_\ell}\,, \end{aligned}$$

where $\boldsymbol{u}_{\textbf{coord}}(t)$, $\boldsymbol{\xi}(t)$, and $\boldsymbol{\chi}_I(t)$ are defined as

$$\begin{aligned} \boldsymbol{u}_{\textbf{coord}}(t) &:= [u_{\text{coord},1}(t), \ldots, u_{\text{coord},n}(t)]^\top && \in \mathbb{R}^n\,, \\ \boldsymbol{\xi}(t) &:= [\xi_1(t), \ldots, \xi_n(t)]^\top && \in \mathbb{R}^n\,, \\ \boldsymbol{\chi}_I(t) &:= [\chi_{I,n_\ell+1}(t), \ldots, \chi_{I,n}(t)]^\top && \in \mathbb{R}^{n-n_\ell}\,, \end{aligned}$$

$\boldsymbol{C}^\top := [\,\mathbf{0}\quad \mathbb{I}_{n-n_\ell}\,] \in \mathbb{R}^{(n-n_\ell)\times n}$, and $\boldsymbol{L}(t)$ is the Laplacian of the undirected graph $\Gamma(t)$ that captures the underlying bidirectional communications network topology of the fleet at time t. It is well known that the Laplacian of an undirected graph is symmetric, $\boldsymbol{L}^\top(t) = \boldsymbol{L}(t)$, and positive semi-definite, $\boldsymbol{L}(t) \geq 0$, $\lambda_1(\boldsymbol{L}(t)) = 0$ is an eigenvalue with eigenvector $\mathbf{1}_n$, $\boldsymbol{L}(t)\mathbf{1}_n = \mathbf{0}$, and the second smallest eigenvalue of $\boldsymbol{L}(t)$ is positive if and only if the graph $\Gamma(t)$ is connected.

At the kinematic level, the coordination control law (4.6) yields the closed-loop coordination system

$$\begin{aligned} \dot{\boldsymbol{\xi}} &= -k_P \boldsymbol{L}(t)\,\boldsymbol{\xi} + \begin{bmatrix} \mathbf{1}_{n_\ell} \\ \boldsymbol{\chi}_I \end{bmatrix}, && \boldsymbol{\xi}(0) = \boldsymbol{\xi}_0\,, \\ \dot{\boldsymbol{\chi}}_I &= -k_I\, \boldsymbol{C}^\top \boldsymbol{L}(t)\,\boldsymbol{\xi}\,, && \boldsymbol{\chi}_I(0) = \mathbf{1}_{n-n_\ell}\,. \end{aligned} \tag{4.7}$$

To analyze the dynamics of this feedback system, we reformulate the coordination problem stated above as a stabilization problem for a related one. To this end, we define the *coordination projection matrix* $\boldsymbol{\Pi}_{\xi}$ as

$$\boldsymbol{\Pi}_{\xi} := \mathbb{I}_n - \tfrac{1}{n}\mathbf{1}_n\mathbf{1}_n^\top ,$$

and we note that $\boldsymbol{\Pi}_{\xi} = \boldsymbol{\Pi}_{\xi}^\top = \boldsymbol{\Pi}_{\xi}^2$ and also that $\boldsymbol{Q}^\top \boldsymbol{Q} = \boldsymbol{\Pi}_{\xi}$, where $\boldsymbol{Q}$ is the $(n-1)\times n$ matrix introduced in (2.8).[1] Moreover, we have that $\boldsymbol{L}(t)\,\boldsymbol{\Pi}_{\xi} = \boldsymbol{\Pi}_{\xi}\,\boldsymbol{L}(t) = \boldsymbol{L}(t)$ and the spectrum of the matrix $\bar{\boldsymbol{L}}(t) := \boldsymbol{Q}\,\boldsymbol{L}(t)\,\boldsymbol{Q}^\top$ is equal to the spectrum of $\boldsymbol{L}(t)$ except for the eigenvalue $\lambda_1 = 0$ corresponding to the eigenvector $\mathbf{1}_n$. Finally, we define the coordination error state $\boldsymbol{\zeta}(t) := [\boldsymbol{\zeta}_1^\top(t),\ \boldsymbol{\zeta}_2^\top(t)]^\top$ as

$$\begin{aligned} \boldsymbol{\zeta}_1(t) &:= \boldsymbol{Q}\,\boldsymbol{\xi}(t) && \in \mathbb{R}^{n-1} , \\ \boldsymbol{\zeta}_2(t) &:= \boldsymbol{\chi}_I(t) - \mathbf{1}_{n-n_\ell} && \in \mathbb{R}^{n-n_\ell} . \end{aligned} \tag{4.8}$$

Note that, at the kinematic level, $\boldsymbol{\zeta}(t) = \mathbf{0}$ is equivalent to $\boldsymbol{\xi}(t) \in \mathrm{span}\{\mathbf{1}_n\}$ and $\dot{\boldsymbol{\xi}}(t) = \mathbf{1}_n$, which implies that, if $\boldsymbol{\zeta}(t) = \mathbf{0}$, then at time t all target vehicles are coordinated and travel at the desired speed. With the above notation, the closed-loop coordination dynamics (4.7) can be reformulated as (see Appendix B.2.2)

$$\dot{\boldsymbol{\zeta}} = \boldsymbol{A}_{\zeta}(t)\,\boldsymbol{\zeta} , \qquad \boldsymbol{\zeta}(0) = \boldsymbol{\zeta}_0 , \tag{4.9}$$

where $\boldsymbol{A}_{\zeta}(t)$ is given by

$$\boldsymbol{A}_{\zeta}(t) := \begin{bmatrix} -k_P\bar{\boldsymbol{L}}(t) & \boldsymbol{Q}\boldsymbol{C} \\ -k_I\boldsymbol{C}^\top\boldsymbol{Q}^\top\bar{\boldsymbol{L}}(t) & \mathbf{0} \end{bmatrix} \in \mathbb{R}^{(2n-n_\ell-1)\times(2n-n_\ell-1)} . \tag{4.10}$$

We now show that if the connectivity of the communication graph $\Gamma(t)$ verifies the PE-like condition (2.8), then the coordination control law (4.6) solves the coordination control problem at the kinematic level. The next lemma summarizes this result.

Lemma 4.1. *Assume that the Laplacian* $\mathbf{L}(t)$ *of the graph* $\Gamma(t)$ *that models the communication topology of a network of vehicles satisfies the PE-like condition* (2.8) *for some parameters* $\mu, T > 0$. *Then, there exist coordination control gains* k_P *and* k_I *such that the origin of the kinematic coordination error dynamics* (4.9) *is exponentially stable with guaranteed rate of convergence*

$$\bar{\lambda}_{cd} := \frac{k_P n\mu}{(1+k_P nT)^2}\left(1+\rho_k\frac{n}{n_\ell}\right)^{-1} , \qquad \rho_k \geq 2 . \tag{4.11}$$

Furthermore, the coordination states $\xi_i(t)$ *and their rates of change* $\dot{\xi}_i(t)$ *satisfy*

$$|\xi_i(t) - \xi_j(t)| \leq \kappa_{\xi 0}\|\boldsymbol{\zeta}(0)\|e^{-\bar{\lambda}_{cd}t} , \qquad \textit{for all } i, j \in \{1,\ldots,n\} , \tag{4.12}$$

$$|\dot{\xi}_i(t) - 1| \leq \kappa_{\dot{\xi} 0}\|\boldsymbol{\zeta}(0)\|e^{-\bar{\lambda}_{cd}t} , \qquad \textit{for all } i \in \{1,\ldots,n\} , \tag{4.13}$$

for some constants $\kappa_{\xi 0}, \kappa_{\dot{\xi} 0} \in (0,\infty)$. ◇

Proof. The proof of this result is given in Appendix B.2.5. □

[1] A proof of the equality $\boldsymbol{Q}^\top\boldsymbol{Q} = \boldsymbol{\Pi}_{\xi}$ can be found in Appendix B.1.1.

Remark 4.4. The proof of Lemma 4.1 is constructive and explicitly specifies a particular choice for the coordination control gains k_P and k_I that ensures exponential stability of the kinematic coordination error dynamics; see Eq. (B.21) in Appendix B.2.5. △

Remark 4.5. Lemma 4.1 above shows that the guaranteed rate of convergence of the coordination control loop is limited by the QoS of the network (characterized by parameters T and μ). According to the lemma, for a given QoS of the network, the maximum guaranteed rate of convergence $\bar{\lambda}^*_{cd}$ is achieved by setting $k_P = \frac{1}{Tn}$, which results in

$$\bar{\lambda}^*_{cd} := \tfrac{\mu}{4T}\left(1 + \rho_k \tfrac{n}{n_\ell}\right)^{-1}, \qquad \rho_k \geq 2\,.$$

Note that the convergence rate $\bar{\lambda}^*_{cd}$ scales with the ratio (n_ℓ/n). We also note that, as the parameter T goes to zero (and the graph becomes connected pointwise in time), the convergence rate can be set arbitrarily high by increasing the coordination control gains k_P and k_I. This is consistent with results obtained in previous work on cooperative path-following control; see [1, Lemma 2].

Finally, we notice that

$$\bar{\lambda}^P_{cd} := \frac{k_P n \mu}{(1+k_P n T)^2}$$

represents the (guaranteed) convergence rate for the coordination loop with a proportional control law, rather than a proportional-integral control law (see Appendix B.2.5). It is straightforward to verify that, for a given proportional gain k_P, we obtain $\bar{\lambda}_{cd} < \bar{\lambda}^P_{cd}$, which implies that a proportional control law can provide higher rates of convergence than the proportional-integral control law presented in this chapter. However, as mentioned earlier and proven in [7], the integral term in the coordination control law is important in the current application as it improves the disturbance rejection capabilities at the coordination level. △

4.2.2 Convergence Analysis with Non-ideal Inner-Loop Performance

When the dynamics of the vehicle are included in the problem formulation, the evolution of the ith coordination state with the speed command (4.4) becomes

$$\dot{\xi}_i = u_{\text{coord},i} + \frac{e_{v,i}}{v_{d,i}(\xi_i)}\,\hat{\mathbf{w}}_{1,i}\cdot\hat{\mathbf{t}}_i\,,$$

where $e_{v,i}(t) := v_i(t) - v_{c,i}(t)$ denotes the speed tracking error for the ith vehicle. In this case, the coordination law (4.6) leads to the closed-loop coordination dynamics

$$\begin{aligned}
\dot{\boldsymbol{\xi}} &= -k_P \boldsymbol{L}(t)\,\boldsymbol{\xi} + \begin{bmatrix} \mathbf{1}_{n_\ell} \\ \boldsymbol{\chi}_I \end{bmatrix} + \boldsymbol{e}'_v\,, & \boldsymbol{\xi}(0) &= \boldsymbol{\xi}_0\,,\\
\dot{\boldsymbol{\chi}}_I &= -k_I\,\boldsymbol{C}^\top \boldsymbol{L}(t)\,\boldsymbol{\xi}\,, & \boldsymbol{\chi}_I(0) &= \mathbf{1}_{n-n_\ell}\,,
\end{aligned}$$

where $\boldsymbol{e}'_v(t) \in \mathbb{R}^n$ is a vector whose ith component is equal to $e'_{v,i} := \frac{e_{v,i}}{v_{d,i}(\xi_i)}\,\hat{\mathbf{w}}_{1,i}\cdot\hat{\mathbf{t}}_i$.

Similarly to the previous section, we can now derive the closed-loop coordination error dynamics, which become

$$\dot{\boldsymbol{\zeta}} = \boldsymbol{A}_\zeta(t)\,\boldsymbol{\zeta} \;+\; \boldsymbol{B}_\zeta\,\boldsymbol{e}'_v\,, \qquad \boldsymbol{\zeta}(0) = \boldsymbol{\zeta}_0\,, \tag{4.14}$$

where $\boldsymbol{A}_\zeta(t)$ was defined in (4.10) and $\boldsymbol{B}_\zeta$ is given by

$$\boldsymbol{B}_\zeta := \begin{bmatrix} \boldsymbol{Q} \\ \boldsymbol{0} \end{bmatrix} \qquad \in \mathbb{R}^{(2n-n_\ell-1)\times n}\,.$$

The next lemma shows that if the connectivity of the communication graph $\Gamma(t)$ verifies the PE-like condition (2.8), then the coordination control law (4.6) solves the coordination control problem in a practical sense. The lemma also proves that the coordination error vector degrades gracefully with the *size* of the speed tracking error vector $\boldsymbol{e}_v(t) := [e_{v,1}(t), \ldots, e_{v,n}(t)]^\top$.

Lemma 4.2. *Consider the coordination error dynamics* (4.14) *and suppose that the information flow satisfies the PE-like condition* (2.8) *for some parameters* $\mu, T > 0$. *Moreover, assume that the speed tracking error vector* $\boldsymbol{e}_v(t)$ *is a piecewise continuous, bounded function of* t *for all* $t \geq 0$. *Then, there exist coordination control gains* k_P *and* k_I *such that system* (4.14) *is input-to-state stable (ISS) with respect to* $\boldsymbol{e}_v(t)$, *satisfying*

$$\|\boldsymbol{\zeta}(t)\| \leq \kappa_{\zeta 0}\|\boldsymbol{\zeta}(0)\|\mathrm{e}^{-\lambda_{cd}t} + \kappa_{\zeta 1}\sup_{\tau\in[0,t)}\|\boldsymbol{e}_v(\tau)\|\,, \qquad \text{for all } t \geq 0\,, \tag{4.15}$$

for some constants $\kappa_{\zeta 0}, \kappa_{\zeta 1} \in (0,\infty)$, *and with* $\lambda_{cd} := \bar{\lambda}_{cd}(1-\theta_\lambda)$, *where* $\bar{\lambda}_{cd}$ *was defined in* (4.11) *and* θ_λ *is a constant verifying* $0 < \theta_\lambda < 1$. *Further, the coordination states* $\xi_i(t)$ *and their rates of change* $\dot{\xi}_i(t)$ *satisfy*

$$|\xi_i(t) - \xi_j(t)| \leq \kappa_{\xi 0}\|\boldsymbol{\zeta}(0)\|\mathrm{e}^{-\lambda_{cd}t} + \kappa_{\xi 1}\sup_{\tau\in[0,t)}\|\boldsymbol{e}_v(\tau)\|\,, \quad \text{for all } t \geq 0\,, \quad \text{and all } i,j \in \{1,\ldots,n\}\,, \tag{4.16}$$

$$|\dot{\xi}_i(t) - 1| \leq \kappa_{\dot{\xi} 0}\|\boldsymbol{\zeta}(0)\|\mathrm{e}^{-\lambda_{cd}t} + \kappa_{\dot{\xi} 1}\sup_{\tau\in[0,t)}\|\boldsymbol{e}_v(\tau)\|\,, \quad \text{for all } t \geq 0\,, \quad \text{and all } i \in \{1,\ldots,n\}\,, \tag{4.17}$$

for some constants $\kappa_{\xi 0}, \kappa_{\xi 1}, \kappa_{\dot{\xi} 0}, \kappa_{\dot{\xi} 1} \in (0,\infty)$. ◇

Proof. The proof of this result is given in Appendix B.2.6. □

Remark 4.6. If the desired speed profiles of all vehicles are constant along the corresponding paths, that is, $v_{d,i}(t_d) = v_{d,i}$ for all $t_d \in [0, t_d^{\mathrm{f}}]$ and all $i \in \{1,\ldots,n\}$, we have shown in Remark 4.1 that the normalized curvilinear abscissas $\ell'_i(t)$ can be used as coordination states. In this case, the dynamics of the ith coordination state can be written as

$$\dot{\ell}'_i = \frac{\left(v_i\,\hat{\mathbf{w}}_{1,i} + k_\ell\,\boldsymbol{p}_{F,i}\right)\cdot\hat{\mathbf{t}}_i}{\ell_{fi}}\,, \qquad i = 1,\ldots,n\,,$$

which implies that the speed commands $v_{c,i}(t)$ can be generated as

$$v_{c,i} := \frac{u_{\text{coord},i}\,\ell_{fi} - k_\ell\, \mathbf{p}_{F,i} \cdot \hat{\mathbf{t}}_i}{\hat{\mathbf{w}}_{\mathbf{1},i} \cdot \hat{\mathbf{t}}_i}, \qquad i = 1, \ldots, n,$$

where $u_{\text{coord},i}(t)$ is now given by

$$u_{\text{coord},i}(t) = -k_P \sum_{j \in \mathcal{N}_i(t)} (\ell_i'(t) - \ell_j'(t)) + \frac{v_{d,i}}{\ell_{fi}}, \qquad i = 1, \ldots, n_\ell,$$

$$\begin{aligned} u_{\text{coord},i}(t) &= -k_P \sum_{j \in \mathcal{N}_i(t)} (\ell_i'(t) - \ell_j'(t)) + \chi_{I,i}(t) \\ \dot{\chi}_{I,i}(t) &= -k_I \sum_{j \in \mathcal{N}_i(t)} (\ell_i'(t) - \ell_j'(t)), \qquad \chi_{I,i}(0) = \frac{v_{d,i}}{\ell_{fi}}, \end{aligned} \qquad i = n_\ell + 1, \ldots, n.$$

The analysis done in Lemmas 4.1 and 4.2 carries through to this setup, yielding analogous results. △

Remark 4.7. To effectively solve the coordination problem in the presence of winds and gusts, it is beneficial to implement the fleet leaders as virtual agents with "uncertainty-free dynamics" (in our case, no groundspeed tracking error). This fact was illustrated through simulation in [6], where we also proposed a methodology to add these virtual agents and derived a lower bound on the QoS of the resulting extended communications network. In the setup adopted in [6], the virtual agents (along with the corresponding coordination control law) act as local controllers that ensure that the fleet of vehicles reaches the desired agreement. △

4.3 Combined Path Following and Time Coordination

Chapter 3 and Section 4.2 have shown that, under an appropriate set of assumptions, the path-following and coordination control laws ensure stability of the path-following and time-coordination dynamics when treated separately. In particular, the solution developed for the path-following problem assumes that the speed of the vehicle is bounded above and below, while the control law designed for vehicle coordination relies on the assumption that the angle between the vehicle's velocity vector and the tangent direction to the path is less than 90 deg (see Eqs. (3.9) and (4.3)). This section addresses the convergence properties of the combined cooperation and path-following systems, and derives design constraints for the inner-loop tracking performance bounds that guarantee stability of the complete system. The overall cooperative path-following control architecture for the ith vehicle is presented in Fig. 4.1.

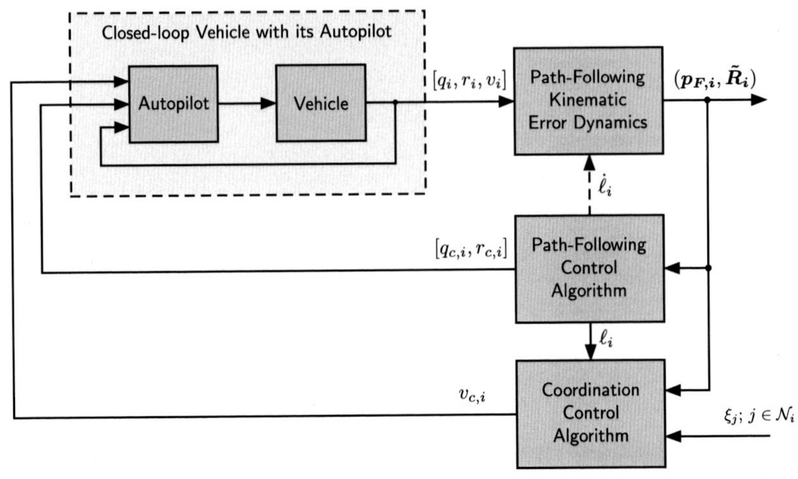

Figure 4.1 Time-critical cooperative path-following closed-loop system. The figure illustrates the control architecture for one of the vehicles involved in the mission, which exchanges information with a group of neighboring vehicles $\mathcal{N}_i$.

4.3.1 Stability Analysis at the Kinematic Level

We start by analyzing the stability of the cooperative path-following closed-loop system at the kinematic level. To this effect, we consider the path-following kinematic error dynamics (3.6) and the coordination-state dynamics (4.5), and show that the rate commands (3.8) and the speed command (4.4) with the coordination control law (4.6) solve the time-critical cooperative path-following problem with guaranteed rates of convergence. The next theorem states this result formally for the case of spatially deconflicted trajectories (see Chapter 2, Section 2.2). Stability conditions for the case of time-deconflicted trajectories are given in Remark 4.8.

Theorem 4.1. *Consider a fleet of n vehicles supported by a communications network that verifies the PE-like condition* (2.8)*, and a set of desired spatially deconflicted 3D time-trajectories with spatial clearance E. Let c and c_1 be positive constants satisfying $c < \frac{1}{\sqrt{2}}$ and $c_1 < \frac{E}{2c}$ and, for each UAV, choose the path-following control parameters k_ℓ, $k_{\tilde{R}}$, and d such that*

$$d > \frac{2(1-c^2)^{\frac{1}{2}}}{1-2c^2} c c_1 , \qquad k_{\tilde{R}} \tilde{k}_\ell > \frac{v_{\max}^2}{c_1^2(1-2c^2)^2} , \tag{4.18}$$

where $\tilde{k}_\ell$ was defined in (3.11)*. Also, let $\rho_k \geq 2$ and set the coordination control gains k_P and k_I such that*

$$k_P > 0 , \qquad \frac{k_I}{k_P} = \frac{k_P n \mu}{(1+k_P n T)^2} \frac{\rho_k \frac{n}{n_\ell}}{1+\rho_k \frac{n}{n_\ell}} . \tag{4.19}$$

Then, for all initial conditions

$$(\boldsymbol{p}_{F,i}(0), \tilde{\boldsymbol{R}}_i(0)) \in \Omega_{pf}\,, \qquad i = 1, \ldots, n\,, \tag{4.20}$$

$$\|\boldsymbol{\zeta}(0)\| \leq \frac{1}{\kappa_{\zeta 0}\kappa_1} \min\left\{ \left(1 - \frac{v_{\min} + k_\ell c c_1}{v_{d\,\min}}\right), \left(\frac{v_{\max} c_2 - k_\ell c c_1}{v_{d\,\max}} - 1\right) \right\}, \tag{4.21}$$

where $\kappa_1 = 2k_P\left(\frac{(n-1)^3}{n}\right)^{\frac{1}{2}} + 1$ *and* $c_2 = \frac{(1-2c^2)d - 2c(1-c^2)^{\frac{1}{2}} c c_1}{(d^2 + (cc_1)^2)^{\frac{1}{2}}}$, *the progression law* (3.7), *the angular-rate commands* (3.8), *and the speed commands* (4.4) *with the coordination control law* (4.6) *ensure, first, that the speed of each vehicle satisfies*

$$v_{\min} \leq v_i(t) \leq v_{\max}\,, \qquad \textit{for all } t \geq 0\,, \quad \textit{and all } i \in \{1, \ldots, n\}\,, \tag{4.22}$$

and, second, that the origin of the path-following kinematic error dynamics (3.6) *and the origin of the kinematic coordination error dynamics* (4.9) *are exponentially stable with guaranteed rates of convergence* $\bar{\lambda}_{pf}$ *and* $\bar{\lambda}_{cd}$, *respectively.* ◇

Proof. The proof of this result is given in Appendix B.2.7. □

Remark 4.8. In the case of time-deconflicted trajectories (with spatial clearance E), the initial conditions of the kinematic coordination error dynamics must satisfy the additional inequality

$$\|\boldsymbol{\zeta}(0)\| < \frac{E - 2cc_1}{\kappa_{\xi 0} v_{d\,\max}}$$

that ensures that no two vehicles are at the same place at the same time. △

4.3.2 Stability Analysis with Inner-Loop Autopilots

Next, we analyze the stability of the cooperative path-following dynamics assuming that each UAV is equipped with an onboard autopilot designed to provide angular-rate as well as speed tracking capabilities. In particular, in this section, we make the assumption that each vehicle is able to track bounded pitch-rate, yaw-rate, as well as speed commands with the performance bounds (2.9) and (2.11).

At this point, we note that, while the pitch-rate and yaw-rate commands (3.8) are continuous in time, the same cannot be said about the speed command (4.4). In fact, due to the time-varying nature of the network topology, the coordination law (4.6) is discontinuous, which implies that the speed command $v_{c,i}(t)$ is also discontinuous. Note that, if the bound

$$|v_{c,i}(t) - v_i(t)| \leq \gamma_v \tag{4.23}$$

holds for all $t \geq 0$ and all vehicles, implying that $\sup_{t\geq 0} \|\boldsymbol{e}_v(t)\| \leq \sqrt{n}\,\gamma_v$, then an upper bound $\Delta v_{c,i}$ on jumps in the speed command $v_{c,i}(t)$ can be derived from (4.4),

(4.6), and the results of Lemma 4.2, and is given by

$$\Delta v_{c,i} = \frac{k_P(n-1)\left(\kappa_{\xi 0}\|\boldsymbol{\zeta}(0)\| + \kappa_{\xi 1}\sqrt{n}\,\gamma_v\right) v_{d\,\max}}{c_2}\,, \qquad i = 1, \ldots, n\,.$$

A necessary (but by no means sufficient!) condition for the bound in (4.23) to hold is thus:

$$\Delta v_{c,i} < \gamma_v\,, \qquad i = 1, \ldots, n\,.$$

The above condition limits the choice of the coordination control gains, which in particular need to satisfy the following inequality:

$$k_P(n-1)\kappa_{\xi 1}\sqrt{n} v_{d\,\max} < c_2\,.$$

The derivation of sufficient conditions ensuring that the bound in (2.11) holds for all $t \geq 0$ requires, however, assumptions on vehicle dynamics and autopilot design, and is thus beyond the scope of this book. Hence, for the subsequent developments, we make the assumption that the bound in (2.11) holds, provided the speed command $v_{c,i}(t)$ satisfies the bounds in (2.10), and derive design constraints for this inner-loop tracking performance bound that ensure that the overall time-critical cooperative path-following control system is stable and has desired convergence properties.

The next theorem summarizes the stability and convergence properties of the time-critical cooperative path-following control system for the case of spatially deconflicted trajectories.

Theorem 4.2. *Consider a fleet of n vehicles supported by a communications network that verifies the PE-like condition* (2.8)*, and a set of desired spatially deconflicted 3D time-trajectories. For given positive constants c and c_1 satisfying $c < \frac{1}{\sqrt{2}}$ and $c_1 < \frac{E}{2c}$, choose the path-following control parameters k_ℓ, $k_{\tilde{R}}$, and d according to the design constraints* (4.18)*. Also, let $\rho_k \geq 2$, and set the coordination control gains k_P and k_I as in* (4.19)*. Further, let $\lambda_{pf} := \bar{\lambda}_{pf}(1-\delta_\lambda)$ and $\lambda_{cd} := \bar{\lambda}_{cd}(1-\theta_\lambda)$, where δ_λ and θ_λ are positive constants verifying $0 < \delta_\lambda, \theta_\lambda < 1$. If the performance bounds γ_q, γ_r, and γ_v satisfy*

$$\left(\gamma_q^2 + \gamma_r^2\right)^{\frac{1}{2}} < \frac{2c}{(1-c^2)^{\frac{1}{2}}}\,\bar{\lambda}_{pf}\delta_\lambda\,, \tag{4.24}$$

$$\gamma_v < \min\left\{\frac{v_{d\,\min} - (v_{\min} + k_\ell c c_1)}{1 + \kappa_1\kappa_{\zeta 1}\sqrt{n} v_{d\,\min}}\,, \frac{(v_{\max}c_2 - k_\ell c c_1) - v_{d\,\max}}{c_2 + \kappa_1\kappa_{\zeta 1}\sqrt{n} v_{d\,\max}}\right\}, \tag{4.25}$$

then, for all initial conditions

$$(\boldsymbol{p}_{F,i}(0), \tilde{\boldsymbol{R}}_i(0)) \in \Omega_{pf}, \qquad i = 1, \ldots, n, \tag{4.26}$$

$$\|\boldsymbol{\zeta}(0)\| \leq \frac{1}{\kappa_{\zeta 0}\kappa_1} \min\left\{\left(1 - \frac{v_{c\,\min} + k_\ell c c_1}{v_{d\,\min}}\right), \left(\frac{v_{c\,\max} c_2 - k_\ell c c_1}{v_{d\,\max}} - 1\right)\right\} - \frac{\kappa_{\zeta 1}}{\kappa_{\zeta 0}}\sqrt{n}\gamma_v, \tag{4.27}$$

the progression law (3.7) *for the virtual target vehicle, the rate commands* (3.8), *and the speed commands* (4.4) *with the coordination control law* (4.6) *ensure, first, that the speed of each vehicle satisfies*

$$v_{\min} \leq v_i(t) \leq v_{\max}, \qquad \textit{for all } t \geq 0, \quad \textit{and all } i \in \{1, \ldots, n\}, \tag{4.28}$$

and, second, that there exist times $T_{b,i} \geq 0$ *such that the path-following errors* $\boldsymbol{p}_{F,i}(t)$ *and* $\boldsymbol{e}_{\tilde{R},i}(t)$, $i = 1, \ldots, n$, *satisfy*

$$\|\boldsymbol{e}_{\tilde{R},i}(t)\|^2 + \frac{1}{c_1^2}\|\boldsymbol{p}_{F,i}(t)\|^2 \leq \left(\frac{1}{1-c^2}\|\boldsymbol{e}_{\tilde{R},i}(0)\|^2 + \frac{1}{c_1^2}\|\boldsymbol{p}_{F,i}(0)\|^2\right) \mathrm{e}^{-2\lambda_{pf} t}, \qquad \textit{for all } 0 \leq t < T_{b,i}, \tag{4.29a}$$

$$\|\boldsymbol{e}_{\tilde{R},i}(t)\|^2 + \frac{1}{c_1^2}\|\boldsymbol{p}_{F,i}(t)\|^2 \leq \frac{(1-c^2)\,\gamma_\omega^2}{4\bar{\lambda}_{pf}^2 \delta_\lambda^2}, \qquad \textit{for all } t \geq T_{b,i}, \tag{4.29b}$$

while the coordination error state $\boldsymbol{\zeta}(t)$ *satisfies*

$$\|\boldsymbol{\zeta}(t)\| \leq \kappa_{\zeta 0}\|\boldsymbol{\zeta}(0)\|\mathrm{e}^{-\lambda_{cd} t} + \kappa_{\zeta 1}\sqrt{n}\,\gamma_v, \qquad \textit{for all } t \geq 0. \tag{4.30}$$

◇

Proof. The proof of this result is given in Appendix B.2.8. □

Remark 4.9. In the case of time-deconflicted trajectories, the initial conditions of the coordination error dynamics must satisfy the additional inequality

$$\|\boldsymbol{\zeta}(0)\| < \frac{E - 2cc_1}{\kappa_{\xi 0} v_{d\,\max}} - \frac{\kappa_{\zeta 1}}{\kappa_{\zeta 0}}\sqrt{n}\gamma_v,$$

which ensures that no two vehicles are at the same place at the same time. △

4.4 Implementation Details

In this section, similarly to what was done in Section 3.3 of Chapter 3, we discuss briefly some details about the practical implementation of the coordination control law proposed in this chapter:

- The strategy for time coordination adopted in this chapter requires the ground-speed of each vehicle to be adjusted based on coordination information exchanged among the vehicles. However, typical commercial off-the-shelf autopilots accept airspeed commands. These two speeds can differ significantly, especially for the case of small UAVs operating in high winds, which implies that the speed command (4.4) needs to be transformed to an airspeed command before being sent to the autopilot. This transformation requires the integration of a wind estimator into the coordination control architecture. Note that some commercial autopilots, such as the Piccolo Plus autopilot,[2] are capable of generating a rough estimate of winds aloft.
- For safety reasons, the airspeed command sent to the autopilot is first limited between a minimum and a maximum value. Anti-windup compensation is thus needed to prevent the integral term of the coordination control law from winding up, which could lead to highly oscillatory speed commands or even closed-loop instability. A possible solution is to feed back the difference between the saturated airspeed command and the unsaturated airspeed command to the input of the integrator:

$$\dot{\chi}_{I,i}(t) = -k_I \sum_{j \in \mathcal{N}_i(t)} (\xi_i(t) - \xi_j(t)) \; + \; k_{\mathrm{aw}} \left[G_{\mathrm{aw}}(s)\right] \left(\mathrm{sat}\left(v_{c,i}(t)\right) - v_{c,i}(t)\right) ,$$
$$\chi_{I,i}(0) = 1 ,$$

  where k_{aw} is the anti-windup feedback gain and $G_{\mathrm{aw}}(s)$ is a stable, low-pass transfer function.
- In addition, to prevent division by zero when computing the speed command (4.4), the quantity $\hat{\mathbf{w}}_{1,i} \cdot \hat{\mathbf{t}}_i$ must be saturated below a certain (small) positive value.

4.5 Simulation Examples

This section presents simulation results of two cooperative multi-vehicle mission scenarios that show the efficacy of the cooperative framework proposed in this book. In the first mission, three UAVs are requested to execute a coordinated maneuver to arrive at predefined positions at the same time. We then consider a second mission in which three UAVs must execute sequential auto-landing while maintaining a pre-specified safe-guarding separation along the glide slope. Similarly to the results presented in Chapter 3, the simulations that follow are based on the kinematic model of the UAV in (3.16) along with a simplified, decoupled linear model describing the roll, pitch, yaw, and speed dynamics of the closed-loop UAV with its autopilot.

In this set of simulations, the path-following controller gains are selected as follows:

$$k_\ell = 0.20\ [1/\mathrm{s}] , \qquad k_{\tilde{R}} = 0.50\ [1/\mathrm{s}] , \qquad d = 125\ [\mathrm{m}] ,$$

[2] Information available online at http://www.cloudcaptech.com/piccolo_system.shtm [Online; accessed 22 January 2017].

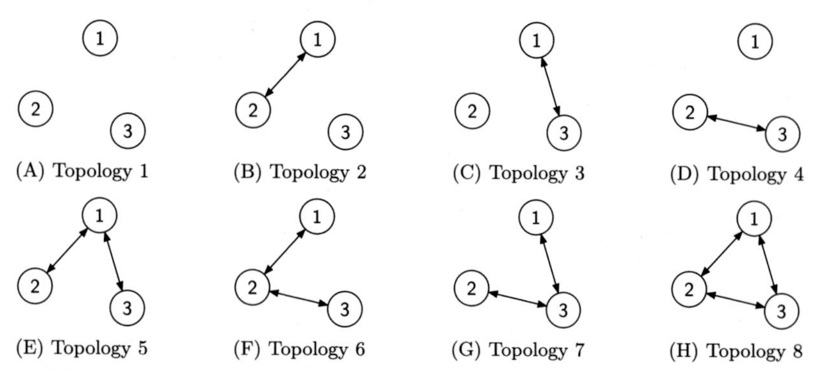

Figure 4.2 Network topologies. At any given time t, the dynamic information flow is characterized by one of these graphs.

while the coordination control gains k_P and k_I, and the anti-windup compensation are set to

$$k_P = 1.0\ 10^{-1}\ [1/\mathrm{s}]\,, \qquad k_I = 1.0\ 10^{-2}\ [1/\mathrm{s}^2]\,,$$

$$k_{\mathrm{aw}} = 2.0\ 10^{-3}\ [1/\mathrm{m}]\,, \qquad G_{\mathrm{aw}}(s) = \frac{1}{s+1}\,.$$

In all of the simulations, vehicle 1 is elected as the single leader of the fleet. The angular-rate commands are saturated to ± 0.3 rad/s, and the speed commands are saturated between 18 m/s and 32 m/s. To achieve coordination, the UAVs rely on a supporting communications network. The information flow is assumed to be time-varying and, at any given time t, is characterized by one of the graphs in Fig. 4.2.

4.5.1 Path Following with Simultaneous Arrival

In this mission scenario, three UAVs are tasked to converge to and follow three spatially deconflicted paths and arrive at their final destinations at the same time. A representative example of such a mission is coordinated rendezvous. Note that this mission imposes only *relative*, but not absolute temporal constraints on the arrival of the UAVs.

Fig. 4.3 shows the three paths with the parallel transport frames as well as the corresponding desired speed profiles, which assume a final desired speed of 20 m/s for all UAVs. The beginning of each path is indicated in this figure with a circle. The figure also shows the coordination maps η_i relating the desired normalized curvilinear abscissa $\ell'_{d,i}$ to the desired mission time t_d. The paths have lengths $\ell_{f1} = 2084.8$ m, $\ell_{f2} = 1806.4$ m, and $\ell_{f3} = 2221.0$ m, and the desired time of arrival is $t_d^{\mathrm{f}} = 85.0$ s. Note that this time of arrival is not a strict requirement to be enforced by the coordination algorithm; in fact, the emphasis is placed on having the vehicles observe the relative temporal assignments of the mission. Fig. 4.4 presents the path separations, which show a minimum spatial clearance between paths of approximately 125 m, and the desired inter-vehicle separations for this particular mission.

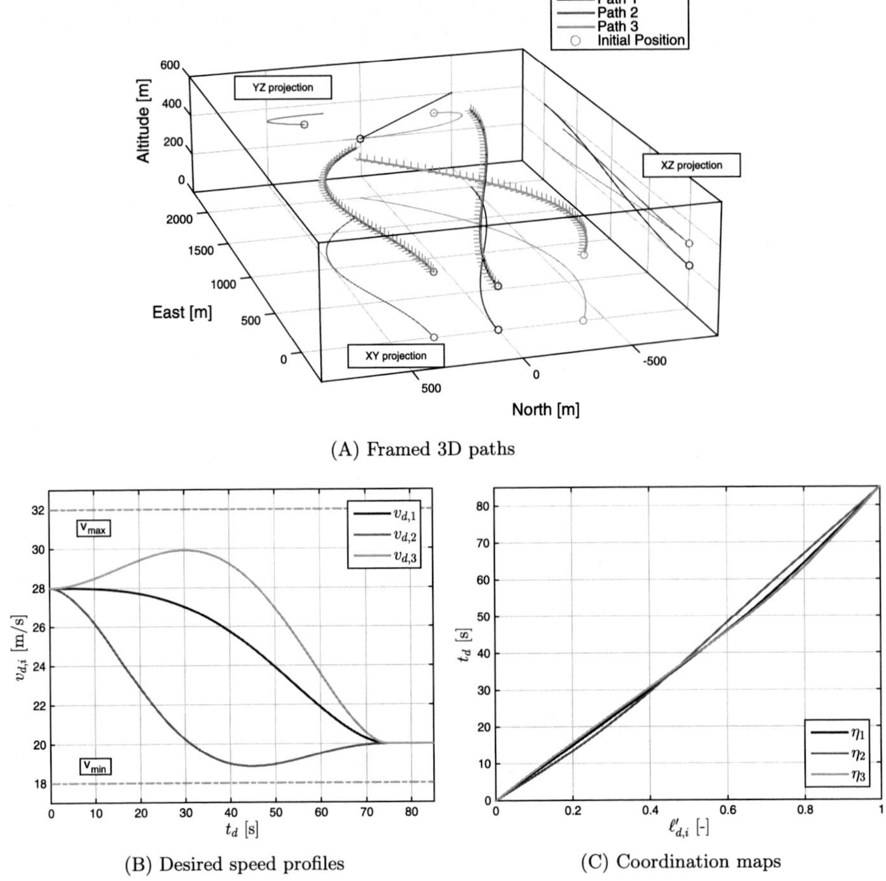

(A) Framed 3D paths

(B) Desired speed profiles (C) Coordination maps

Figure 4.3 Simultaneous arrival. Framed 3D spatial paths along with the corresponding desired speed profiles and coordination maps.

Simulation results are presented next. Fig. 4.5 illustrates the evolution of the UAVs as well as the virtual targets moving along the paths. This figure also includes the $\{\mathcal{W}\}$ frame attached to each UAV as well as the $\{\mathcal{F}\}$ frame attached to the virtual targets. The UAVs start the mission with an initial offset in both position and attitude with respect to the beginning of the framed paths. As can be seen in the figure, the path-following algorithm eliminates this initial offset and steers the UAVs along the corresponding paths, while the coordination algorithm ensures simultaneous arrival at the end of the path at $t = 84.1$ s.

Details about the performance of the path-following algorithm are shown in Fig. 4.6; the path-following attitude and position errors, $\Psi(\tilde{\boldsymbol{R}}_i(t))$ and $\boldsymbol{p}_{F,i}(t)$, converge to a neighborhood of zero within 10 s and 30 s, respectively. The figure also presents the angular-rate commands, $q_{c,i}(t)$ and $r_{c,i}(t)$, in dashed lines, the actual an-

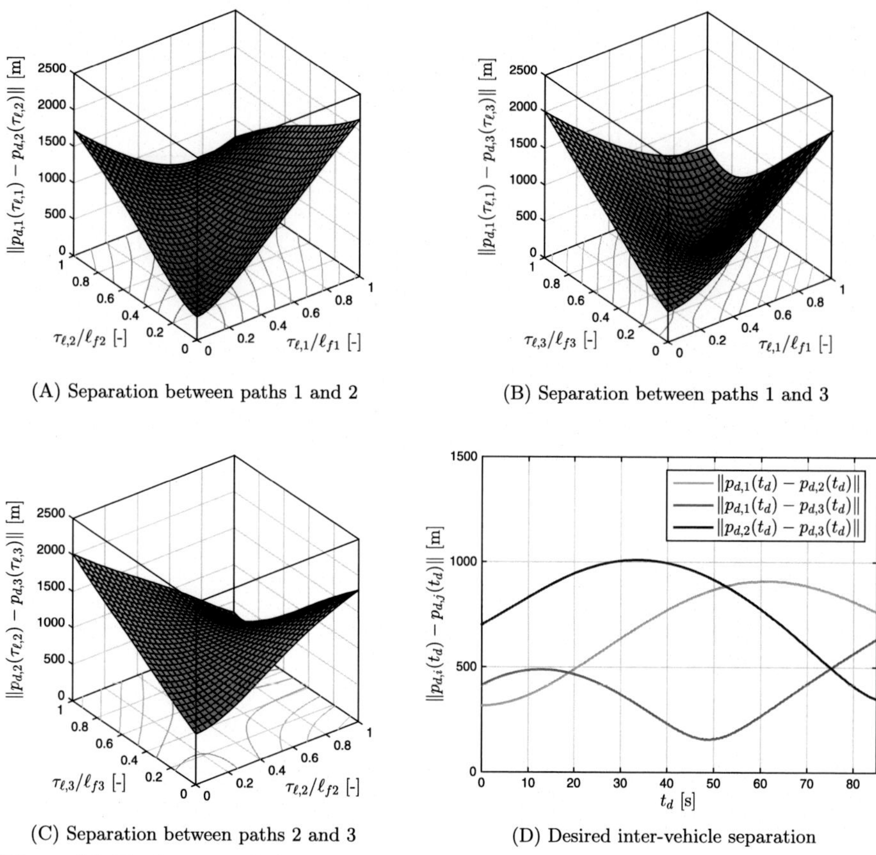

(A) Separation between paths 1 and 2

(B) Separation between paths 1 and 3

(C) Separation between paths 2 and 3

(D) Desired inter-vehicle separation

Figure 4.4 Simultaneous arrival. Path separation and desired inter-vehicle separation; the three paths are spatially deconflicted with a minimum clearance of 125 m.

gular rates, $q_i(t)$ and $r_i(t)$, in solid lines, as well as the rate of progression $\dot{\ell}_i(t)$ of the virtual targets along the path.

The evolution of both the coordination errors $(\xi_i(t) - \xi_j(t))$ and the rate of change of the coordination states $\dot{\xi}_i(t)$ is illustrated in Fig. 4.7, along with the resulting UAV speeds and the integral states implemented on the follower vehicles. The figure shows that the coordination errors converge to a neighborhood of zero, while the rate of change of the coordination states converges to the desired rate of 1 s/s. In particular, Fig. 4.7B illustrates how the vehicles adjust their speeds (solid thick lines) with respect to the desired speed profile (dashed lines) to achieve coordination. The figure also shows that, as a result of the switching nature of the network topology, the speed commands (solid thin lines) of the three vehicles are discontinuous. Finally, Fig. 4.8 describes the evolution of the information flow as the mission unfolds, and presents an estimate of the QoS of the network, com-

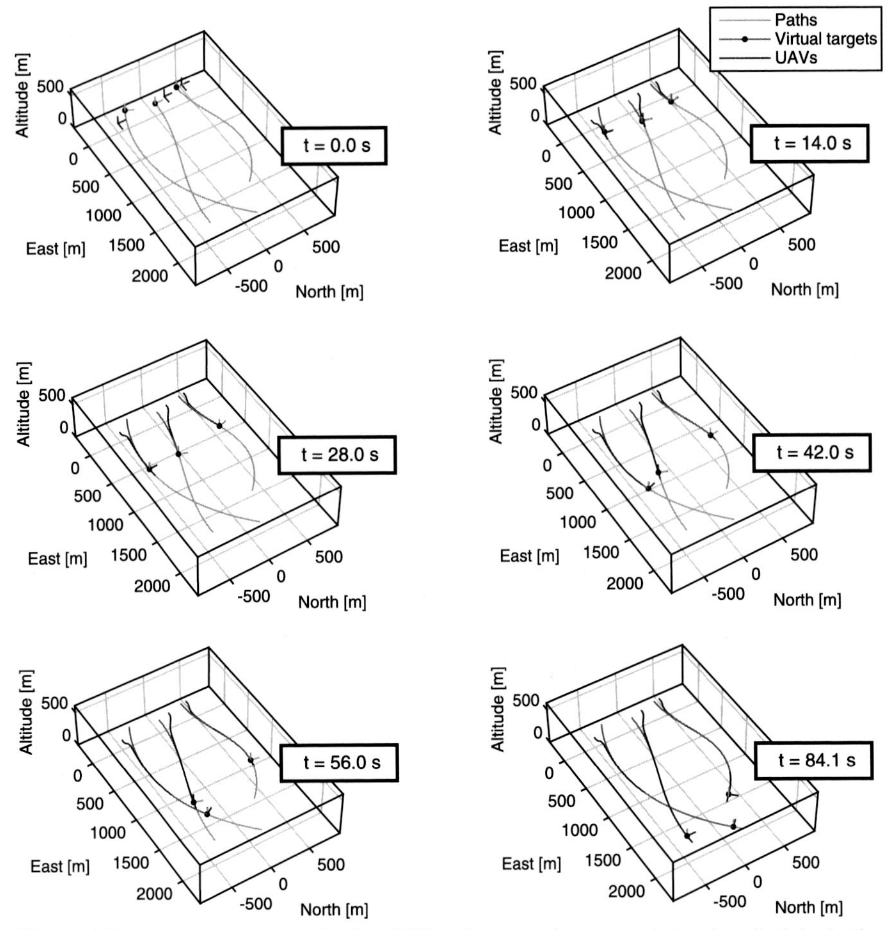

Figure 4.5 Simultaneous arrival. The three UAVs achieve simultaneous arrival at their final destinations at $t = 84.1$ s.

puted as

$$\hat{\mu}(t) := \lambda_{\min}\left(\frac{1}{3}\frac{1}{T}\int_{t-T}^{t} \boldsymbol{Q}_3 \boldsymbol{L}(\tau)\boldsymbol{Q}_3^\top d\tau\right), \qquad t \geq T, \tag{4.31}$$

with $T = 5$ s. The network topology changes every 0.5 s and, as can be seen in the figure, the QoS estimate is always greater than 0.15.

4.5.2 Sequential Auto-Landing

In this mission scenario, three UAVs must arrive at the assigned glide slope separated by pre-specified safe-guarding time-intervals, and then follow the glide path at a constant approach speed while maintaining the safe-guarding separation. To this end,

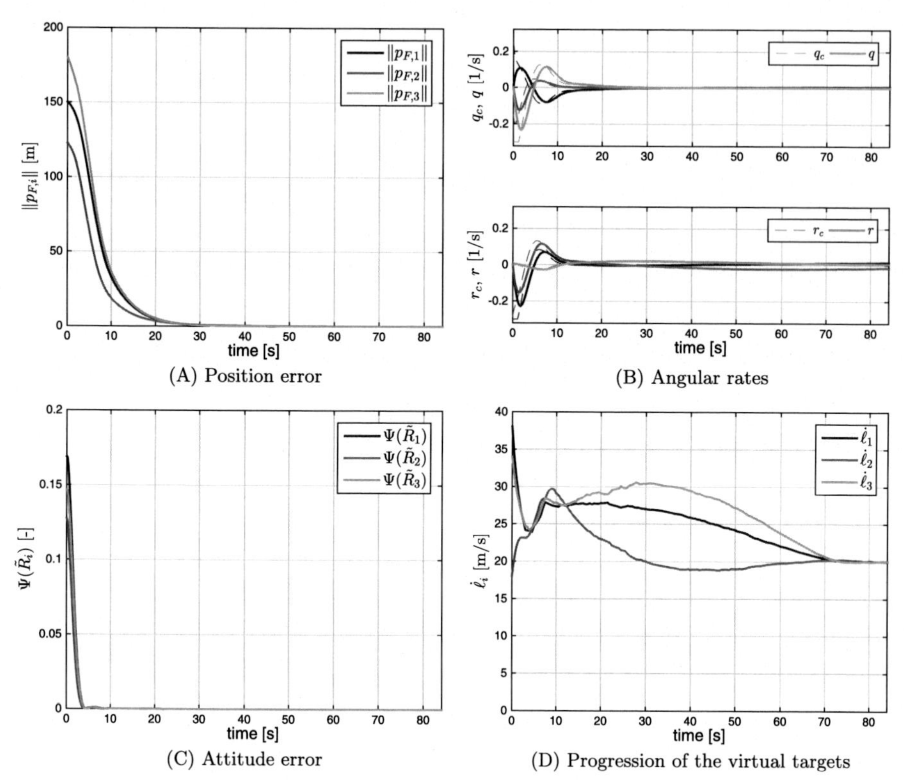

(A) Position error

(B) Angular rates

(C) Attitude error

(D) Progression of the virtual targets

Figure 4.6 Simultaneous arrival. The path-following algorithm drives the path-following position and attitude errors to a neighborhood of zero.

time-deconflicted transition trajectories are generated from pre-specified initial conditions to the beginning of the glide path, satisfying the desired inter-vehicle arrival schedule and taking the UAVs to the desired approach speed. Again, this mission imposes only *relative* temporal constraints on the arrival of the UAVs.

Fig. 4.9 shows the three transition paths with the parallel transport frames as well as the framed 3-deg glide path. The beginning of each transition path is indicated with a circle, while the beginning of the glide path is indicated with a triangle. The figure also presents the desired speed profiles for the initial transition phase that ensure a desired safe-guarding arrival separation of 30 s, trajectory deconfliction, as well as a final approach speed of 20 m/s. The transition coordination maps are shown in Fig. 4.9C. Finally, the figure also includes the desired speed profile for the approach along the glide slope as well as the corresponding coordination map. The transition paths have lengths $\ell_{f1} = 1609.0$ m, $\ell_{f2} = 1962.7$ m, and $\ell_{f3} = 2836.7$ m, and the nominal times of arrival at the glide slope are $t_{d1}^{\mathrm{f}} = 65.0$ s, $t_{d2}^{\mathrm{f}} = 95.0$ s, and $t_{d3}^{\mathrm{f}} = 125.0$ s. Again, as in the previous mission scenario, these nominal times of arrival are not to be enforced by the coordination algorithm, as the vehicles are only required to observe the desired 30 s inter-vehicle schedule. Fig. 4.10 presents the path separations,

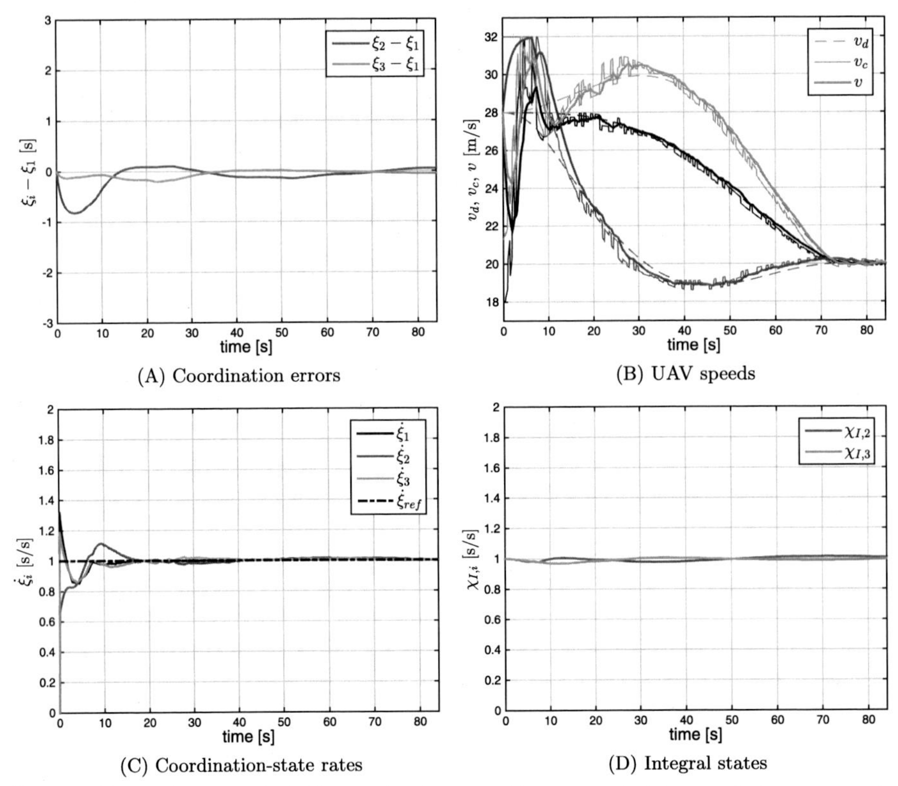

(A) Coordination errors (B) UAV speeds

(C) Coordination-state rates (D) Integral states

Figure 4.7 Simultaneous arrival. The coordination control law ensures that the coordination errors converge to a neighborhood of zero and also that the rate of change of the coordination states evolves at about the desired rate of 1 s/s.

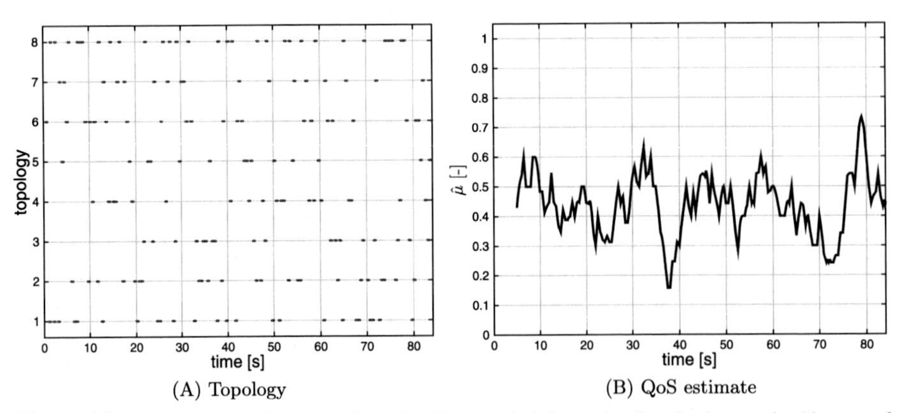

(A) Topology (B) QoS estimate

Figure 4.8 Simultaneous arrival. At a given time instant, the information flow is characterized by one of the eight topologies in Fig. 4.2. The resulting graph is only connected in an integral sense, and not pointwise in time.

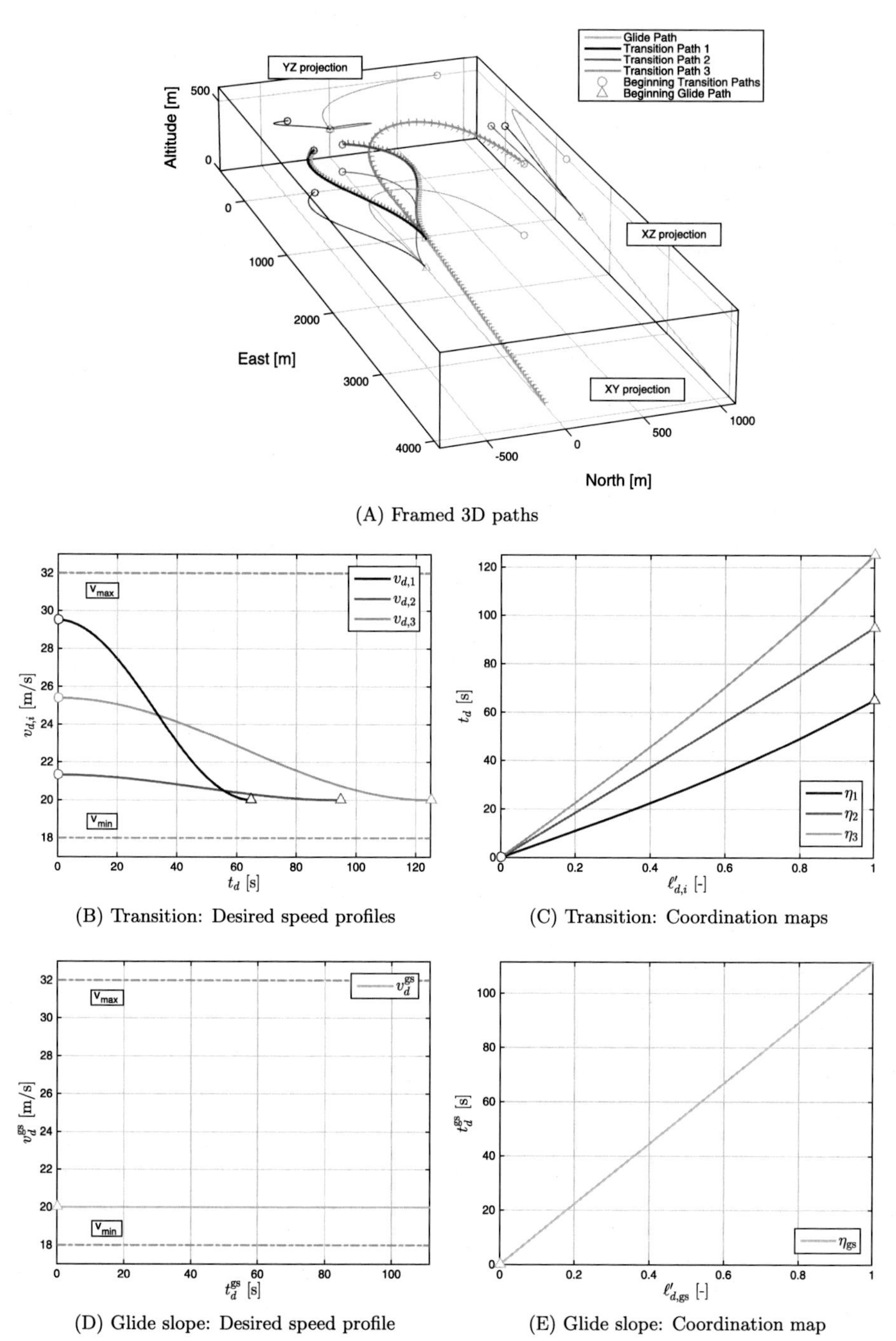

(A) Framed 3D paths

(B) Transition: Desired speed profiles

(C) Transition: Coordination maps

(D) Glide slope: Desired speed profile

(E) Glide slope: Coordination map

Figure 4.9 Sequential auto-landing. Framed 3D paths along with the corresponding desired speed profiles and coordination maps for both the transition trajectories and the glide slope.

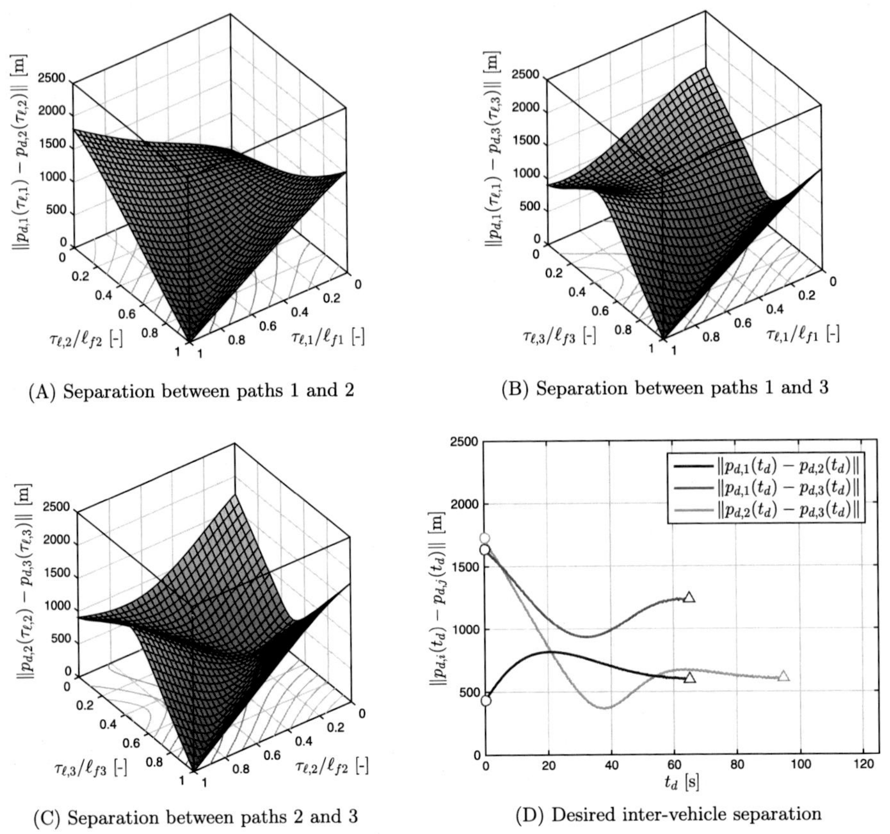

(A) Separation between paths 1 and 2

(B) Separation between paths 1 and 3

(C) Separation between paths 2 and 3

(D) Desired inter-vehicle separation

Figure 4.10 Sequential auto-landing. Path separation and desired inter-vehicle separation during the transition phase; the speed profiles ensure deconfliction of the three desired trajectories with a minimum clearance of 350 m.

which show that the three transition paths meet at their end positions (beginning of the glide slope), whereas the desired inter-vehicle separations for this particular mission are never less than 350 m.

The cooperative motion-control algorithms described in this chapter can be used to solve this sequential auto-landing problem. In this case, however, since the UAVs are required to maintain a safe-guarding separation during the approach along the glide path, the coordination states have to be redefined as the vehicles reach the glide slope. Hence, while the ith UAV is flying along its transition path, its coordination state is defined as

$$\xi_i(t) = \eta_i(\ell'_i(t)) \,, \qquad\qquad i = 1, 2, 3 \,,$$

where $\ell'_i(t)$ is the normalized curvilinear abscissa of the ith virtual target along the corresponding transition path. When the UAV reaches the beginning of the glide path,

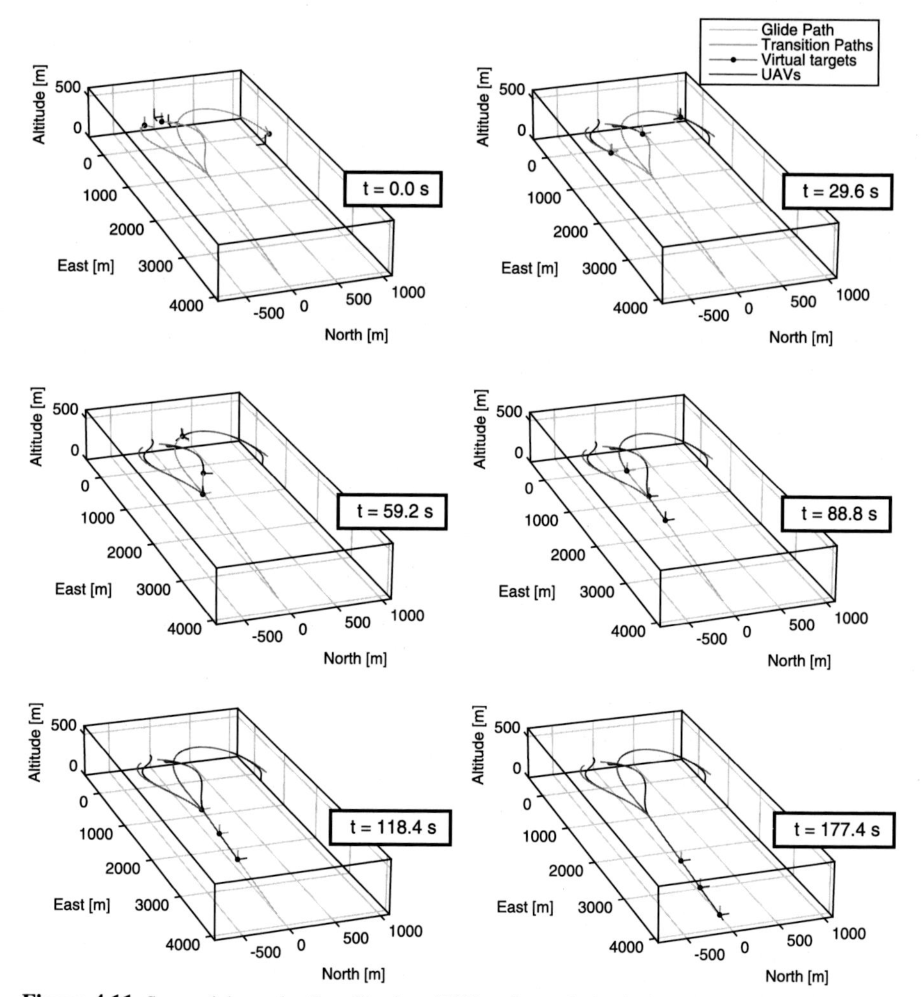

Figure 4.11 Sequential auto-landing. The three UAVs arrive at the beginning of the glide path separated by approximately 30 s and maintain this safe-guarding separation as they fly along the glide slope.

then its coordination state is redefined as

$$\xi_i(t) = \eta_{\text{gs}}(\ell_i'(t)) + t_{di}^{\text{f}}\,, \qquad\qquad i = 1, 2, 3\,,$$

where $\ell_i'(t)$ is now the normalized curvilinear abscissa of the ith virtual target along the glide path, and t_{di}^{f} is the desired time of arrival of the ith UAV at the beginning of the glide slope. Note that, with the above definitions, the coordination states $\xi_i(t)$ are continuous, as $\eta_i(1) = t_{di}^{\text{f}}$ and $\eta_{\text{gs}}(0) = 0$.

Next, we present simulation results for this mission scenario. Fig. 4.11 illustrates the evolution of the UAVs as well as the virtual targets moving along the paths. Similarly to the previous scenario, the UAVs start the mission with an initial offset in both

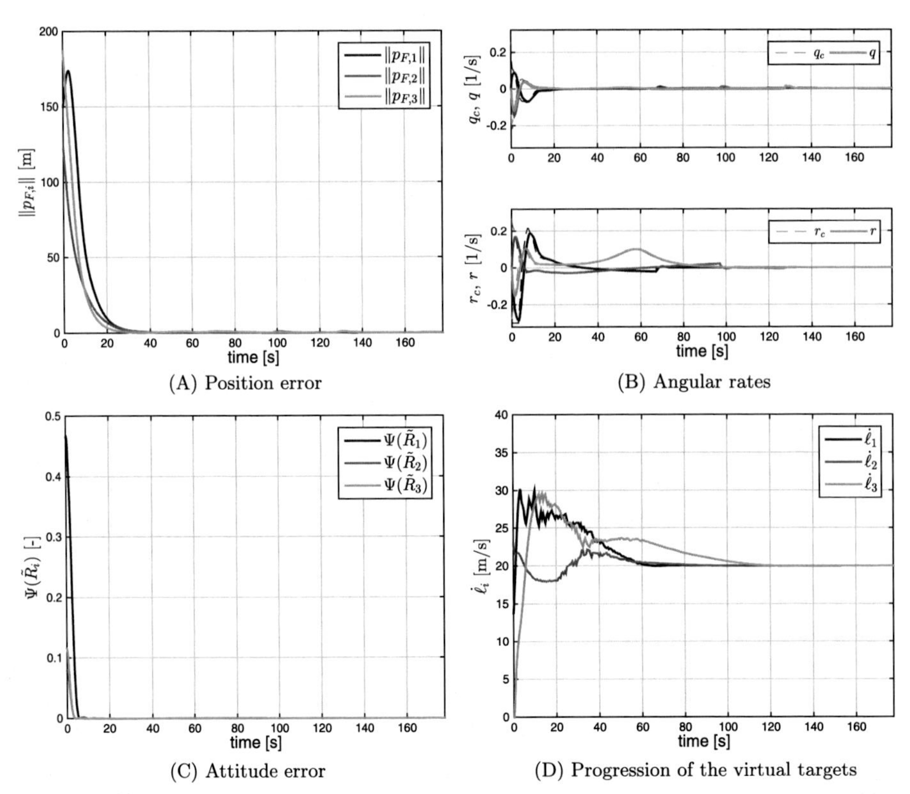

Figure 4.12 Sequential auto-landing. The path-following algorithm drives the path-following position and attitude errors to a neighborhood of zero.

position and attitude with respect to the beginning of the transition paths. As can be seen in the figure, the path-following algorithm eliminates this initial offset and steers the UAVs along the corresponding transition paths, while the coordination algorithm ensures that the UAVs reach the glide slope separated by the desired time-interval. The UAVs reach the glide slope at $t = 67.2$ s, $t = 97.1$ s, and $t = 127.1$ s, approximately meeting the desired 30 s inter-vehicle separation. After reaching the glide slope, the path-following algorithm ensures that the UAVs stay on the glide path as the coordination algorithm maintains the safe-guarding separation. The simulation is stopped when the first UAV reaches the end of the glide path.

Fig. 4.12 shows the path-following position and attitude errors, $\boldsymbol{p}_{F,i}(t)$ and $\Psi(\tilde{\boldsymbol{R}}_i(t))$, the angular-rate commands, $q_{c,i}(t)$ and $r_{c,i}(t)$, the actual angular rates, $q_i(t)$ and $r_i(t)$, and the rate of progression of the virtual targets $\dot{\ell}_i(t)$. The path-following errors converge to a neighborhood of zero within 40 s.

The coordination errors $(\xi_i(t) - \xi_j(t))$ also converge to a neighborhood of zero, while the rate of change of the coordination states $\dot{\xi}_i(t)$ converges to a neighborhood of the desired rate of 1 s/s; see Fig. 4.13. This figure also shows the UAV speeds and the integral states implemented on the follower vehicles. In particular, Fig. 4.13B

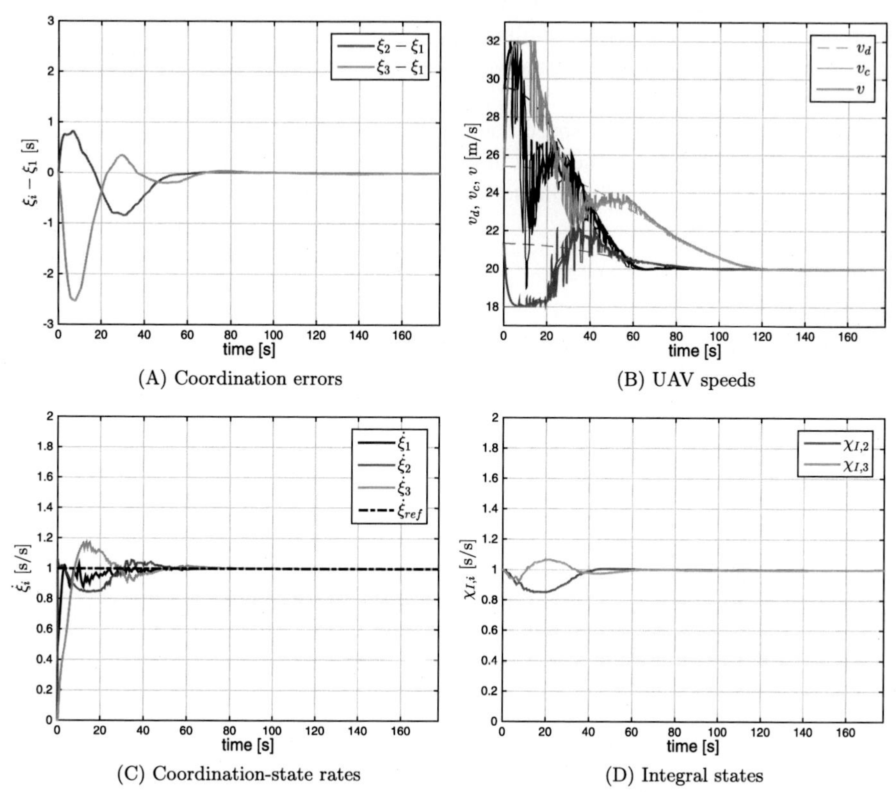

(A) Coordination errors (B) UAV speeds

(C) Coordination-state rates (D) Integral states

Figure 4.13 Sequential auto-landing. The coordination control law ensures that the coordination errors converge to a neighborhood of zero, thus ensuring trajectory deconfliction, and also that the rate of change of the coordination states evolves at about the desired rate of 1 s/s.

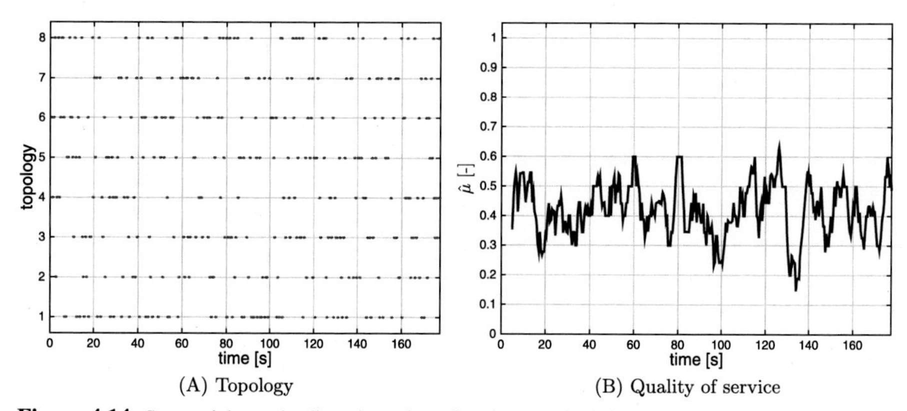

(A) Topology (B) Quality of service

Figure 4.14 Sequential auto-landing. At a given time instant, the information flow is characterized by one of the eight topologies in Fig. 4.2. The resulting graph is only connected in an integral sense, and not pointwise in time.

shows that, after a transient caused by the initial path-following errors as well as the speed corrections introduced by the coordination control law, the speed of each UAV converges to its desired speed and, as the vehicles enter the glide path, their speeds converge to the desired approach speed of 20 m/s. Again, as a result of the switching nature of the network topology, the speed commands of the three vehicles are discontinuous. Finally, Fig. 4.14 shows the evolution of the time-varying network topology along with an estimate of the QoS of the network, computed as in (4.31).

References

[1] A.P. Aguiar, I. Kaminer, R. Ghabcheloo, A.M. Pascoal, E. Xargay, N. Hovakimyan, C. Cao, V. Dobrokhodov, Time-coordinated path following of multiple UAVs over time-varying networks using $\mathcal{L}_1$ adaptation, in: AIAA Guidance, Navigation and Control Conference, Honolulu, HI, August 2008, AIAA 2008-7131.

[2] R. Ghabcheloo, A.P. Aguiar, A.M. Pascoal, C. Silvestre, I. Kaminer, J.P. Hespanha, Coordinated path-following in the presence of communication losses and delays, SIAM Journal on Control and Optimization 48 (2009) 234–265.

[3] I. Kaminer, O.A. Yakimenko, A.M. Pascoal, R. Ghabcheloo, Path generation, path following and coordinated control for time-critical missions of multiple UAVs, in: American Control Conference, Minneapolis, MN, June 2006, pp. 4906–4913.

[4] L. Lapierre, D. Soetanto, A.M. Pascoal, Coordinated motion control of marine robots, in: IFAC Conference on Manoeuvering and Control of Marine Craft, Girona, Spain, September 2003.

[5] E. Xargay, Time-Critical Cooperative Path-Following Control of Multiple Unmanned Aerial Vehicles, PhD thesis, University of Illinois at Urbana-Champaign, Urbana, IL, United States, May 2013.

[6] E. Xargay, R. Choe, N. Hovakimyan, I. Kaminer, Convergence of a PI coordination protocol in networks with switching topology and quantized measurements, in: IEEE Conference on Decision and Control, Maui, HI, December 2012, pp. 6107–6112.

[7] E. Xargay, R. Choe, N. Hovakimyan, I. Kaminer, Multi-leader coordination algorithm for networks with switching topology and quantized measurements, Automatica 50 (2014) 841–851.

[8] E. Xargay, I. Kaminer, A.M. Pascoal, N. Hovakimyan, V. Dobrokhodov, V. Cichella, A.P. Aguiar, R. Ghabcheloo, Time-critical cooperative path following of multiple unmanned aerial vehicles over time-varying networks, Journal of Guidance, Control and Dynamics 36 (2013) 499–516.

Meeting Absolute Temporal Specifications

5

The coordination control law described in Chapter 4 ensures that the vehicles meet the *relative* temporal assignments of a cooperative mission. As discussed in Chapter 1, however, some missions also impose *absolute* temporal constraints on the trajectories of the vehicles, in addition to relative temporal assignments. Here, we extend the coordination control law introduced in Section 4.2.1 of Chapter 4 to accommodate absolute temporal specifications. To this end, we reformulate the consensus problem as a collective tracking problem, and propose a distributed coordination control law that (i) uses a (virtual) clock vehicle to set the pace of the mission, and (ii) relies on dynamic link weights to coordinate the vehicles with this clock vehicle. The chapter also analyzes the stability and convergence properties of the closed-loop coordination dynamics in the case of cooperative missions that require the vehicles to strictly observe the absolute temporal specifications as planned by the trajectory-generation algorithm. The results presented in this chapter are the outcome of joint work with J. Puig-Navarro; see [3,4].

5.1 Strict and Loose Absolute Temporal Constraints

We start by formally defining the temporal specifications that will be considered in this chapter. In particular, we will propose coordination control laws for two types of absolute temporal constraints, which we will refer to as *strict* and *loose* assignments. 'Strict temporal constraints' are those that impose a precise timing schedule that all of the vehicles must observe punctually. On the other hand, similarly to [2], the term 'loose' is used here to denote temporal assignments that allow the vehicles to deviate from the desired timing plan within a pre-specified time window. Mathematically, we say the ith vehicle strictly meets the absolute temporal assignments of a mission if the following equality holds:

$$\xi_i(t) - t_d = 0\,, \qquad \text{for all } t_d \in [0, t_d^{\mathrm{f}}]\,.$$

Alternatively, loose absolute temporal constraints are met if the coordination variable of the vehicle satisfies the following bound:

$$|\xi_i(t) - t_d| \leq \Delta_d\,, \qquad \text{for all } t_d \in [0, t_d^{\mathrm{f}}]\,,$$

where Δ_d is a positive parameter that characterizes the width of the predefined time window.

The two temporal specifications above relate to distinct mission profiles. For example, loose temporal constraints are common in landing scenarios, such as the one described in Chapter 1. Strict temporal constraints are relevant in missions that require

Time-Critical Cooperative Control of Autonomous Air Vehicles, DOI: 10.1016/B978-0-12-809946-9.00007-5

synchronization with an external agent with a fixed schedule, such as the calibration of satellite instruments for Earth remote sensing. In this mission scenario, a formation of aircraft must cooperatively fly a set of predefined trajectories at different levels within the atmosphere, while coordinating their maneuvers with an Earth-observing satellite; the fleet of aircraft is tasked to collect measurements of some particular features of interest, which are then compared to the data obtained by the satellite and, if necessary, used for calibration of its instruments. As illustrated in Fig. 5.1, this strategy (flown manually) was used for the calibration of the radar altimeter onboard the CryoSat-2 in support of NASA's IceBridge program [5].

In what follows, we present coordination control laws to support the execution of time-critical cooperative missions that impose either strict or loose absolute temporal assignments. In particular, strict temporal specifications are addressed in Section 5.2, while loose temporal constraints are dealt with in Section 5.3.

5.2 Coordinating with a Virtual Clock Vehicle

5.2.1 Coordination Control Law

To solve the coordination problem with both strict absolute and relative temporal constraints, we first introduce a *clock reference vehicle* with dynamics

$$\dot{\xi}_R(t) = 1\,, \qquad \xi_R(0) = 0\,, \tag{5.1}$$

and propose the following modification of the distributed coordination control law (4.6):

$$u_{\text{coord},i}(t) = -k_P \sum_{j \in \mathcal{N}_i(t)} (\xi_i(t) - \xi_j(t)) - k_R(\xi_i(t) - \xi_R(t)) + 1\,, \qquad i = 1, \ldots, n_\ell\,,$$

$$\left.\begin{aligned} u_{\text{coord},i}(t) &= -k_P \sum_{j \in \mathcal{N}_i(t)} (\xi_i(t) - \xi_j(t)) + \chi_{I,i}(t) \\ \dot{\chi}_{I,i}(t) &= -k_I \sum_{j \in \mathcal{N}_i(t)} (\xi_i(t) - \xi_j(t))\,, \qquad \chi_{I,i}(0) = 1 \end{aligned}\right\}, \qquad i = n_\ell + 1, \ldots, n\,. \tag{5.2}$$

The control law above differs from the original coordination control law in that, now, the speed commands for the leader vehicles are adjusted based on the state of the clock vehicle (5.1), in addition to the coordination states of the neighboring vehicles. We note that, unlike the leaders, the clock vehicle runs as an isolated agent. Moreover, the clock vehicle need not be an actual vehicle and can, in fact, be separately implemented as a virtual agent running onboard the leader vehicles.

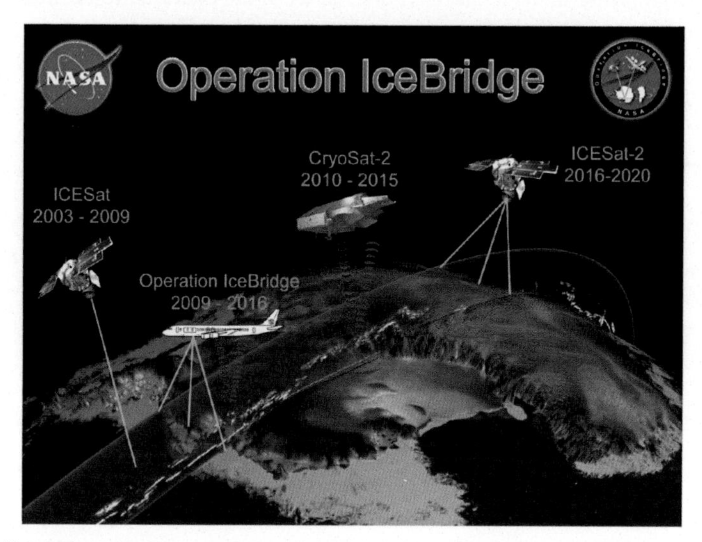

(A) NASA's IceBridge program uses satellite and aircraft measurements for polar-ice observation. *(Image is courtesy of Dr. Michael Studinger/NASA.)*

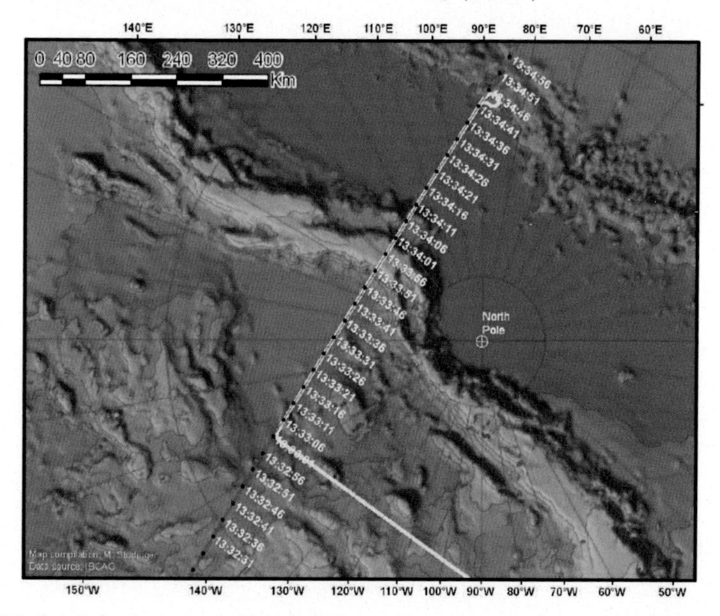

(B) Coordinated trajectories of IceBridge's DC-8 aircraft (yellow) and the CryoSat-2 (orange) with the times of the satellite over-pass indicated. *(Image is courtesy of Dr. Michael Studinger/NASA and IBCAO.)*

Figure 5.1 As part of NASA's IceBridge program, low-altitude flights were conducted to support the calibration and validation of the radar altimeter onboard the CryoSat-2, a European Space Agency environmental research satellite. To compensate for the drifting sea ice pack, the flights had to be both spatially and temporally coordinated with the over-pass of the satellite [5].

Following steps similar to those outlined in Chapter 4, the coordination control law can be rewritten in compact form as

$$\begin{aligned} \boldsymbol{u}_{\mathbf{coord}}(t) &= -k_P \boldsymbol{L}(t)\,\boldsymbol{\xi}(t) - k_R \begin{bmatrix} \boldsymbol{C}_\ell^\top \\ \boldsymbol{0} \end{bmatrix} (\boldsymbol{\xi}(t) - \xi_R(t)\mathbf{1}_n) + \begin{bmatrix} \mathbf{1}_{n_\ell} \\ \boldsymbol{\chi}_I(t) \end{bmatrix}, \\ \dot{\boldsymbol{\chi}}_I(t) &= -k_I\, \boldsymbol{C}_f^\top \boldsymbol{L}(t)\,\boldsymbol{\xi}(t), \qquad\qquad \boldsymbol{\chi}_I(0) = \mathbf{1}_{n-n_\ell}, \end{aligned}$$

where the matrices $\boldsymbol{C}_\ell$ and $\boldsymbol{C}_f$ are defined as $\boldsymbol{C}_\ell^\top := [\ \mathbb{I}_{n_\ell} \quad \boldsymbol{0}\ ] \in \mathbb{R}^{n_\ell \times n}$ and $\boldsymbol{C}_f^\top := [\ \boldsymbol{0} \quad \mathbb{I}_{n-n_\ell}\ ] \in \mathbb{R}^{(n-n_\ell)\times n}$, respectively. This coordination control law leads to the closed-loop coordination dynamics

$$\begin{aligned} \dot{\xi}_R(t) &= 1, & \xi_R(0) &= 0, \\ \dot{\boldsymbol{\xi}}(t) &= -k_P \boldsymbol{L}(t)\,\boldsymbol{\xi}(t) - k_R \begin{bmatrix} \boldsymbol{C}_\ell^\top \\ \boldsymbol{0} \end{bmatrix} (\boldsymbol{\xi}(t) - \xi_R(t)\mathbf{1}_n) + \begin{bmatrix} \mathbf{1}_{n_\ell} \\ \boldsymbol{\chi}_I(t) \end{bmatrix}, & \boldsymbol{\xi}(0) &= \boldsymbol{\xi}_0, \\ \dot{\boldsymbol{\chi}}_I(t) &= -k_I\, \boldsymbol{C}_f^\top \boldsymbol{L}(t)\,\boldsymbol{\xi}(t), & \boldsymbol{\chi}_I(0) &= \mathbf{1}_{n-n_\ell}. \end{aligned} \tag{5.3}$$

Finally, similarly to Section 4.2 in Chapter 4, we redefine the coordination error state $\boldsymbol{\zeta}(t) := [\zeta_R(t),\ \boldsymbol{\zeta}_1^\top(t),\ \boldsymbol{\zeta}_2^\top(t)]^\top$ as

$$\begin{aligned} \zeta_R(t) &:= \sqrt{n}\left(\xi_R(t) - \tfrac{1}{n}\mathbf{1}_n^\top \boldsymbol{\xi}(t)\right) && \in \mathbb{R}, \\ \boldsymbol{\zeta}_1(t) &:= \boldsymbol{Q}\,\boldsymbol{\xi}(t) && \in \mathbb{R}^{n-1}, \\ \boldsymbol{\zeta}_2(t) &:= \boldsymbol{\chi}(t) - \mathbf{1}_{n-n_\ell} && \in \mathbb{R}^{n-n_\ell}. \end{aligned}$$

Note that, at the kinematic level, $\boldsymbol{\zeta}(t) = \boldsymbol{0}$ is equivalent to $\boldsymbol{\xi}(t) = \xi_R(t)\mathbf{1}_n$ and $\dot{\boldsymbol{\xi}}(t) = \mathbf{1}_n$, which implies that, if $\boldsymbol{\zeta}(t) = \boldsymbol{0}$, then at time t all target vehicles are coordinated, travel at the desired speed, and meet the absolute temporal assignments as specified in the trajectory-generation step.

With the above notation, the closed-loop coordination dynamics (5.3) can now be reformulated as (see Appendix A in [3])

$$\dot{\boldsymbol{\zeta}}(t) = \boldsymbol{A}_\zeta(t)\,\boldsymbol{\zeta}(t), \qquad \boldsymbol{\zeta}(0) = \boldsymbol{\zeta}_0, \tag{5.4}$$

where $\boldsymbol{A}_\zeta(t)$ is given by

$$\boldsymbol{A}_\zeta(t) := \begin{bmatrix} -k_R \frac{n_\ell}{n} & k_R \frac{1}{\sqrt{n}} \begin{bmatrix} \mathbf{1}_{n_\ell} \\ \boldsymbol{0} \end{bmatrix}^\top \boldsymbol{Q}^\top & -\frac{1}{\sqrt{n}} \mathbf{1}_{n_\ell}^\top \\ k_R \frac{1}{\sqrt{n}} \boldsymbol{Q} \begin{bmatrix} \mathbf{1}_{n_\ell} \\ \boldsymbol{0} \end{bmatrix} & -k_P \tilde{\boldsymbol{L}}(t) & \boldsymbol{Q}\boldsymbol{C}_f \\ \boldsymbol{0} & -k_I \boldsymbol{C}_f^\top \boldsymbol{Q}^\top \bar{\boldsymbol{L}}(t) & \boldsymbol{0} \end{bmatrix} \in \mathbb{R}^{(2n-n_\ell)\times(2n-n_\ell)},$$

with $\tilde{\boldsymbol{L}}(t)$ being defined as $\tilde{\boldsymbol{L}}(t) := \bar{\boldsymbol{L}}(t) + \frac{k_R}{k_P} \boldsymbol{Q}\boldsymbol{C}_\ell {\boldsymbol{C}_\ell}^\top \boldsymbol{Q}^\top \in \mathbb{R}^{(n-1)\times(n-1)}$.

5.2.2 Stability Analysis at the Kinematic Level

Next we show that, if the connectivity of graph $\Gamma(t)$ verifies the PE-like condition (2.8), then the control law (5.1)-(5.2) solves —at the kinematic level— the coordination control problem and ensures that the fleet of vehicles meets the absolute temporal assignments of the mission as specified by the trajectory-generation algorithm. The next theorem summarizes this result.

Theorem 5.1. *Consider the coordination error dynamics* (5.4) *and assume that the information flow satisfies the PE-like condition* (2.8) *for some parameters* $\mu, T > 0$. *Then, there exist coordination control gains* k_P, k_R, *and* k_I *such that the origin of the error dynamics* (5.4) *is exponentially stable with guaranteed rate of convergence*

$$\bar{\lambda}_{cd}^R := \frac{k_P n \mu}{\left(1 + k_P\left(n + \frac{k_R}{k_P}\right)T\right)^2}\left(1 + \left(1 + \tfrac{\rho}{1-\rho}\nu\right)\rho\frac{n}{n_\ell}\right)^{-1},$$

where $\rho \geq 2$ *is a design parameter, and* $\nu := \nu\left(k_R\frac{k_P}{k_I}, \frac{n}{n_\ell}\right)$ *is a known nonlinear function with* $\nu\left(0, \frac{n}{n_\ell}\right) = 0$. *Furthermore, the coordination states* $\xi_i(t)$ *and their rates of change* $\dot{\xi}_i(t)$ *satisfy*

$$\begin{aligned}
|\xi_i(t) - \xi_j(t)| &\leq \kappa_{\xi_r}\|\boldsymbol{\zeta}(0)\|\,\mathrm{e}^{-\bar{\lambda}_{cd}^R t}, &\quad& \forall\, i, j \in \{1,\ldots,n\},\\
|\dot{\xi}_i(t) - 1| &\leq \kappa_{\dot{\xi}}\|\boldsymbol{\zeta}(0)\|\,\mathrm{e}^{-\bar{\lambda}_{cd}^R t}, && \forall\, i \in \{1,\ldots,n\},\\
|\xi_i(t) - t| &\leq \kappa_{\xi}\|\boldsymbol{\zeta}(0)\|\,\mathrm{e}^{-\bar{\lambda}_{cd}^R t}, && \forall\, i \in \{1,\ldots,n\},
\end{aligned}$$

for some constants $\kappa_{\xi_r}, \kappa_{\dot{\xi}}, \kappa_{\xi} \in (0, \infty)$. ◇

Proof. The proof of this result is given in [3]. □

Remark 5.1. Similarly to the proof of Lemma 4.1 in Chapter 4, the proof of Theorem 5.1 above is constructive and explicitly specifies a particular choice for the control gains k_P, k_R, and k_I that ensures exponential stability of the closed-loop coordination dynamics; see Appendix B in [3]. △

Remark 5.2. If the coordination control gain k_R is set to zero, the coordination control (5.1)-(5.2) reduces to the original proportional-integral control law in (4.6) for missions with only relative temporal constraints. In this case, one recovers the rate of exponential convergence derived in Lemma 4.1:

$$\bar{\lambda}_{cd} = \bar{\lambda}_{cd}^R\Big|_{k_R=0} = \frac{k_P n \mu}{(1 + k_P n T)^2}\left(1 + \rho\frac{n}{n_\ell}\right)^{-1}, \qquad \rho \geq 2. \quad △$$

5.3 Coordination with Loose Absolute Temporal Constraints

The coordination control law proposed in the previous section ensures that the fleet of vehicles strictly observes the absolute temporal constraints of the mission as planned by the trajectory-generation algorithm. As discussed earlier in this chapter, however, some mission scenarios require that the fleet of cooperating vehicles meet loose absolute temporal constraints (such as arrival within a desired temporal window), while meeting at the same time a desired inter-vehicle schedule. Next, we present a modification of the coordination control law (5.2) that, without strictly enforcing a desired timing plan, confines temporal deviations with respect to this plan within a prespecified window and, at the same time, still allows the vehicles to coordinate their positions along the paths so as to meet the relative temporal assignments of the mission.

To this end, we let the clock reference vehicle evolve with the same dynamics as in (5.1)

$$\dot{\xi}_R(t) = 1\,, \qquad \xi_R(0) = 0\,,$$

and adopt the following distributed control law for vehicle coordination:

$$\begin{aligned}
u_{\text{coord},i}(t) &= -k_P \sum_{j\in\mathcal{N}_i(t)} (\xi_i(t) - \xi_j(t)) - \alpha_i^R(t) k_R (\xi_i(t) - \xi_R(t)) + 1\,, \\
&\qquad\qquad\qquad\qquad\qquad\qquad\qquad\qquad i = 1, \ldots, n_\ell\,, \\
\left.\begin{aligned}
u_{\text{coord},i}(t) &= -k_P \sum_{j\in\mathcal{N}_i(t)} (\xi_i(t) - \xi_j(t)) + \chi_{I,i}(t) \\
\dot{\chi}_{I,i}(t) &= -k_I \sum_{j\in\mathcal{N}_i(t)} (\xi_i(t) - \xi_j(t))\,, \qquad \chi_{I,i}(0) = 1
\end{aligned}\right.&\,, \qquad i = n_\ell + 1, \ldots, n\,.
\end{aligned} \tag{5.5}$$

In the above control law, the influence of the clock vehicle on the coordination control input of the leaders is adjusted through the time-varying link weights $\alpha_i^R(t)$, $i \in \{1, \ldots, n_\ell\}$. Various strategies can be adopted to design the dynamics of the weights $\alpha_i^R(t)$. Here, we propose the following hybrid approach:

$$\alpha_i^R(t) = \begin{cases} 0\,, & \text{if } |\xi_i(t) - \xi_R(t)| \leq \Delta_d\,, \\ 1\,, & \text{if } |\xi_i(t) - \xi_R(t)| > \Delta_d\,, \end{cases} \qquad \forall\, i \in \{1, \ldots, n_\ell\}\,, \tag{5.6}$$

with conveniently selected dwell times between activation/deactivation events of a particular weight. With this strategy, the link between the ith leader and the clock agent is activated only if the coordination state of this leader vehicle, $\xi_i(t)$, falls outside the reference temporal window $\tau_R(t) := [\xi_R(t) - \Delta_d, \xi_R(t) + \Delta_d]$. Clearly, activation of the link weight $\alpha_i^R(t)$ draws the ith leader towards the reference window $\tau_R(t)$. However, unlike the coordination control law (5.2), the control law above does not necessarily enforce coordination of the fleet of vehicles with the clock reference vehicle. In fact, if at a given time t all of the coordination states of the leader vehicles are such that $\xi_i(t) \in \tau_R(t)$, $i \in \{1, \ldots, n_\ell\}$, then the control law above reduces to the proportional-integral coordination control law in Chapter 4, which is only designed to enforce the relative temporal assignments of the mission. The dwell times between activation/deactivation events prevent infinitely fast switching [1].

The coordination control law described in this section with the hybrid strategy for the activation of the weights $\alpha_i^R(t)$ in (5.6) leads to linear closed-loop coordination error dynamics with state-dependent switching. Two of the individual subsystems in the family of systems defining these switched dynamics correspond to (i) the system in (5.4), which arises when all link weights are equal to one, and (ii) system (4.9) in Chapter 4, resulting from the coordination control law for relative temporal constraints. While these two error subsystems have been proven to be exponentially stable (and it can be easily proven that the other error subsystems in the family are also stable), the stability of the resulting switched system under state-dependent switching cannot be guaranteed. The derivation of conditions under which these closed-loop coordination error dynamics are stable is an open problem and will be the subject of future research. In the next section, we present simulation results that illustrate the efficacy of the proposed control law for loose absolute temporal constraints, and compare its performance with other coordination control laws presented in the book.

5.4 Illustrative Example: Sequential Auto-Landing with Predefined Arrival Windows

In this section, we consider a sequential auto-landing mission scenario with five small tactical UAVs, and implement the coordination control laws for absolute temporal constraints presented earlier. Similarly to the results presented in Chapters 3 and 4, the simulations in this section are based on the kinematic model of the UAV in (3.16) along with a simplified, decoupled linear model of the UAV with its autopilot.

In this set of simulations, the path-following controller gains are selected as follows:

$$k_\ell = 0.20\,[1/\mathrm{s}]\,, \qquad k_{\tilde{R}} = 0.50\,[1/\mathrm{s}]\,, \qquad d = 125\,[\mathrm{m}]\,,$$

while the control gains for the coordination algorithm are chosen as

$$k_P = 1.0\;10^{-1}\,[1/\mathrm{s}]\,, \qquad k_R = 6.0\;10^{-2}\,[1/\mathrm{s}]\,, \qquad k_I = 1.0\;10^{-2}\,[1/\mathrm{s}^2]\,,$$

$$k_{\mathrm{aw}} = 2.0\;10^{-3}\,[1/\mathrm{m}]\,, \qquad G_{\mathrm{aw}}(s) = \frac{1}{s+1}\,.$$

In all of the simulations in this section, vehicles 1 and 5 are elected as leaders of the fleet. The angular-rate commands are saturated to ± 0.3 rad/s, and the speed commands are saturated between 18 m/s and 32 m/s. The information flow is assumed to be bidirectional, time-varying, and the underlying communications graph is not necessarily connected pointwise in time.

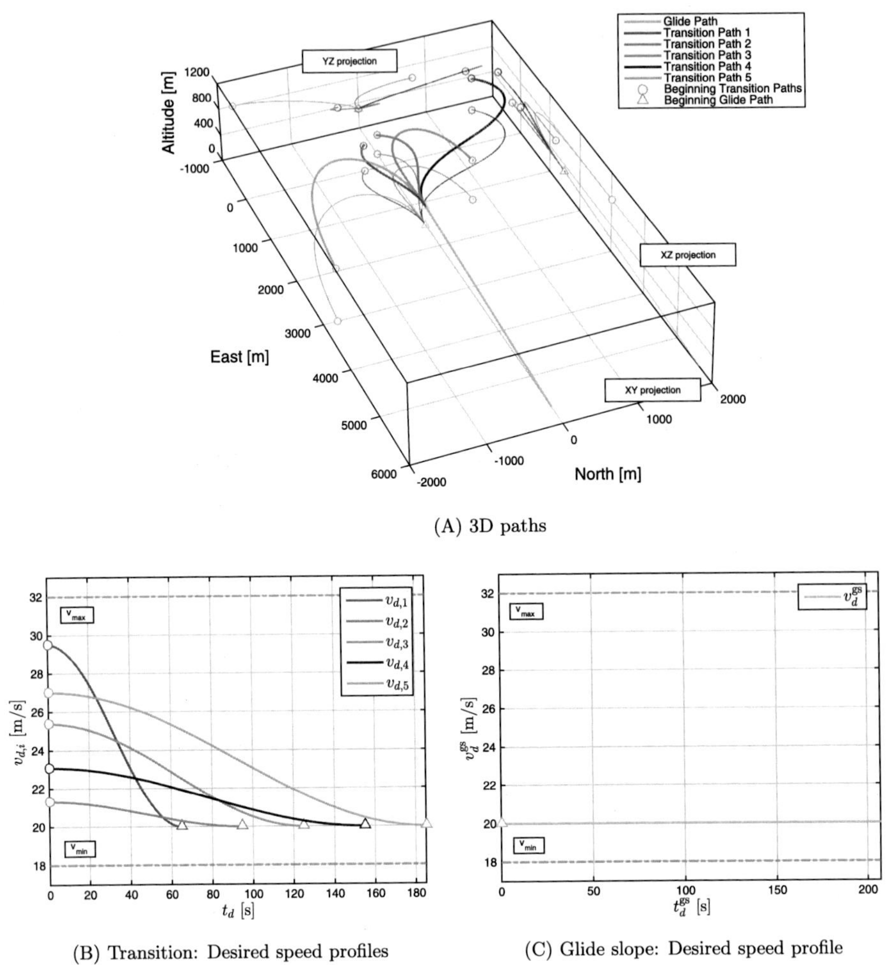

(A) 3D paths

(B) Transition: Desired speed profiles

(C) Glide slope: Desired speed profile

Figure 5.2 Sequential auto-landing. 3D paths along with the corresponding desired speed profiles for both the transition trajectories and the glide slope.

5.4.1 *Transition Trajectories and Glide Slope*

In this scenario, five UAVs are requested to autonomously land at a pre-specified airfield, by following a designated glide slope and separated by a pre-specified safe-guarding time-interval of 30 s. To this end, time-deconflicted transition trajectories are generated from pre-specified initial conditions to the beginning of the glide path, satisfying the desired inter-vehicle arrival schedule, and taking the UAVs to the desired approach speed of 20 m/s. Fig. 5.2 shows the five transition paths as well as the 3-deg glide slope. The beginning of each transition path is indicated with a circle, while the beginning of the glide slope is indicated with a triangle. The figure also presents the desired speed profiles for the initial transition phase that satisfy the desired inter-vehicle arrival schedule, ensure a minimal inter-vehicle separation of 300 m, and comply with the desired final approach speed. Finally, the figure also includes the desired speed profile for the approach along the glide slope. The transition paths have lengths $\ell_{f1} = 1609.0$ m, $\ell_{f2} = 1,962.7$ m, $\ell_{f3} = 2836.7$ m, $\ell_{f4} = 3338.3$ m, and $\ell_{f5} = 4348.4$ m, while the desired times of arrival at the glide slope are $t_{d1}^{\mathrm{f}} = 65.0$ s, $t_{d2}^{\mathrm{f}} = 95.0$ s, $t_{d3}^{\mathrm{f}} = 125.0$ s, $t_{d4}^{\mathrm{f}} = 155.0$ s, and $t_{d5}^{\mathrm{f}} = 185.0$ s. Fig. 5.3 presents the path and inter-vehicle separations, which show that the five transition paths meet at their end positions (beginning of the glide slope), while the planned inter-vehicle separations satisfy the desired clearance of 300 m.

5.4.2 Mission Execution

Loose Absolute Temporal Constraints

In this first set of simulations, the vehicles are allowed to deviate from the desired timing plan by at most ± 5 s, and must observe the desired inter-vehicle separation of 30 s as they execute the sequential auto-landing. To solve this coordination problem, we use the control law for loose absolute temporal constraints in (5.5) with the hybrid strategy for the link weights $\alpha_i^R(t)$. The parameter Δ_d is set to 5 s, while the dwell times for activation and deactivation of these links are set to 3 s and 1 s, respectively.

Fig. 5.4 illustrates the evolution of the UAVs moving along the paths. The UAVs start the mission with an initial offset in both position and attitude with respect to the beginning of the transition paths. As can be seen in the figure, the path-following algorithm eliminates this initial offset and steers the UAVs along the corresponding transition paths, while the coordination algorithm ensures that the UAVs reach the glide slope within the pre-specified arrival windows and separated by the desired time-interval of 30 s. In fact, the UAVs reach the glide slope at $t = 68.2$ s, $t = 97.6$ s, $t = 127.5$ s, $t = 157.4$ s, and $t = 187.4$ s, thus meeting the loose absolute temporal specifications as well as the desired inter-vehicle separation. After reaching the glide slope, the path-following algorithm ensures that the UAVs stay on the glide path as the coordination algorithm maintains the safe-guarding separation. The simulation is stopped when the first UAV reaches the end of the glide path.

Details about the performance of the path-following control algorithm are shown in Fig. 5.5. As illustrated in Figs. 5.5A and 5.5B, the path-following position and attitude errors, $\boldsymbol{p}_{F,i}(t)$ and $\Psi(\tilde{\boldsymbol{R}}_i(t))$, respectively converge to a neighborhood of zero within 40 s and 10 s. Figs. 5.5C and 5.5D present the angular-rate commands, $q_{c,i}(t)$ and $r_{c,i}(t)$, in dashed lines, the actual angular rates, $q_i(t)$ and $r_i(t)$, in solid lines, as well as the rate of progression $\dot{\ell}_i(t)$ of the virtual targets along the paths.

Fig. 5.6 illustrates the performance of the coordination control law. Fig. 5.6A shows that the coordination errors $(\xi_i(t) - \xi_R(t))$ converge to and stay within the desired ± 5 s temporal window, while the rate of change of the coordination states $\dot{\xi}_i(t)$ converges to a neighborhood of the desired rate of 1 s/s. This figure also shows the UAV speeds and the integral states implemented on the follower vehicles. In particular, Fig. 5.6B shows that, after a transient caused by the initial path-following errors

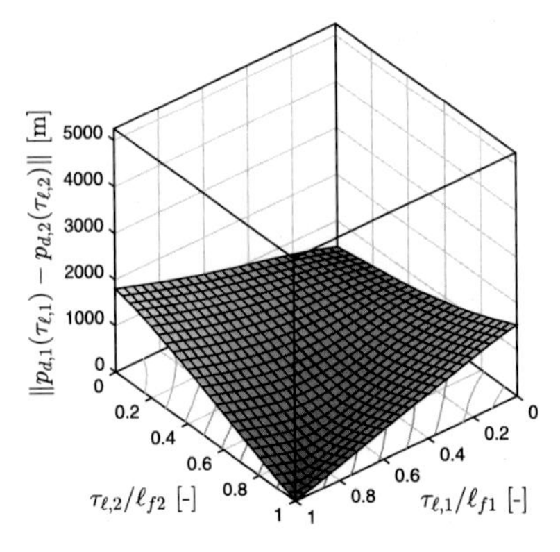

(A) Separation between paths 1 and 2

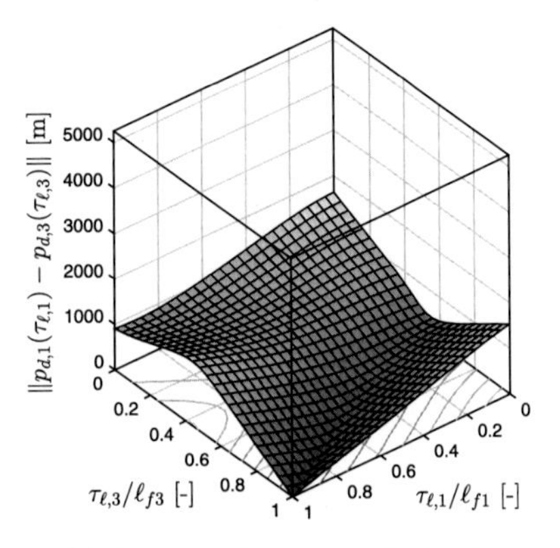

(B) Separation between paths 1 and 3

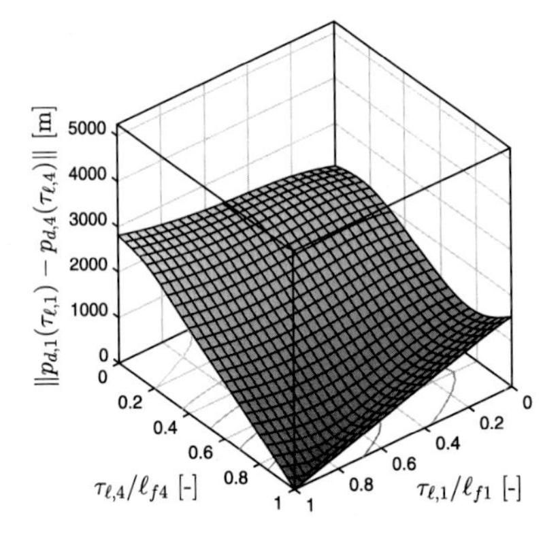

(C) Separation between paths 1 and 4

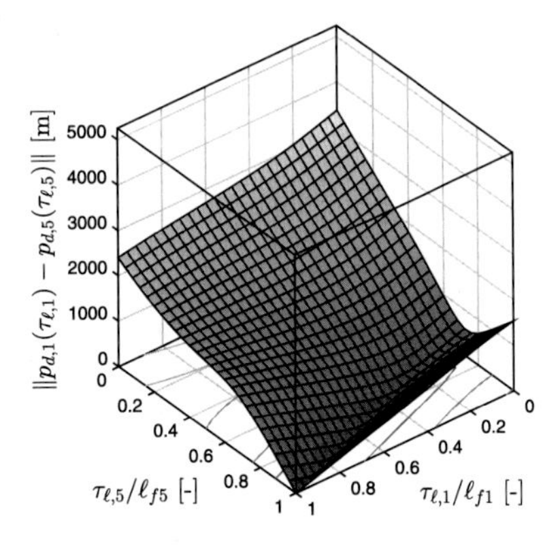

(D) Separation between paths 1 and 5

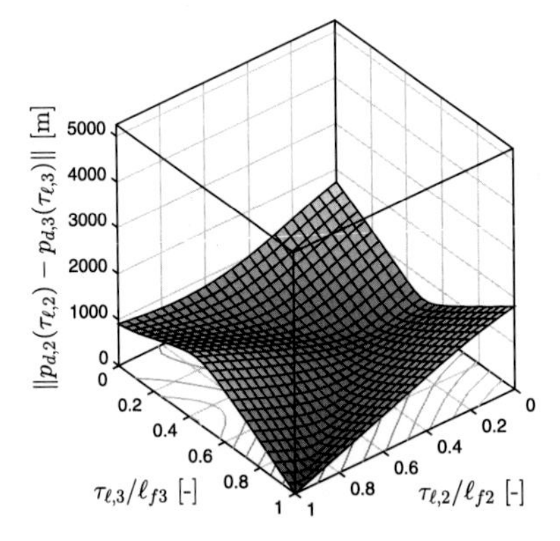

(E) Separation between paths 2 and 3

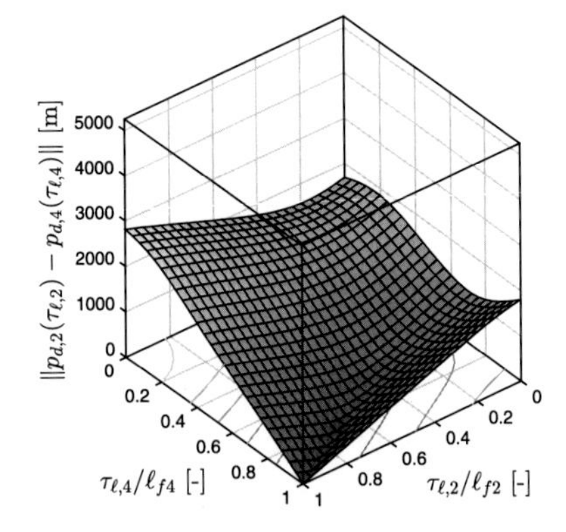

(F) Separation between paths 2 and 4

Figure 5.3 Sequential auto-landing. Path separation and inter-vehicle separation during the transition phase; the speed profiles ensure deconfliction of the five planned trajectories with a minimum clearance of 300 m.

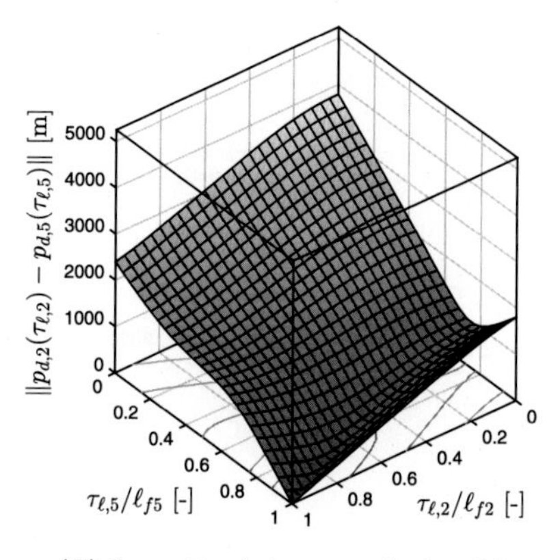

(G) Separation between paths 2 and 5

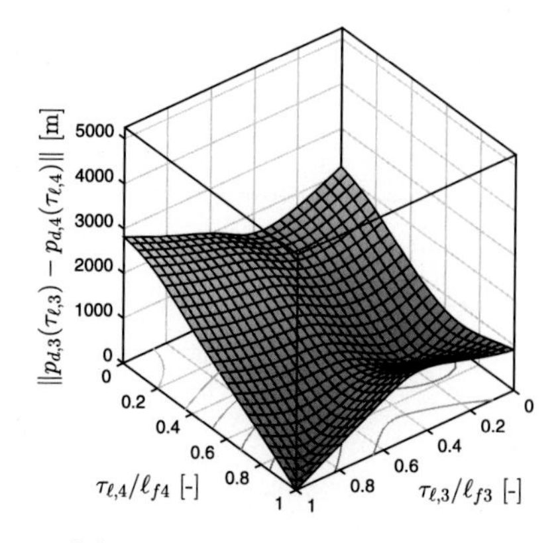

(H) Separation between paths 3 and 4

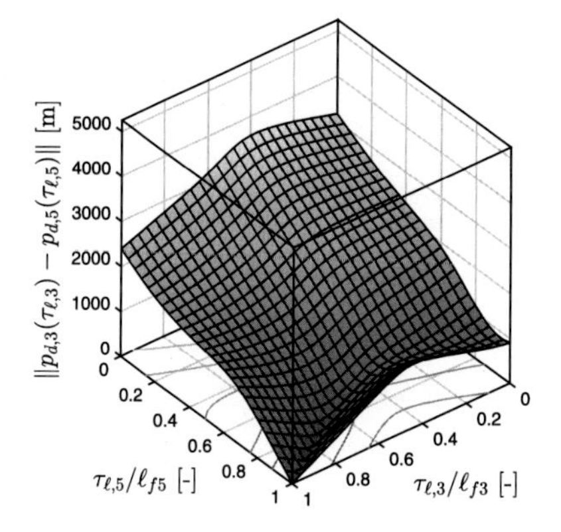

(I) Separation between paths 3 and 5

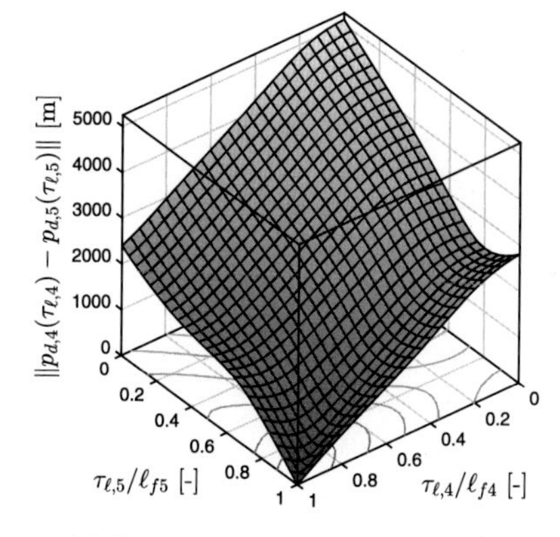

(J) Separation between paths 4 and 5

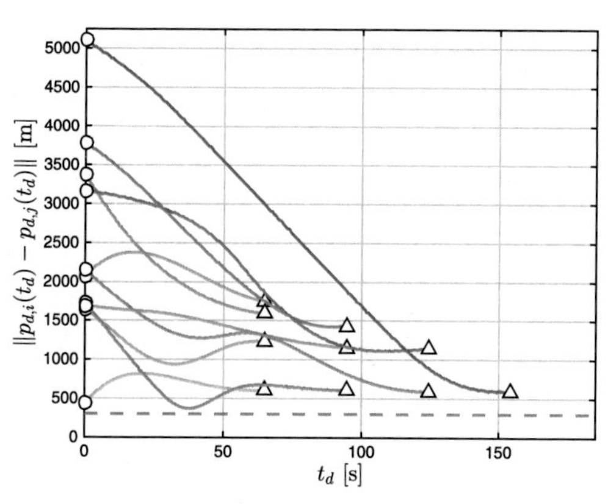

(K) Desired inter-vehicle separation

Figure 5.3 *(continued)*

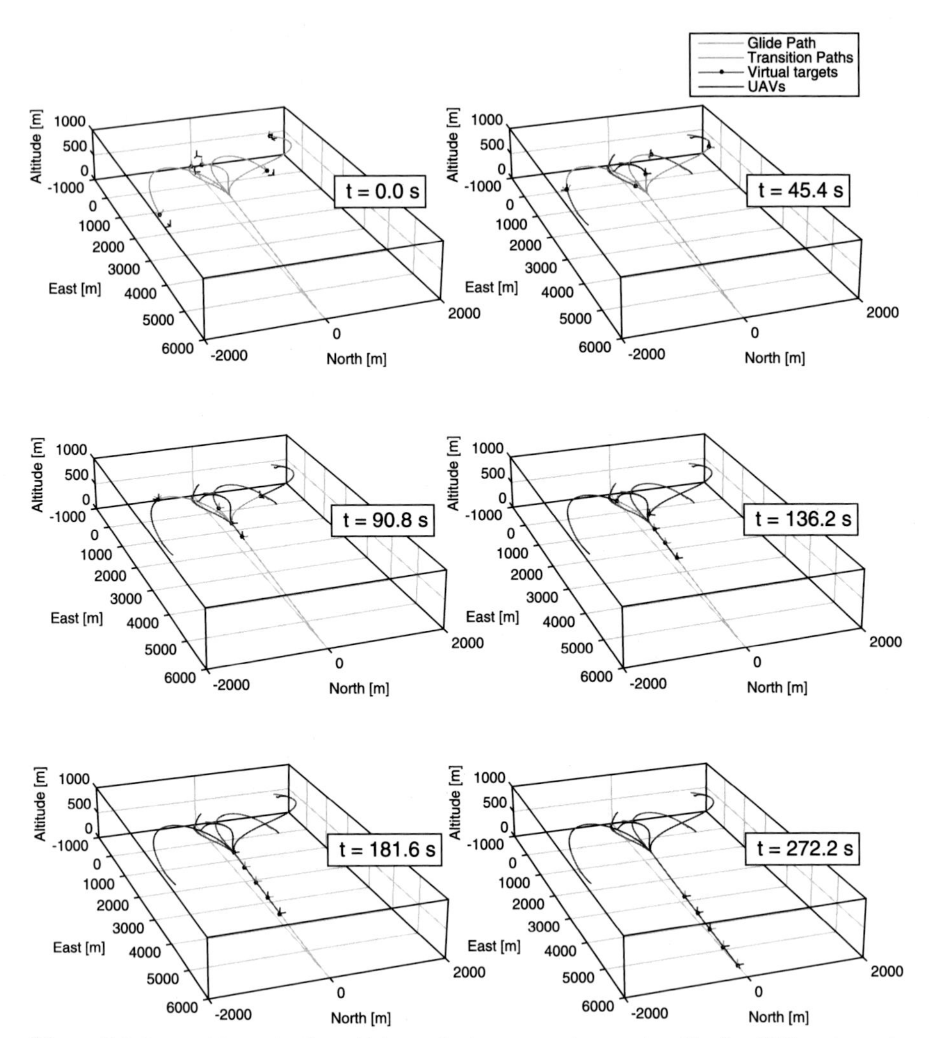

Figure 5.4 Sequential auto-landing with loose absolute temporal constraints. The five UAVs arrive at the beginning of the glide path within the pre-specified arrival windows and separated by approximately 30 s, and maintain this safe-guarding separation as they fly along the glide slope.

as well as the speed corrections introduced by the coordination control law, the speed of each UAV (solid thick lines) converges to its desired speed (dashed lines) and, as the vehicles enter the glide path, their speeds converge to the desired approach speed of 20 m/s. Notice that, as a result of the switching nature of the network topology, the speed commands (solid thin lines) of the five vehicles are discontinuous. In all of these figures, small triangles indicate the arrival of the UAVs at the glide slope. The coordination control law also guarantees that the vehicles always maintain the desired inter-vehicle separation of 300 m; see Fig. 5.7. Finally, Fig. 5.8 presents an estimate

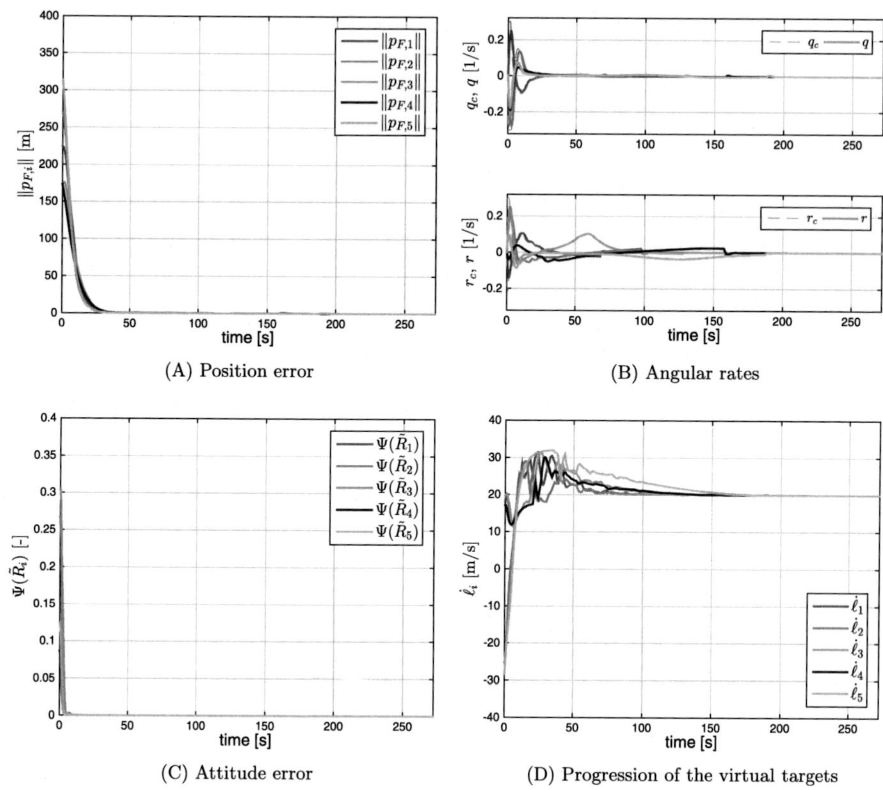

(A) Position error

(B) Angular rates

(C) Attitude error

(D) Progression of the virtual targets

Figure 5.5 Sequential auto-landing with loose absolute temporal constraints. The path-following algorithm drives the path-following position and attitude errors to a neighborhood of zero.

of the QoS of the network, computed as

$$\hat{\mu}(t) := \lambda_{\min}\left(\frac{1}{5}\frac{1}{T}\int_{t-T}^{t} \boldsymbol{Q}_5 \boldsymbol{L}(\tau)\boldsymbol{Q}_5^\top d\tau\right), \qquad t \geq T,$$

with $T = 10$ s. The network topology changes every 2 s and, as can be seen in the figure, the QoS estimate is always greater than 0.3.

Strict Absolute Temporal Constraints

Next we show simulation results with the coordination control law for strict absolute temporal constraints in (5.2). Only the results for the coordination dynamics are reported here, as there are no significant differences regarding path-following performance.

As can be seen in Fig. 5.9, the coordination control law ensures, on the one hand, that the vehicles coordinate their relative positions along the paths so as to observe the desired inter-vehicle separation of 30 s and, on the other hand, that they coordinate their absolute positions with the (virtual) clock reference vehicle. In fact, the

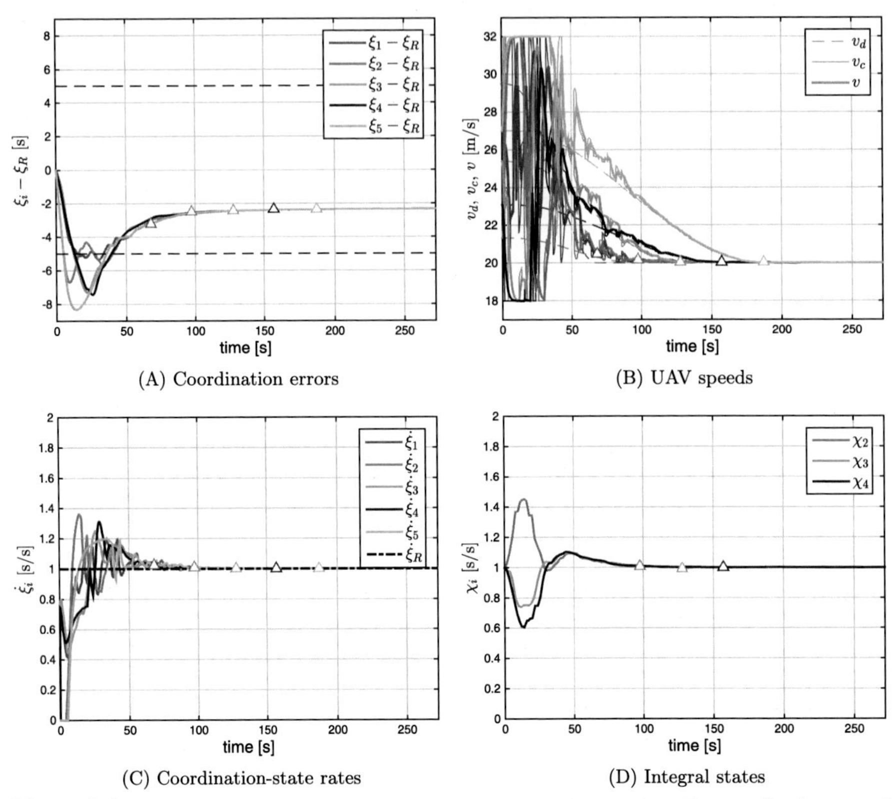

(A) Coordination errors (B) UAV speeds

(C) Coordination-state rates (D) Integral states

Figure 5.6 Sequential auto-landing with loose absolute temporal constraints. The coordination control law ensures that the coordination errors converge to and stay within the desired ± 5 s temporal window and also that the rate of change of the coordination states evolves at about the desired rate of 1 s/s.

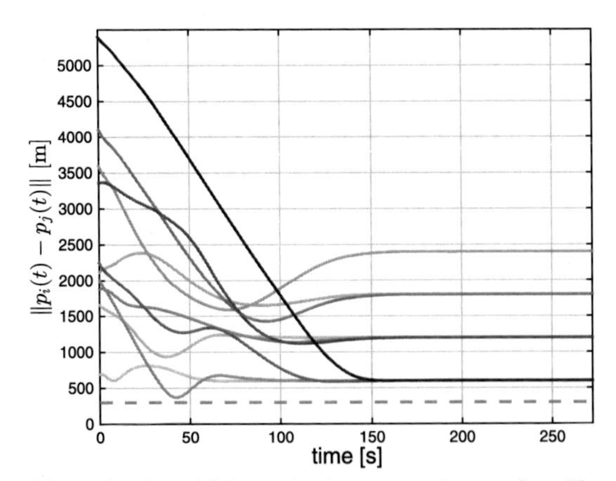

Figure 5.7 Sequential auto-landing with loose absolute temporal constraints. The coordination control law guarantees that the UAVs always maintain the desired inter-vehicle separation of 300 m.

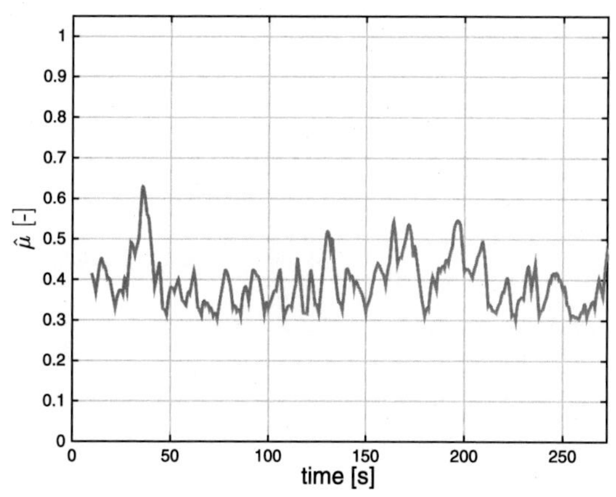

Figure 5.8 Sequential auto-landing with loose absolute temporal constraints. The information flow among UAVs is time-varying, bidirectional, and is only connected in an integral sense.

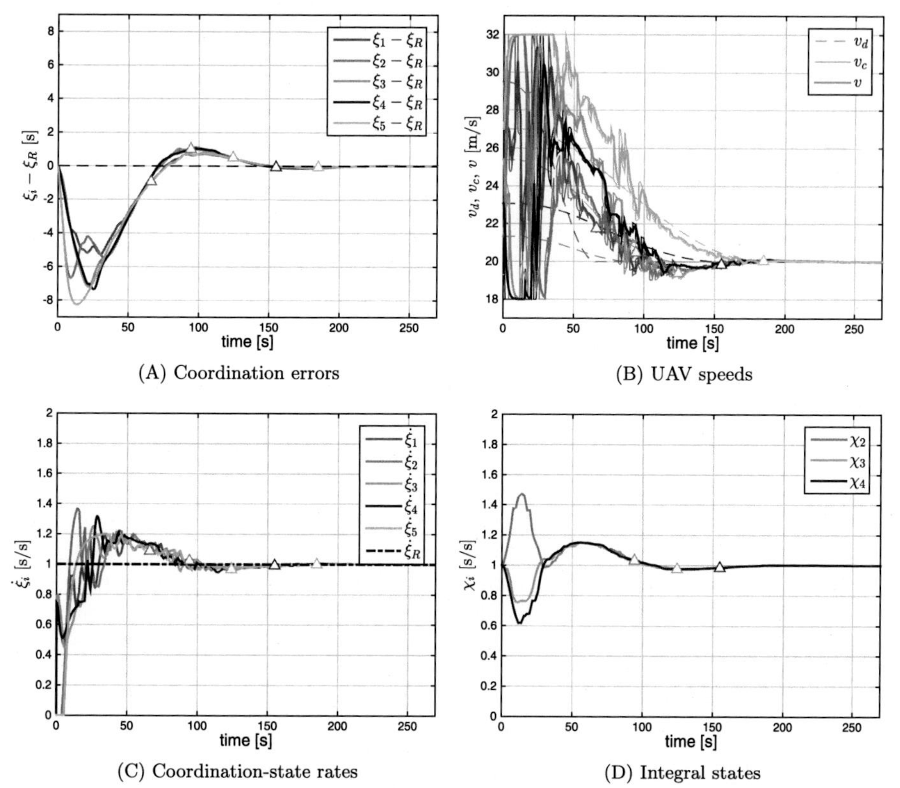

(A) Coordination errors

(B) UAV speeds

(C) Coordination-state rates

(D) Integral states

Figure 5.9 Sequential auto-landing with strict absolute temporal constraints. The coordination control law ensures that the vehicles maintain the desired inter-vehicle separation and, at the same time, meet the absolute temporal assignments of the mission, as specified in the trajectory-generation step.

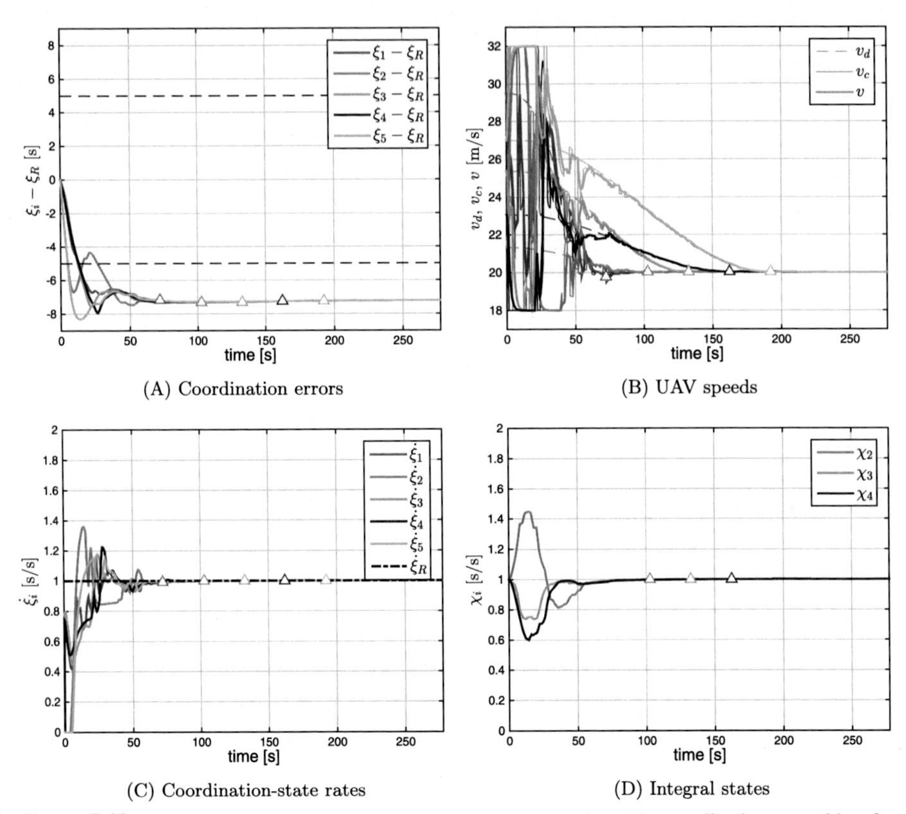

(A) Coordination errors

(B) UAV speeds

(C) Coordination-state rates

(D) Integral states

Figure 5.10 Sequential auto-landing with relative temporal constraints. The coordination control law for relative temporal constraints proposed in Chapter 4 cannot guarantee that the vehicles meet the absolute temporal assignments of the mission.

figure shows that the coordination errors $(\xi_i(t) - \xi_R(t))$ converge to a neighborhood of zero, while the rate of change of the coordination states $\dot{\xi}_i(t)$ converges to a neighborhood of the desired rate of 1 s/s. The UAVs reach the beginning of the glide slope at $t = 66.0$ s, $t = 94.0$ s, $t = 124.5$ s, $t = 155.1$ s, and $t = 185.1$ s, and complete their synchronization with the clock vehicle as they fly along the glide slope. Notice that the selection of control gains is such that coordination among UAVs is achieved significantly faster than coordination with the clock vehicle. This is particularly important to prevent collisions between vehicles, as the desired trajectories are separated in time, but not in space.

Relative Temporal Constraints

Finally, Fig. 5.10 shows simulation results obtained with the coordination control law for relative temporal constraints in Chapter 4. The vehicles coordinate their relative positions along the paths so as to meet the desired inter-vehicle schedule and guarantee collision-free maneuvers. In addition, after an initial transient phase, the speed of each UAV converges to its desired speed profile. However, as can be seen in

Fig. 5.10A, the fleet fails to meet the absolute temporal specifications of the mission. In fact, the UAVs reach the beginning of the glide slope at $t = 72.2$ s, $t = 102.4$ s, $t = 132.3$ s, $t = 162.3$ s, and $t = 192.3$ s, with a delay of more than 2 seconds with respect to the desired arrival windows.

References

[1] D. Liberzon, Switching in Systems and Control, Birkhäuser, Boston, MA, 2003.
[2] T.W. McLain, R.W. Beard, Coordination variables, coordination functions, and cooperative timing missions, Journal of Guidance, Control and Dynamics 28 (2005) 150–161.
[3] J. Puig-Navarro, Time-Critical Coordination Control of Multiple Unmanned Aerial Vehicles with Absolute Temporal Constraints, Master's thesis, Universitat Politècnica de València, València, Spain, January 2015.
[4] J. Puig-Navarro, E. Xargay, R. Choe, N. Hovakimyan, I. Kaminer, Time-critical coordination control of multiple UAVs with absolute temporal constraints, in: AIAA Guidance, Navigation and Control Conference, Kissimmee, FL, January 2015, AIAA 2015-0595.
[5] M. Studinger, Operation IceBridge – Personal Blog, http://blogs.nasa.gov/icebridge/tag/studinger/ [Online; accessed 19 April 2016].

Time Coordination Under Quantization

6

When vehicles communicate over a network with finite-rate communication links, the exchanged information must be quantized into an appropriate finite number of bits. In this chapter, we analyze the effect of quantization on the stability and convergence properties of the closed-loop coordination dynamics. In particular, we restrict the analysis to the problem of time-critical coordination with relative temporal specifications at the kinematic level; stability of the combined coordination and path-following systems with vehicle dynamics can be investigated following an approach similar to that in Chapter 4. Also, for the sake of simplicity, we consider only the case of *uniform quantization* with step size Δ. The results in this chapter show that, under sufficiently fine quantization, the (quantized) coordination control law solves the coordination control problem in a practical sense, and the coordination error degrades gracefully with the value of the quantizer step size. In addition, we also prove that, depending on the design of the quantized coordination control law, the closed-loop kinematic coordination error dynamics may have undesirable "zero-speed" attractors. From a mathematical point of view, a consequence of quantization is that the resulting dynamics are not guaranteed to admit solutions either in the classical sense or in the sense of Carathéodory. This implies that a weaker concept of solution has to be considered for the derivations in this chapter. The results presented in this chapter are the outcome of joint work with R. Choe; see [3].

6.1 Convergence with Quantized Information

6.1.1 Coordination Control Law and Coordination Dynamics

When only quantized information is exchanged among different vehicles, the distributed coordination control law for relative temporal specifications in (4.6) becomes

$$\begin{aligned} u_{\mathrm{coord}} &= -k_P\big(\boldsymbol{D}(t)\,\boldsymbol{\xi} - \boldsymbol{A}(t)\;\mathbf{q}(\boldsymbol{\xi})\big) + \begin{bmatrix} \mathbf{1}_{n_\ell} \\ \chi_I \end{bmatrix}, \\ \dot{\chi}_I &= -k_I\,\boldsymbol{C}^\top\big(\boldsymbol{D}(t)\,\boldsymbol{\xi} - \boldsymbol{A}(t)\;\mathbf{q}(\boldsymbol{\xi})\big), \qquad \chi_I(0) = \mathbf{1}_{n-n_\ell}, \end{aligned} \tag{6.1}$$

where the time-varying matrices $\boldsymbol{D}(t)$ and $\boldsymbol{A}(t)$ are respectively the *degree* and *adjacency matrices* of $\boldsymbol{L}(t)$, while $\mathbf{q}(\boldsymbol{\xi}(t)) \in \Delta\mathbb{Z}^n$ is the quantized coordination state

$$\mathbf{q}(\boldsymbol{\xi}(t)) := [\,\mathrm{q}_\Delta(\xi_1(t)), \ldots, \mathrm{q}_\Delta(\xi_n(t))]^\top,$$

with $\mathrm{q}_\Delta(\cdot) : \mathbb{R} \to \Delta\mathbb{Z}$ defined as

$$\mathrm{q}_\Delta(\phi) := \mathrm{sgn}(\phi)\Delta\left\lfloor \tfrac{|\phi|}{\Delta} + \tfrac{1}{2} \right\rfloor, \qquad \phi \in \mathbb{R}.$$

Time-Critical Cooperative Control of Autonomous Air Vehicles, DOI: 10.1016/B978-0-12-809946-9.00008-7

Here, $\Delta\mathbb{Z}$ denotes the set of the (integer) multiples of the quantization step size Δ, while the notation $\lfloor\cdot\rfloor$ represents the floor function. Then, the closed-loop kinematic coordination dynamics can be written as

$$\begin{aligned} \dot{\boldsymbol{\xi}} &= -k_P \left(\boldsymbol{D}(t)\boldsymbol{\xi} - \boldsymbol{A}(t)\,\mathbf{q}(\boldsymbol{\xi})\right) + \begin{bmatrix} \mathbf{1}_{n_\ell} \\ \boldsymbol{\chi}_I \end{bmatrix}, & \boldsymbol{\xi}(0) &= \boldsymbol{\xi}_0, \\ \dot{\boldsymbol{\chi}}_I &= -k_I\, \boldsymbol{C}^\top \left(\boldsymbol{D}(t)\boldsymbol{\xi} - \boldsymbol{A}(t)\,\mathbf{q}(\boldsymbol{\xi})\right), & \boldsymbol{\chi}_I(0) &= \mathbf{1}_{n-n_\ell}. \end{aligned} \tag{6.2}$$

In terms of the coordination error state $\boldsymbol{\zeta}(t)$, and noting that $\boldsymbol{L}(t) = \boldsymbol{D}(t) - \boldsymbol{A}(t)$, the closed-loop coordination error dynamics can be expressed as

$$\dot{\boldsymbol{\zeta}} = \boldsymbol{A}_\zeta(t)\,\boldsymbol{\zeta} + \boldsymbol{f}_{\mathbf{q}}, \qquad \boldsymbol{\zeta}(0) = \boldsymbol{\zeta}_0, \tag{6.3}$$

where $\boldsymbol{A}_\zeta(t)$ was defined in (4.10) and $\boldsymbol{f}_{\mathbf{q}}(t)$ is given by

$$\boldsymbol{f}_{\mathbf{q}}(t) := \begin{bmatrix} k_P\, \boldsymbol{Q}\boldsymbol{A}(t)\, \boldsymbol{e}_\xi(t) \\ k_I\, \boldsymbol{C}^\top \boldsymbol{A}(t)\, \boldsymbol{e}_\xi(t) \end{bmatrix} \qquad \in \mathbb{R}^{2n-n_\ell-1},$$

with $\boldsymbol{e}_\xi(t) := \mathbf{q}(\boldsymbol{\xi}(t)) - \boldsymbol{\xi}(t)$ being the *quantization error vector.*

Note that, in this case, the right-hand side of the kinematic coordination error dynamics (6.3) is discontinuous not only due to the switching network topology, but also due to the presence of quantized states. As proven in [1], Carathéodory solutions may not exist for quantized consensus problems, implying that a weaker concept of solution has to be considered. Inspired by the results in [1], in this chapter we will consider solutions *in the sense of Krasovskii*, which we define next.

Definition 6.1 (Krasovskii solution [2]). Let $\boldsymbol{\phi} : J \to \mathbb{R}^n$ (J is an interval in $\mathbb{R}$) be an absolutely continuous function on each compact subinterval of J. Then, $\boldsymbol{\phi}$ is called a *Krasovskii solution* of the vector differential equation $\dot{\boldsymbol{\phi}} = \boldsymbol{f}(t, \boldsymbol{\phi})$ if

$$\dot{\boldsymbol{\phi}}(t) \in \mathrm{K}\left(\boldsymbol{f}(t, \boldsymbol{\phi}(t))\right) \qquad \text{almost everywhere in } J,$$

where the operator $\mathrm{K}(\cdot)$ is defined as

$$\mathrm{K}\left(\boldsymbol{f}(t, \boldsymbol{\phi})\right) := \bigcap_{\epsilon > 0} \overline{\mathrm{co}}\, \boldsymbol{f}(t, \boldsymbol{\phi} + \epsilon\mathcal{B}),$$

with $\mathcal{B}$ being the open unit ball in $\mathbb{R}^n$. In the above definition, $\overline{\mathrm{co}}$ denotes the closure of the convex hull. ♠

To show that Krasovskii solutions to (6.3) exist (at least) locally, we note that, during continuous evolution of the system between "quantization jumps", the quantized coordination error dynamics (6.3) are linear, with the quantized state $\mathbf{q}(\boldsymbol{\xi}(t))$ acting as a bounded exogenous input. This implies that the solutions $\boldsymbol{\xi}(t)$ are locally bounded (no *finite escape time* occurs). Then, local existence of Krasovskii solutions is guaranteed by the fact that the right-hand side of (6.3) is measurable and locally bounded [2]. At this point, however, we cannot claim that Krasovskii solutions to (6.3) are complete; for this, we will need to prove that solutions are bounded (see Theorem 6.1, Section 6.1.3).

6.1.2 Krasovskii Equilibria

Before investigating the convergence properties of the quantized coordination error dynamics (6.3), in this section we analyze the existence of equilibria for these dynamics. In particular, we show that (*i*) unlike the unquantized case, $\boldsymbol{\zeta}_{\mathbf{eq}} = \mathbf{0}$ is not an equilibrium point of the quantized kinematic coordination error dynamics; and (*ii*) other (undesirable) equilibria might exist, depending on the step size of the quantizers. The first result follows easily from the error dynamics (6.3) and recalling that, at the kinematic level, $\boldsymbol{\zeta}(t) = \mathbf{0}$ is equivalent to $\boldsymbol{\xi}(t) \in \mathrm{span}\{\mathbf{1}_n\}$ and $\dot{\boldsymbol{\xi}}(t) = \mathbf{1}_n$. The proof of the second result is, instead, more involved, and we only show it here for the case of static and connected network topologies.

We start by noting that $\dot{\boldsymbol{\zeta}}(t) \equiv \mathbf{0}$ is equivalent to $\dot{\boldsymbol{\xi}}(t) \in \mathrm{span}\{\mathbf{1}_n\}$ and $\dot{\boldsymbol{\chi}}_I(t) \equiv \mathbf{0}$ holding simultaneously. This implies that $\boldsymbol{\zeta}_{\mathbf{eq}} = [\boldsymbol{\zeta}_{\mathbf{1eq}}^\top, \boldsymbol{\zeta}_{\mathbf{2eq}}^\top]^\top$ is an equilibrium of (6.3) if the following inclusions hold for all $t \geq 0$:

$$\beta(t)\mathbf{1}_n \in \mathrm{K}\left(-k_P\left(\boldsymbol{D}\boldsymbol{\xi}_{\mathbf{eq}}(t) - \boldsymbol{A}\,\mathbf{q}(\boldsymbol{\xi}_{\mathbf{eq}}(t))\right) + \begin{bmatrix} \mathbf{1}_{n_\ell} \\ \boldsymbol{\chi}_{I,\mathrm{eq}} \end{bmatrix}\right),$$

$$\mathbf{0} \in \mathrm{K}\left(-k_I\boldsymbol{C}^\top\left(\boldsymbol{D}\boldsymbol{\xi}_{\mathbf{eq}}(t) - \boldsymbol{A}\,\mathbf{q}(\boldsymbol{\xi}_{\mathbf{eq}}(t))\right)\right),$$

where $\beta(t) \in \mathbb{R}$ is an arbitrary time-varying signal, $\boldsymbol{\xi}_{\mathbf{eq}}(t)$ is a continuous coordination-state trajectory satisfying $\boldsymbol{\zeta}_{\mathbf{1eq}} = \boldsymbol{Q}\boldsymbol{\xi}_{\mathbf{eq}}(t)$, while $\boldsymbol{\chi}_{I,\mathbf{eq}} = \boldsymbol{\zeta}_{\mathbf{2eq}} + \mathbf{1}_{n-n_\ell}$. The second inclusion above and the continuity of $\boldsymbol{\xi}_{\mathbf{eq}}(t)$, along with the fact that the network is assumed to be static and connected, preclude the existence of equilibria involving time-varying coordination-state trajectories, that is, $\beta(t) \equiv 0$ (or equivalently $\dot{\boldsymbol{\xi}}_{\mathbf{eq}}(t) \equiv 0$). Then, the set of (Krasovskii) equilibrium points of the error dynamics (6.3) can be defined as

$$\Theta := \Bigg\{(\boldsymbol{\zeta}_{\mathbf{1eq}}, \boldsymbol{\zeta}_{\mathbf{2eq}}) \in \mathbb{R}^{n-1} \times \mathbb{R}^{n-n_\ell} :$$

$$\boldsymbol{\zeta}_{\mathbf{1eq}} = \boldsymbol{Q}\boldsymbol{\xi}_{\mathbf{eq}},\ \boldsymbol{\zeta}_{\mathbf{2eq}} = \boldsymbol{\chi}_{I,\mathbf{eq}} - \mathbf{1}_{n-n_\ell},\ \mathbf{0} \in \mathrm{K}\left(\begin{bmatrix} -k_P\left(\boldsymbol{D}\boldsymbol{\xi}_{\mathbf{eq}} - \boldsymbol{A}\,\mathbf{q}(\boldsymbol{\xi}_{\mathbf{eq}})\right) + \begin{bmatrix} \mathbf{1}_{n_\ell} \\ \boldsymbol{\chi}_{I,\mathrm{eq}} \end{bmatrix} \\ -k_I\boldsymbol{C}^\top\left(\boldsymbol{D}\boldsymbol{\xi}_{\mathbf{eq}} - \boldsymbol{A}\,\mathbf{q}(\boldsymbol{\xi}_{\mathbf{eq}})\right) \end{bmatrix}\right)\Bigg\}. \tag{6.4}$$

Next, we show that, under sufficiently fine quantization, the set Θ is empty. We also prove, on the contrary, the existence of other (undesirable) "zero-speed" equilibrium points in the presence of coarse quantization.

Lemma 6.1. *Consider the quantized kinematic coordination error dynamics* (6.3), *and assume that the network topology is static and connected. If the step size of the quantizers satisfies*

$$\Delta < \frac{2n_\ell}{(3n - 2n_\ell)(n-1)}\frac{1}{k_P}, \tag{6.5}$$

then the set of equilibrium points Θ is empty. ◇

Proof. The proof of this result is given in Appendix B.2.9. □

The next corollary follows from the proof of Lemma 6.1.

Corollary 6.1. *Consider the quantized kinematic coordination dynamics* (6.2), *and assume that the network topology is static and connected. If the step size of the quantizers is such that*

$$\left\| \frac{1}{k_P} \boldsymbol{D}^{-1} \begin{bmatrix} \mathbf{1}_{n_\ell} \\ \mathbf{0} \end{bmatrix} \right\|_\infty < \frac{\Delta}{2}, \tag{6.6}$$

then, for any $k \in \mathbb{Z}$, the point

$$(\hat{\boldsymbol{\xi}}, \hat{\boldsymbol{\chi}}_I) = \left(k\Delta \mathbf{1}_n + \frac{1}{k_P} \boldsymbol{D}^{-1} \begin{bmatrix} \mathbf{1}_{n_\ell} \\ \mathbf{0} \end{bmatrix}, \mathbf{0} \right) \tag{6.7}$$

is a "zero-speed" equilibrium point of system (6.2). ◇

Proposition 6.1. *Consider the quantized kinematic coordination dynamics* (6.2), *and assume that the network topology is static and connected. Further, assume that the step size of the quantizers satisfies inequality* (6.6). *Then, the "zero-speed" equilibrium points defined in* (6.7) *are locally asymptotically stable.* ◇

Proof. The proof of this result is given in Appendix B.2.10. □

Remark 6.1. The equilibrium points characterized by (6.7) correspond to solutions in which the vehicles have zero groundspeed and, therefore, are to be avoided. Unfortunately, as shown in Proposition 6.1, these equilibrium points are asymptotically stable, which implies that the quantizers must be designed to preclude the existence of such equilibria. The bound in (6.5) should thus be understood as a design constraint for the quantizers that prevents the existence of such undesirable equilibria. △

Remark 6.2. From inequality (6.6), it follows that all of the components of $\hat{\boldsymbol{\xi}}$ fall into the same quantization level, with the components corresponding to the follower vehicles at its center. Moreover, note that all of the equilibrium points characterized by (6.7) map to the same coordination error state

$$(\hat{\boldsymbol{\zeta}}_1, \hat{\boldsymbol{\zeta}}_2) = \left(\frac{1}{k_P} \boldsymbol{D}^{-1} \begin{bmatrix} \mathbf{1}_{n_\ell} \\ \mathbf{0} \end{bmatrix}, -\mathbf{1}_{n-n_\ell} \right).$$

△

6.1.3 Stability Analysis at the Kinematic Level

Next we show that, if the connectivity of graph $\Gamma(t)$ verifies the PE-like condition (2.8), then the coordination control law (6.1) solves —at the kinematic level— the coordination control problem in a practical sense and, in addition, the coordination error vector degrades gracefully with the value of the quantizer step size. The next theorem summarizes this result.

Theorem 6.1. *Consider the closed-loop kinematic coordination error dynamics* (6.3) *and suppose that the information flow satisfies the PE-like condition* (2.8) *for some parameters* $\mu, T > 0$. *Then, there exist coordination control gains* k_P *and* k_I *such that the solution of the quantized coordination error dynamics* (6.3) *satisfies*

$$\|\boldsymbol{\zeta}(t)\| \leq \kappa'_{\zeta 0}\|\boldsymbol{\zeta}(0)\|\mathrm{e}^{-\lambda'_{cd}t} + \kappa'_{\zeta 1}\Delta\,, \qquad \text{for all } t \geq 0\,, \tag{6.8}$$

for some constants $\kappa'_{\zeta 0}, \kappa'_{\zeta 1} \in (0, \infty)$, *and with* $\lambda'_{cd} := \bar{\lambda}_{cd}(1 - \theta'_{\lambda})$, *where* $\bar{\lambda}_{cd}$ *was defined in* (4.11) *and* θ'_{λ} *is a constant verifying* $0 < \theta'_{\lambda} < 1$. ◇

Proof. The proof of this result is given in Appendix B.2.11. □

Remark 6.3. The theorem above admits a slightly stronger version. In fact, from the proof of the theorem it follows that the coordination error state $\boldsymbol{\zeta}(t)$ is uniformly bounded for all $t \geq 0$ and uniformly ultimately bounded with ultimate bound proportional to the step size of the quantizers. However, we prefer to present here an ISS-type result similar to the one in Lemma 4.2, for such a bound is more convenient when proving stability of the overall cooperative path-following system. △

Remark 6.4. As mentioned earlier, in this chapter we only investigate the stability and convergence properties of the time-coordination problem at the kinematic level. Stability of the overall cooperative control architecture, including vehicle dynamics, is not addressed here and can be analyzed following an approach similar to that in Chapter 4 (Sections 4.2.2 and 4.3). This analysis should yield additional design constraints for the step size of the quantizers, similar to the constraints on the performance bounds in Theorem 4.2. △

6.1.4 Coordination with Fully Quantized Information

In this section we propose a modification of the coordination control law (6.1) that retains $\boldsymbol{\zeta}_{\mathbf{eq}} = \mathbf{0}$ as an equilibrium point of the resulting quantized kinematic coordination error dynamics. In addition, we will show that, for the case of connected network topologies and sufficiently fine quantization, $\boldsymbol{\zeta}_{\mathbf{eq}} = \mathbf{0}$ is the only equilibrium point.

To this end, consider the distributed control law

$$\begin{aligned} u_{\mathbf{coord}} &= -k_P \boldsymbol{L}(t)\,\mathbf{q}(\boldsymbol{\xi}) + \begin{bmatrix} \mathbf{1}_{n_\ell} \\ \boldsymbol{\chi}_I \end{bmatrix}, \\ \dot{\boldsymbol{\chi}}_I &= -k_I\, \boldsymbol{C}^\top \boldsymbol{L}(t)\,\mathbf{q}(\boldsymbol{\xi})\,, \qquad \boldsymbol{\chi}_I(0) = \mathbf{1}_{n-n_\ell}\,, \end{aligned} \tag{6.9}$$

which, unlike control law (6.1), uses only quantized information. The kinematic coordination dynamics can now be written as

$$\begin{aligned} \dot{\boldsymbol{\xi}} &= -k_P \boldsymbol{L}(t)\,\mathbf{q}(\boldsymbol{\xi}) + \begin{bmatrix} \mathbf{1}_{n_\ell} \\ \boldsymbol{\chi}_I \end{bmatrix}, \qquad \boldsymbol{\xi}(0) = \boldsymbol{\xi}_0\,, \\ \dot{\boldsymbol{\chi}}_I &= -k_I \boldsymbol{C}^\top \boldsymbol{L}(t)\,\mathbf{q}(\boldsymbol{\xi})\,, \qquad \boldsymbol{\chi}_I(0) = \mathbf{1}_{n-n_\ell}\,, \end{aligned} \tag{6.10}$$

leading to the quantized kinematic coordination error dynamics

$$\dot{\boldsymbol{\zeta}} = \boldsymbol{A}_{\zeta}(t)\boldsymbol{\zeta} + \boldsymbol{f}'_{\mathbf{q}}, \qquad \boldsymbol{\zeta}(0) = \boldsymbol{\zeta}_{\mathbf{0}}, \tag{6.11}$$

where $\boldsymbol{A}_{\zeta}(t)$ was defined in (4.10) and $\boldsymbol{f}'_{\mathbf{q}}(t)$ is given by

$$\boldsymbol{f}'_{\mathbf{q}}(t) := \begin{bmatrix} k_P\, \boldsymbol{Q}\boldsymbol{L}(t)\, \boldsymbol{e}_{\xi}(t) \\ k_I\, \boldsymbol{C}^{\top}\boldsymbol{L}(t)\, \boldsymbol{e}_{\xi}(t) \end{bmatrix} \in \mathbb{R}^{2n-n_{\ell}-1}.$$

In this case, it can be shown that the set of (Krasovskii) equilibria of (6.11) is characterized by:

$$\begin{aligned}\Theta' := \Bigg\{ &(\boldsymbol{\zeta}_{\mathbf{1eq}}, \boldsymbol{\zeta}_{\mathbf{2eq}}) \in \mathbb{R}^{n-1} \times \mathbb{R}^{n-n_{\ell}} : \\ &\boldsymbol{\zeta}_{\mathbf{1eq}} = \boldsymbol{Q}\boldsymbol{\xi}_{\mathbf{eq}}(t),\ \boldsymbol{\zeta}_{\mathbf{2eq}} = \boldsymbol{\chi}_{I,\mathbf{eq}} - \mathbf{1}_{n-n_{\ell}}, \\ &\begin{bmatrix} \beta(t)\mathbf{1}_n \\ \mathbf{0} \end{bmatrix} \in \mathrm{K}\left(\begin{bmatrix} -k_P\boldsymbol{L}(t)\,\mathbf{q}(\boldsymbol{\xi}_{\mathbf{eq}}(t)) + \begin{bmatrix} \mathbf{1}_{n_{\ell}} \\ \boldsymbol{\chi}_{I,\mathbf{eq}} \end{bmatrix} \\ -k_I\boldsymbol{C}^{\top}\boldsymbol{L}(t)\,\mathbf{q}(\boldsymbol{\xi}_{\mathbf{eq}}(t)) \end{bmatrix} \right) \Bigg\},\end{aligned}$$

where $\beta(t) \in \mathbb{R}$ is an arbitrary time-varying signal. Next, we provide some insights into the set of equilibrium points Θ' defined above. Moreover, similarly to Theorem 6.1, we show that the coordination error state degrades gracefully with the value of the quantizer step size.

Lemma 6.2. *Consider the quantized kinematic coordination error dynamics* (6.11). *The following results apply:*

(*i*) $\boldsymbol{\zeta}_{\mathbf{eq}} = \mathbf{0}$ *is an equilibrium point, independently of the quantizer resolution and the information flow;*

(*ii*) *for the case of connected (undirected) network topologies, if the step size of the quantizers satisfies*

$$\Delta < \frac{1}{2(n-n_{\ell})}\frac{1}{k_P}, \tag{6.12}$$

then $\boldsymbol{\zeta}_{\mathbf{eq}} = \mathbf{0}$ *is the only equilibrium point, that is* $\Theta' = \{(\mathbf{0}, \mathbf{0})\}$. ◇

Proof. The proof is given in Appendix B.2.12. □

Remark 6.5. The bound in (6.12) should be understood as a design constraint for the quantizers that prevents the existence of equilibria other than $\boldsymbol{\zeta}_{\mathbf{eq}} = \mathbf{0}$. △

Theorem 6.2. *Consider the closed-loop kinematic coordination error dynamics* (6.11) *and suppose that the information flow satisfies the PE-like condition* (2.8) *for some parameters* $\mu, T > 0$. *Then, there exist coordination control gains* k_P *and* k_I *such that the solution of the quantized coordination error dynamics* (6.11) *satisfies*

$$\|\boldsymbol{\zeta}(t)\| \leq \kappa''_{\zeta 0}\|\boldsymbol{\zeta}(0)\|\mathrm{e}^{-\lambda''_{cd}t} + \kappa''_{\zeta 1}\Delta, \qquad \textit{for all } t \geq 0,$$

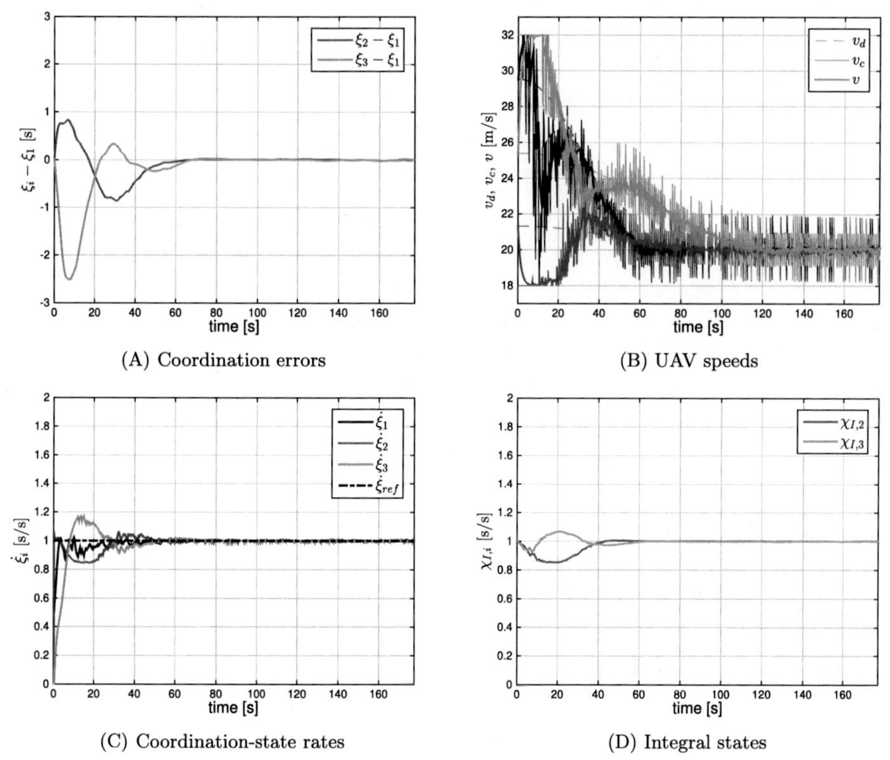

(A) Coordination errors

(B) UAV speeds

(C) Coordination-state rates

(D) Integral states

Figure 6.1 Sequential auto-landing. Closed-loop coordination dynamics for the partially quantized control law under fine quantization.

for some constants $\kappa''_{\zeta 0}, \kappa''_{\zeta 1} \in (0, \infty)$, *and with* $\lambda''_{cd} := \bar{\lambda}_{cd}(1 - \theta''_{\lambda})$, *where* $\bar{\lambda}_{cd}$ *was defined in* (4.11) *and* θ''_{λ} *is a constant verifying* $0 < \theta''_{\lambda} < 1$. ◇

Proof. The proof of this result is similar to the proof of Theorem 6.1, and is therefore omitted. □

6.2 Simulation Example: Sequential Auto-Landing with Quantized Information

We now present simulation results that illustrate the performance of the two coordination control laws introduced in this chapter: the *partially quantized* control law (6.1) and the *fully quantized* control law (6.9). We consider the sequential auto-landing mission scenario described in Section 4.5.2 of Chapter 4, and assume that the information exchanged among UAVs is quantized. Similarly to the results presented in previous chapters, the simulations in this section are based on the kinematic model of the UAV in (3.16) along with a simplified, decoupled linear model of the UAV with its autopilot.

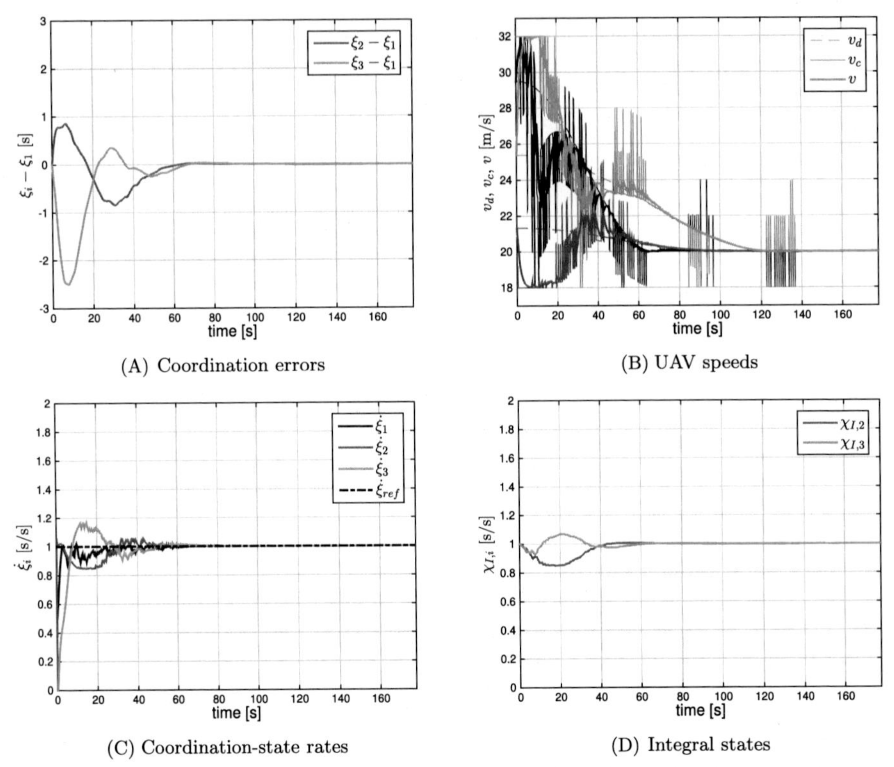

(A) Coordination errors (B) UAV speeds

(C) Coordination-state rates (D) Integral states

Figure 6.2 Sequential auto-landing. Closed-loop coordination dynamics for the fully quantized control law under fine quantization.

Figs. 6.1 and 6.2 present the computed evolution of the quantized coordination dynamics for the two control laws with quantizer step size $\Delta = 1$ s (note that this step size verifies inequalities (6.5) and (6.12)). In particular, the figures show the time-evolution of the coordination errors $(\xi_i(t) - \xi_j(t))$, the rate of change of the coordination states $\dot{\xi}_i(t)$, the UAV speeds $v_i(t)$, and the integral states $\chi_{I,i}(t)$ implemented on the follower vehicles. As can be seen in the figure, for this resolution of the quantizers, the two control laws lead to similar results, with comparable levels of performance in terms of vehicle coordination. However, note that the partially quantized control law results in a speed command with high-frequency content, whereas the fully quantized control law generates a much smoother speed command with a much less "wiggling" speed response. From this perspective, the fully quantized control law seems to be preferable to the partially quantized control law, in which each vehicle uses its own unquantized coordination state.

Next, we illustrate the behavior of the two protocols in the presence of coarse quantization. We consider the same simulation scenario as in Figs. 6.1 and 6.2, but change the quantizer step size to $\Delta = 5$ s, which does not verify inequalities (6.5) or (6.12). The computed responses of the quantized coordination dynamics for the two con-

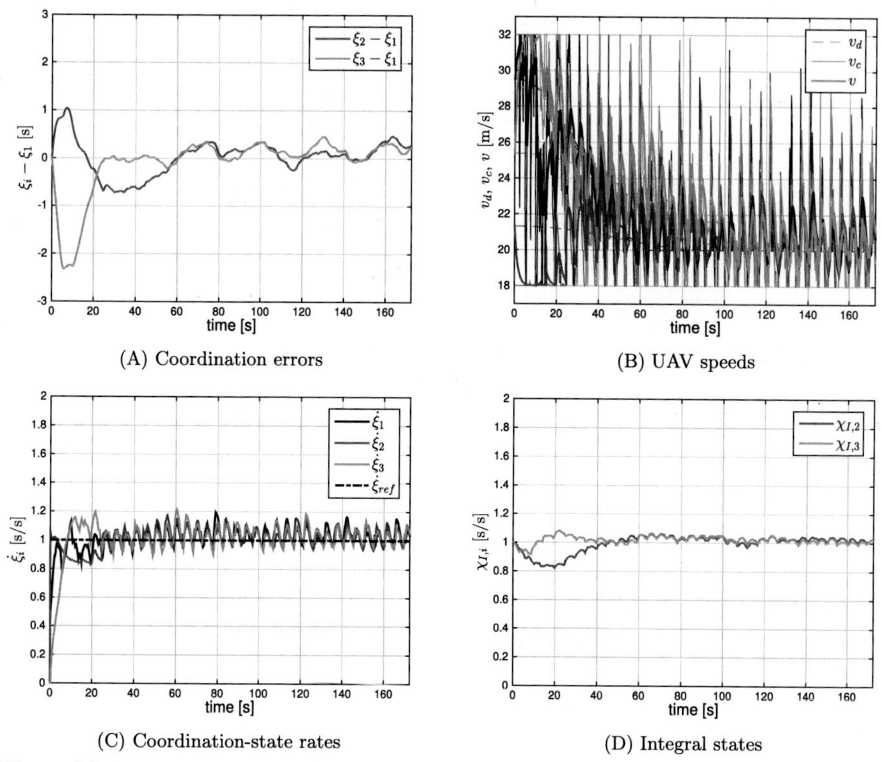

(A) Coordination errors

(B) UAV speeds

(C) Coordination-state rates

(D) Integral states

Figure 6.3 Sequential auto-landing. Closed-loop coordination dynamics for the partially quantized control law under coarse quantization.

trol laws are shown in Figs. 6.3 and 6.4, respectively. In this case, the coordination control law with partially quantized information achieves the desired agreement (in a practical sense); the response is, however, highly oscillatory with speed commands exhibiting large high-frequency content. Meanwhile, when using the control law with fully quantized feedback, the vehicles seem to asymptotically converge to the desired agreement, and the response is again much smoother than the one achieved with the partially quantized control law. Interestingly, note that asymptotic convergence is a much stronger result than that of Theorem 6.2, which instead derives an ISS-type bound for the coordination error state. These simulation results seem to indicate that a stronger version of Theorem 6.2 may hold and, hence, point out to the need to conduct a more in-depth analysis of the fully quantized coordination control law.

Finally, we notice that convergence of the fleet of vehicles to one of the "zero-speed" equilibria characterized in Corollary 6.1 cannot be illustrated by this simulation scenario, as the speed command of each vehicle is saturated below 18 m/s. Simulation results illustrating this undesirable phenomenon can be found in [3].

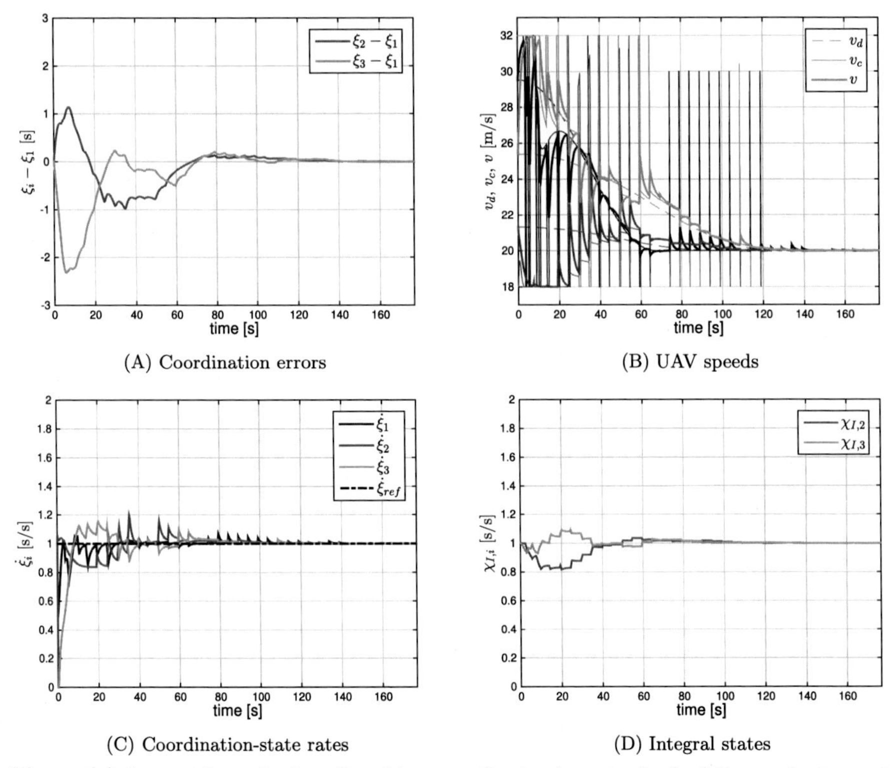

(A) Coordination errors

(B) UAV speeds

(C) Coordination-state rates

(D) Integral states

Figure 6.4 Sequential auto-landing. Closed-loop coordination dynamics for the fully quantized control law under coarse quantization.

References

[1] F. Ceragioli, C. De Persis, P. Frasca, Discontinuities and hysteresis in quantized average consensus, Automatica 47 (2011) 1916–1928.

[2] O. Hájek, Discontinuous differential equations, I, Journal of Differential Equations 32 (1979) 149–170.

[3] E. Xargay, R. Choe, N. Hovakimyan, I. Kaminer, Multi-leader coordination algorithm for networks with switching topology and quantized measurements, Automatica 50 (2014) 841–851.

Time Coordination Under Low Connectivity

7

In Chapter 4 we have proven that the guaranteed rate of convergence of the coordination control loop is limited by the QoS of the communications network (see Lemmas 4.1 and 4.2). This implies that in communication-limited environments, characterized by small parameters μ and large parameters T, long times might be required for the vehicles to reach agreement and coordinate their positions along the paths. In this chapter, we propose a modification of the coordination control law that is intended to improve the convergence rate of the closed-loop coordination dynamics in *low-connectivity scenarios*. The derivations in this chapter build on the coordination control law for missions with relative temporal specifications introduced in Chapter 4 (Section 4.2.1), but can also be easily extended to the case of missions imposing absolute temporal constraints discussed in Chapter 5. While we have —as of now— no theoretical guarantee that the modified coordination algorithm achieves this objective, we provide numerical evidence suggesting that the coordination error state converges to a neighborhood of the origin in a shorter time.

7.1 Local Estimators and Topology Control

The problem of designing a control law that speeds up the convergence of the coordination dynamics under low connectivity can to some extent be seen as dual of that of determining a logic-based communication protocol that is able to reduce the amount of information exchanged over the network while maintaining a desired level of performance. Logic-based protocols use banks of local estimators and communication logic to determine when each node in a network should communicate its own state to the neighboring nodes, and have been shown to significantly reduce the required channel bandwidth [4,5]. Here, instead, we propose the use of local estimators to improve the *knowledge* that each vehicle (or node) has about the coordination states of other vehicles, while continuously broadcasting its own coordination state to its neighbors, as determined by the time-varying communication topology. The states of the local estimators can then be used by the coordination control law to enforce vehicle coordination even during time-intervals when the actual coordination states of other vehicles are not available.

In this approach, the estimators are useful only if vehicles receive "enough" information from the corresponding neighboring vehicles; if this is not the case, then each vehicle is just carrying a "bag" of estimators with worthless information, which may reduce the convergence rate of the coordination error dynamics due to a "large net-

Time-Critical Cooperative Control of Autonomous Air Vehicles, DOI: 10.1016/B978-0-12-809946-9.00009-9

work effect".[1] This implies that precise a priori knowledge about the (local) structure of the network topology would be beneficial for an effective implementation of such estimators. The framework adopted in this book, however, assumes no information about the structure of the network topology, other than the —rather general— PE-like connectivity assumption in (2.8). This means that the vehicles involved in the mission do not know in advance which neighbors they are going to exchange information with or the amount of information received from each neighbor. The question is thus how to design a protocol that can take advantage of the additional information provided by the estimators without experiencing the drawbacks associated with a large extended network.

To address this problem, in this chapter we borrow and expand tools and concepts from control of complex networks, and develop strategies to control the communication links between each vehicle and its estimators. These topology-control strategies are thus responsible for deciding when a vehicle should "listen" to a particular estimator and adjusting the corresponding link weight accordingly. This approach leads to an evolving network, whose topology depends on the local exchange of information among nodes.

7.1.1 Estimator Dynamics

Recalling that, at the kinematic level, the evolution of the coordination states is described by single integrators (see Eq. (4.5)), the estimate of the coordination state of vehicle j implemented at the ith vehicle, denoted here by $\hat{\xi}^i_j(t)$, is obtained from the following dynamics:

$$\dot{\hat{\xi}}^i_j(t) = u^i_j(t)\,, \qquad\qquad \hat{\xi}^i_j(0) = \xi_i(0)\,,$$

with the control law

$$u^i_j(t) = -\hat{k}_P\big(\hat{\xi}^i_j(t) - \xi_i(t)\big) - a_{ij}(t)\hat{l}_P\big(\hat{\xi}^i_j(t) - \xi_j(t)\big) + \hat{\chi}^i_j(t)\,, \tag{7.1a}$$

$$\dot{\hat{\chi}}^i_j(t) = -\hat{k}_I\big(\hat{\xi}^i_j(t) - \xi_i(t)\big) - a_{ij}(t)\hat{l}_I\big(\hat{\xi}^i_j(t) - \xi_j(t)\big)\,, \qquad \hat{\chi}^i_j(0) = 1\,, \tag{7.1b}$$

where $\hat{k}_P, \hat{k}_I > 0$ are consensus gains, $\hat{l}_P, \hat{l}_I > 0$ are learning gains, and $a_{ij}(t)$ is the (i, j) entry of the time-varying adjacency matrix $\boldsymbol{A}(t)$ of graph $\Gamma(t)$. On the one hand, the first term in (7.1a) and (7.1b) ensures that the estimate $\hat{\xi}^i_j(t)$ follows the coordination state of vehicle i in the absence of information from vehicle j. On the other hand, the second term in these equations incorporates the information available from vehicle j into the estimator dynamics, and is thus responsible for learning the evolution of its coordination state. From a practical perspective, the learning gains $\hat{l}_P$ and $\hat{l}_I$ should be larger than the consensus gains $\hat{k}_P$ and $\hat{k}_I$ so that the learning component dominates the estimator dynamics as soon as new information from vehicle j is available.

[1] As will become clear later, the implementation of these estimators in each vehicle creates an extended network, which can have a significantly larger number of nodes and a smaller connectivity degree than the original vehicle network.

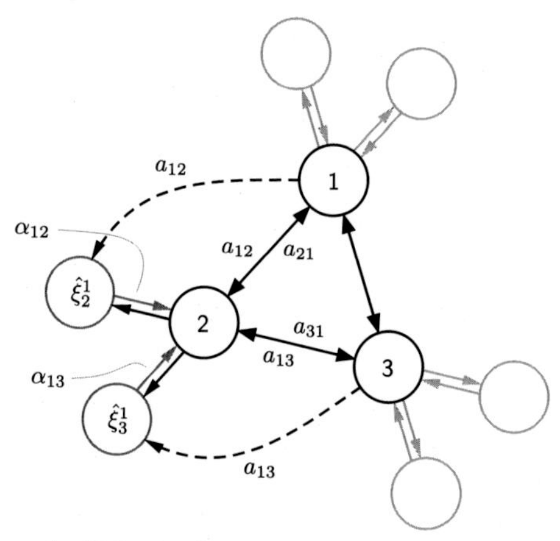

Figure 7.1 Extended network with local estimators. Example of a network of three vehicles, each running two estimators, and the resulting information flow for vehicle 1.

Note, however, that the choice of (large) learning gains is constrained by the presence of channel noise in real-world applications.

7.1.2 Coordination Control Law and Link-Weight Dynamics

To improve the convergence rate of the closed-loop coordination dynamics, we propose the following modification of the distributed coordination control law (4.6):

$$u_{\text{coord},i}(t) = -k_P \sum_{j\in\mathcal{N}_i(t)} (\xi_i(t) - \xi_j(t)) - \hat{k}_P \sum_{j\in\mathcal{S}_i} \alpha_{ij}(t)(\xi_i(t) - \hat{\xi}^i_j(t)) + 1\,, \qquad i = 1,\ldots,n_\ell\,,$$

$$\left.\begin{aligned} u_{\text{coord},i}(t) &= -k_P \sum_{j\in\mathcal{N}_i(t)} (\xi_i(t) - \xi_j(t)) - \hat{k}_P \sum_{j\in\mathcal{S}_i} \alpha_{ij}(t)(\xi_i(t) - \hat{\xi}^i_j(t)) + \chi_{I,i}(t) \\ \dot{\chi}_{I,i}(t) &= -k_I \sum_{j\in\mathcal{N}_i(t)} (\xi_i(t) - \xi_j(t)) - \hat{k}_I \sum_{j\in\mathcal{S}_i} \alpha_{ij}(t)(\xi_i(t) - \hat{\xi}^i_j(t))\,, \quad \chi_{I,i}(0) = 1 \end{aligned}\right\}, \qquad i = n_\ell + 1,\ldots,n\,,$$

where $\mathcal{S}_i$ represents the group of vehicles for which vehicle i runs an estimator, and $\alpha_{ij}(t)$ are time-varying link weights that can be manipulated at will. The control law above adjusts the speed commands for the vehicles based not only on the actual coordination states of the neighboring vehicles (when available), but also on the information provided uninterruptedly by the estimators. This setup leads to an extended network of $\left(n + \sum_i \text{card}(\mathcal{S}_i)\right)$ nodes with a time-varying directed topology of small connectivity degree. Fig. 7.1 presents a simple example illustrating the node configuration and the information flow of this extended network.

Effective use of the estimators requires a careful design of the dynamics of the link weights $\alpha_{ij}(t)$, which should ensure that only the states of the estimators with useful information are included in the coordination control law. The blind addition of coordination-state estimates to the control law, without consideration of the quality of such estimates, is likely to slow down the convergence of the coordination error dynamics. To prevent this undesirable "large network effect", we propose to dynamically adjust the weights $\alpha_{ij}(t)$ as a function of the quality of the local estimates. To this effect, we define the following variables:

$$\hat{\mu}_{ij}(t) := \frac{1}{T_\mu} \int_{t-T_\mu}^{t} a_{ij}(\tau) d\tau \,, \qquad j \in \mathcal{S}_i \,, \quad i = 1, \ldots, n \,,$$

where $T_\mu > 0$ is a constant *characteristic time*. From its definition, it follows that the variable $\hat{\mu}_{ij}(t)$ represents a measure of the quality of the communication link between vehicle i and vehicle j and, hence, can be used to characterize the quality of the coordination-state estimate of the jth vehicle. For example, if $\hat{\mu}_{ij}(t) \approx 0$, then the ith vehicle has received little information from vehicle j in the time-interval $[t - T_\mu, t]$, which implies that the estimate of the coordination state of vehicle j is likely to be inaccurate; instead, if $\hat{\mu}_{ij}(t) \approx 1$, then vehicles i and j have been frequently exchanging information during the interval $[t - T_\mu, t]$, which suggests that the coordination-state estimate of vehicle j is accurate. As a practical note, it is important to emphasize that the variables $\hat{\mu}_{ij}(t)$ can be easily computed locally by the vehicles if each packet exchanged over the network contains the source vehicle's identifier.

At this point, various strategies can be adopted to judiciously design the dynamics of the weight links $\alpha_{ij}(t)$ as a function of the variables $\hat{\mu}_{ij}(t)$. Here, we investigate two different approaches for topology control: (*i*) a *hybrid strategy* in which links switch between different activation/deactivation modes; and (*ii*) a continuous strategy based on *edge snapping* [1,2]. Next, we provide details about these two strategies:

- *Hybrid Strategy:* Letting $\mu_{\min}$ and $\mu_{\max}$ be a priori defined activation thresholds, the link weights $\alpha_{ij}(t)$ can be assigned as follows:

$$\alpha_{ij}(t) = \begin{cases} 0\,, & \text{if } \hat{\mu}_{ij}(t) \le \mu_{\min}\,, \\ 1\,, & \text{if } \mu_{\min} < \hat{\mu}_{ij}(t) \le \mu_{\max}\,, \\ 0\,, & \text{if } \mu_{\max} < \hat{\mu}_{ij}(t)\,. \end{cases}$$

  According to this strategy, vehicle i only "listens" to its jth estimator if the variable $\hat{\mu}_{ij}(t)$ is between the thresholds $\mu_{\min}$ and $\mu_{\max}$. If $\hat{\mu}_{ij}(t) < \mu_{\min}$, it is considered that the estimator does not contain valuable information about the coordination state of vehicle j and, therefore, the corresponding estimate is not included in the control law. If $\hat{\mu}_{ij}(t) > \mu_{\max}$, then it is considered that there is enough communication between vehicles i and j to ensure a fast convergence of the coordination error dynamics, and the information from the estimator can thus be ignored.
- *Edge Snapping:* In this strategy, the evolution of each link weight $\alpha_{ij}(t)$ is modeled as a second-order dynamical system subject to the action of a two-well potential

and driven by an appropriately designed forcing signal. More precisely, this approach sets [1,2]

$$\alpha_{ij}(t) = \sigma_{ij}^2(t)\,, \qquad j \in \mathcal{S}_i\,, \quad i = 1,\ldots,n\,,$$

where $\sigma_{ij}(t)$ is generated through the following dynamics:

$$\ddot{\sigma}_{ij}(t)+d_\sigma\,\dot{\sigma}_{ij}(t)+\frac{\mathrm{d}U(\sigma_{ij})}{\mathrm{d}\sigma_{ij}}(t) = g\big(\hat{\mu}_{ij}(t), \hat{e}^i_j(t)\big)\,, \qquad j \in \mathcal{S}_i\,, \quad i = 1,\ldots,n\,.$$

In the above equation, d_σ is the damping coefficient, $U(\cdot)$ is a two-well potential function, while $g : \mathbb{R} \times \mathbb{R} \to \mathbb{R}$ is the forcing function, which —for the purpose of our problem— we take to depend on the variable $\hat{\mu}_{ij}(t)$ as well as the error $\hat{e}^i_j(t) := \hat{\xi}^i_j(t) - \xi_i(t)$. This strategy yields a bistable dynamical behavior of the link weights, which leads to self-emerging unweighted topologies and, unlike the hybrid approach discussed previously, results in a continuous evolution of the link weights.

Finally, we note that the estimators and topology-control algorithms proposed here do not require any a priori knowledge about the structure of the network topology and can be implemented in a distributed fashion onboard the autonomous vehicles. The main drawback of this approach is an increase in the computational demands of the system, as the algorithms must be run onboard the vehicles in real time. There is, therefore, a trade-off between overall coordination performance and onboard computational requirements. In the next section, we present simulation results that illustrate the benefits of the modified coordination control law proposed here for low-connectivity scenarios.

7.2 Simulation Example: Sequential Auto-Landing Under Severely Limited Communication

We consider again the sequential auto-landing mission scenario described in Section 4.5.2 of Chapter 4, and augment the proportional-integral coordination control law developed in Chapter 4 with the onboard estimators (two per vehicle) as well as the two strategies for topology control and link assignment. Similarly to the results presented in previous chapters, the simulations in this section are based on the kinematic model of the UAV in (3.16) along with a simplified, decoupled linear model of the UAV with its autopilot.

In this set of simulations, the path-following controller gains are selected as follows:

$$k_\ell = 0.20\ [1/\mathrm{s}]\,, \qquad k_{\tilde{R}} = 0.50\ [1/\mathrm{s}]\,, \qquad d = 125\ [\mathrm{m}]\,,$$

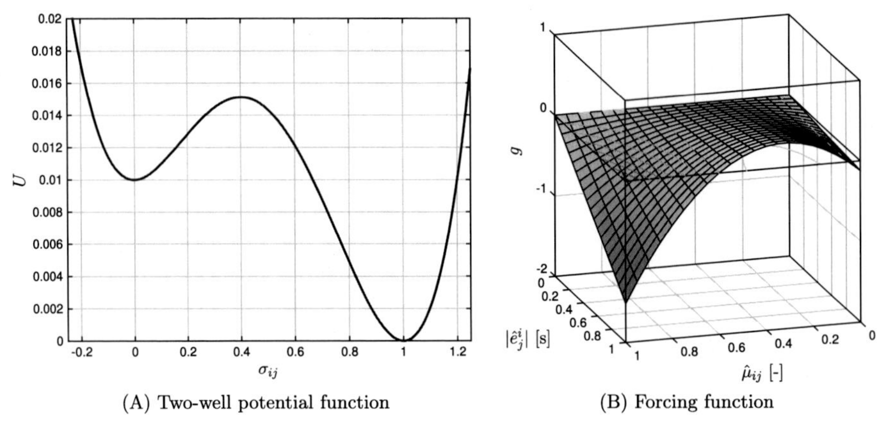

(A) Two-well potential function (B) Forcing function

Figure 7.2 Edge snapping; two-well function and forcing function for link-weight dynamics.

while the control gains for the (extended) coordination algorithm are chosen as

$$k_P = 1.0\,10^{-1}\,[1/\text{s}], \qquad k_I = 1.0\,10^{-2}\,[1/\text{s}^2], \qquad k_{\text{aw}} = 2.0\,10^{-3}\,[1/\text{m}],$$
$$G_{\text{aw}}(s) = \frac{1}{s+1}, \qquad \hat{k}_P = 1.0\,10^{-1}\,[1/\text{s}], \qquad \hat{k}_I = 2.0\,10^{-2}\,[1/\text{s}^2],$$
$$\hat{l}_P = 1.0\,10^{0}\,[1/\text{s}], \qquad \hat{l}_I = 1.0\,10^{-1}\,[1/\text{s}^2],$$

with the following settings for the link-weight dynamics:

- *Hybrid Strategy:* The characteristic time is set to $T_\mu = 5$ s, and the activation thresholds are set to $\mu_{\text{min}} = 0.05$ and $\mu_{\text{max}} = 0.6$.
- *Edge Snapping:* The characteristic time is also set to $T_\mu = 5$ s, and the damping coefficient d_σ, the two-well potential function $U(\cdot)$, and the forcing function $g(\cdot,\cdot)$ are given by

$$d_\sigma = 4, \qquad U(\sigma_{ij}) = 0.01\,(\sigma_{ij} - 1)^2\,(15\sigma_{ij}^2 + 2\sigma_{ij} + 1),$$
$$g(\hat{\mu}_{ij}, \hat{e}^i_j) = 4\,(\mu_{\text{max}} - \hat{\mu}_{ij})(\hat{\mu}_{ij} - \mu_{\text{min}})\,|\hat{e}^i_j|, \qquad \mu_{\text{min}} = 0.05, \quad \mu_{\text{max}} = 0.6.$$

The two-well potential function and the forcing function are illustrated in Fig. 7.2. In all of the simulations, vehicle 1 is elected as the single leader of the fleet. The angular-rate commands are saturated to ± 0.3 rad/s, and the speed commands are saturated between 18 m/s and 32 m/s. The information flow is assumed to be time-varying and, at any given time t, is characterized by one of the first four graphs in Fig. 4.2 (that is, topologies 1 through 4).

To analyze the convergence properties of the proposed algorithms, a set of Monte Carlo simulations is conducted for a total of 1024 simulation configurations, consisting of different initial conditions of the UAVs and different switching topologies. For each configuration (initial condition and switching topology), three simulation runs are performed with the following three coordination algorithms: (i) the

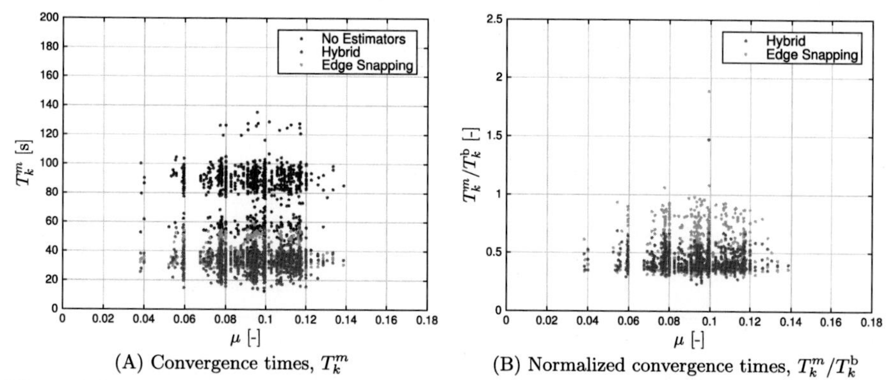

(A) Convergence times, T_k^m (B) Normalized convergence times, T_k^m/T_k^{b}

Figure 7.3 Convergence time of the coordination error dynamics to a 0.1-error band with three different coordination algorithms.

"basic" proportional-integral coordination control law described in Chapter 4; (*ii*) the modified coordination control law with node estimators and binary link weights (hybrid strategy); and (*iii*) the modified coordination control law with node estimators and continuous link weights (edge snapping). For each run we compute the amount of time it takes for the 2-norm of the coordination error state $\boldsymbol{\zeta}(t)$ to converge to a 0.1-error band, which corresponds to a maximum coordination error $(\xi_i(t) - \xi_j(t))$ of approximately 0.2 s (see Eq. (B.26) in Appendix B.2.5). We represent each of these convergence times as T_k^m, $k = 1, \ldots, 1024$ and $m = \{\mathrm{b}, \mathrm{h}, \mathrm{e}\}$, where 'b' stands for *basic*, 'h' stands for *hybrid*, and 'e' stands for *edge snapping*.

The data obtained from these Monte Carlo runs is shown in Fig. 7.3A. To remove the dependency of convergence time on the initial conditions of the UAVs, the figure also presents the convergence times T_k^{h} and T_k^{e} normalized with respect to the corresponding T_k^{b}; see Fig. 7.3B. As can be observed in the figure, all of the normalized convergence times for the hybrid approach are below 1 (except for one, which we will discuss shortly), which suggests that the proposed approach is effective for speeding up coordination in low-connectivity scenarios. The single normalized convergence time above 1 corresponds to sample #546, with convergence times $T_{546}^{\mathrm{b}} = 12.5$ s and $T_{546}^{\mathrm{h}} = 18.4$ s. This sample corresponds thus to a configuration in which the UAVs start the mission almost coordinated, and the modified control law delays convergence of the coordination error dynamics to the 0.1-error band by roughly 6 s for a mission with a desired duration of approximately 175 s. This outlier is, hence, of no particular concern. Similar conclusions can be drawn for the coordination control law with edge snapping. In fact, only three runs present a normalized convergence time above 1, while the remaining 1021 samples are below 1. One of these runs also corresponds to sample #546, with convergence time $T_{546}^{\mathrm{e}} = 23.6$ s. The other two runs correspond to samples #517 and #971, with convergence times $T_{517}^{\mathrm{b}} = 50.8$ s and $T_{517}^{\mathrm{e}} = 54.9$ s, and $T_{971}^{\mathrm{b}} = 44.2$ s and $T_{971}^{\mathrm{e}} = 46.9$ s, respectively, which represent an increase of approximately 6% and 8%. Despite these three outliers, the data seems to support the observation that the edge-snapping strategy is also capable of speeding up coordina-

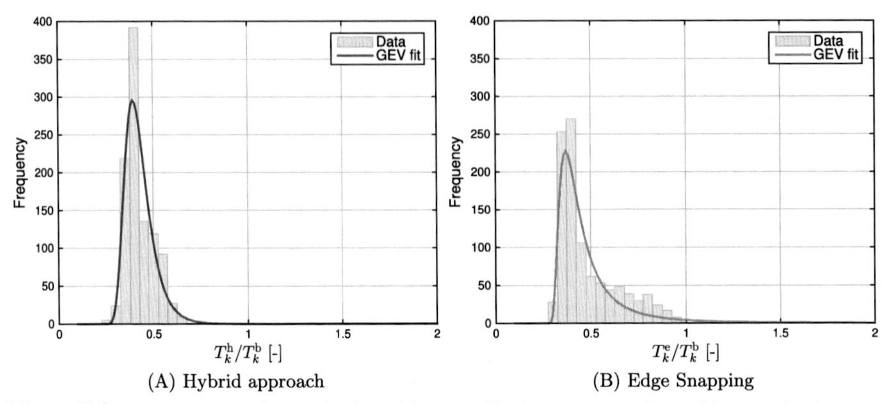

(A) Hybrid approach (B) Edge Snapping

Figure 7.4 Histograms and distribution fits of the normalized convergence time of the coordination error dynamics with both the hybrid approach and the edge-snapping strategy.

Table 7.1 Mean, variance, as well as location, scale, and shape parameters characterizing the two GEV distributions in Fig. 7.4.

	Hybrid		**Edge snapping**	
	Estimate	***95% confidence***	***Estimate***	***95% confidence***
μ	0.3925	(0.3887, 0.3964)	0.3967	(0.3914, 0.4020)
σ	0.0583	(0.0556, 0.0611)	0.0763	(0.0714, 0.0815)
ξ	0.0152	(−0.0139, 0.0442)	0.4656	(0.4072, 0.5239)
Mean	0.4270	–	0.5051	–
Variance	0.0058	–	0.3021	–

tion of the fleet in low-connectivity scenarios, albeit not as effectively as the hybrid approach.

To provide further insight into these results, Fig. 7.4 presents histograms of the normalized convergence time of the coordination error dynamics with both the hybrid approach and the edge-snapping strategy. In addition, the figure also shows *generalized extreme value* (GEV) distribution fits to these histograms [3]. The location, scale, and shape parameters (conventionally denoted in the literature as μ, σ, and ξ) characterizing these two GEV distribution fits are summarized in Table 7.1. The table also contains estimates of the mean of both distributions, which indicate that, on average, the hybrid approach is able to reduce the convergence time to 43% of the original value, while the edge-snapping strategy shortens the convergence time to 51% of this value.

Finally, we present time-history responses of the control algorithms proposed in this chapter. In particular, Figs. 7.5–7.7 show the results obtained for one of the Monte Carlo runs obtained with the hybrid approach, while Figs. 7.8–7.10 illustrate the results obtained with the coordination algorithm with edge snapping for the same scenario. The figures show the time-evolution of the coordination errors $(\xi_i(t) - \xi_j(t))$, the rate of change of the coordination states $\dot{\xi}_i(t)$, the

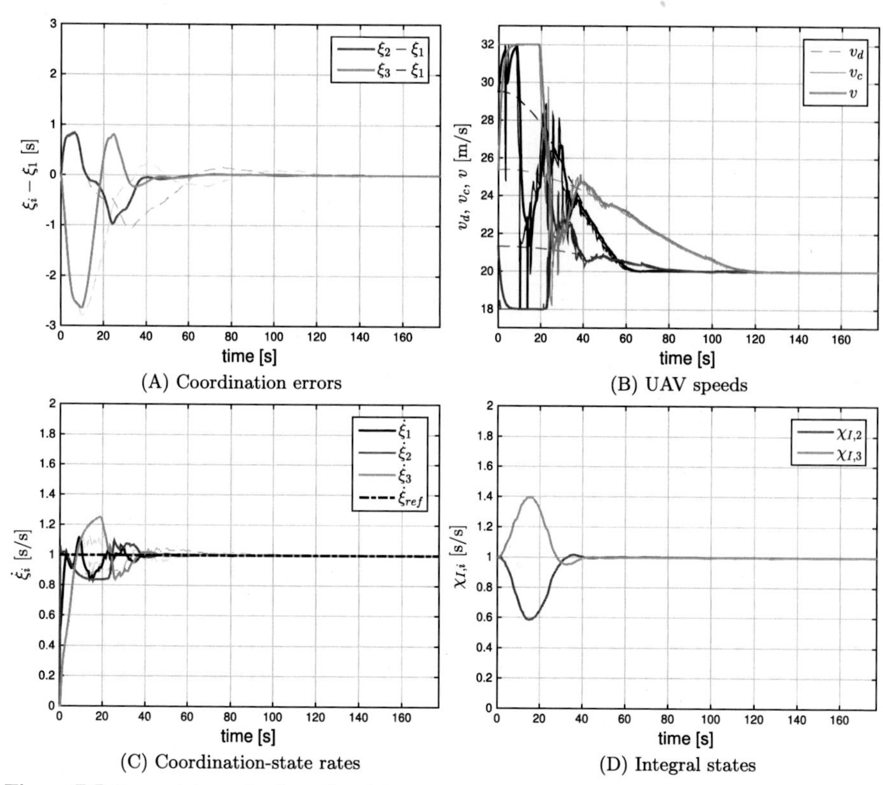

(A) Coordination errors (B) UAV speeds

(C) Coordination-state rates (D) Integral states

Figure 7.5 Sequential auto-landing. Closed-loop coordination dynamics for the modified control law with node estimators and binary link weights (hybrid strategy).

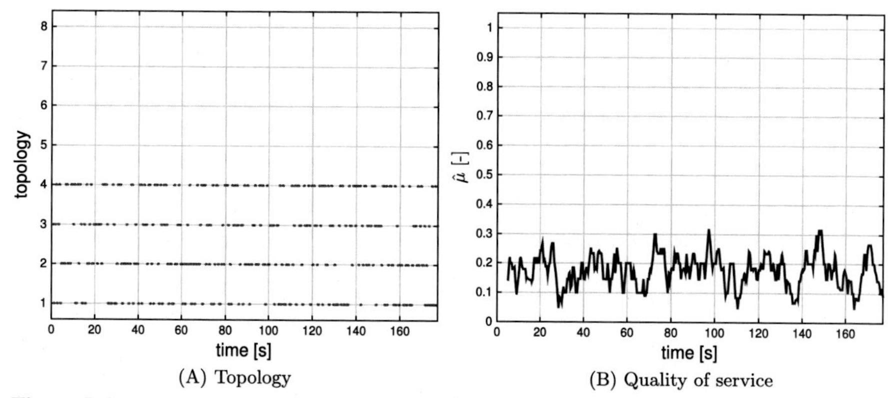

(A) Topology (B) Quality of service

Figure 7.6 Sequential auto-landing. At a given time instant, the information flow is characterized by one of the first four topologies in Fig. 4.2. The resulting graph is only connected in an integral sense, and not pointwise in time.

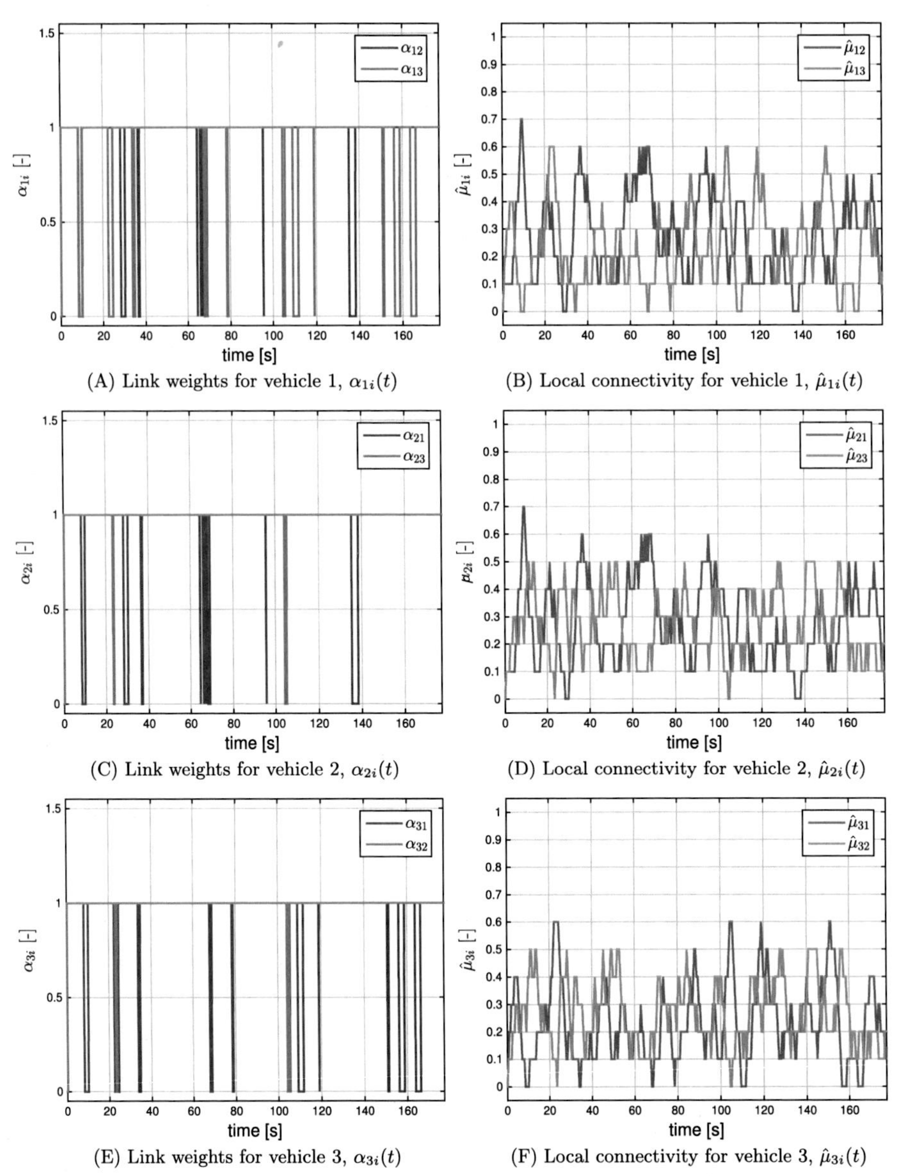

(A) Link weights for vehicle 1, $\alpha_{1i}(t)$

(B) Local connectivity for vehicle 1, $\hat{\mu}_{1i}(t)$

(C) Link weights for vehicle 2, $\alpha_{2i}(t)$

(D) Local connectivity for vehicle 2, $\hat{\mu}_{2i}(t)$

(E) Link weights for vehicle 3, $\alpha_{3i}(t)$

(F) Local connectivity for vehicle 3, $\hat{\mu}_{3i}(t)$

Figure 7.7 Sequential auto-landing. Time-evolution of link weights and local connectivity (hybrid strategy).

UAV speeds $v_i(t)$, the integral states $\chi_{I,i}(t)$ implemented on the follower vehicles, the time-evolution of the link weights between the vehicles and their estimators $\alpha_{ij}(t)$, as well as the variables $\hat{\mu}_{ij}(t)$. As a reference for comparison, the figures also present (in pale dashed lines) the time-responses of the coordination errors $(\xi_i(t) - \xi_j(t))$ as well as the rate of change of the coordination states $\dot{\xi}_i(t)$ obtained with the "basic"

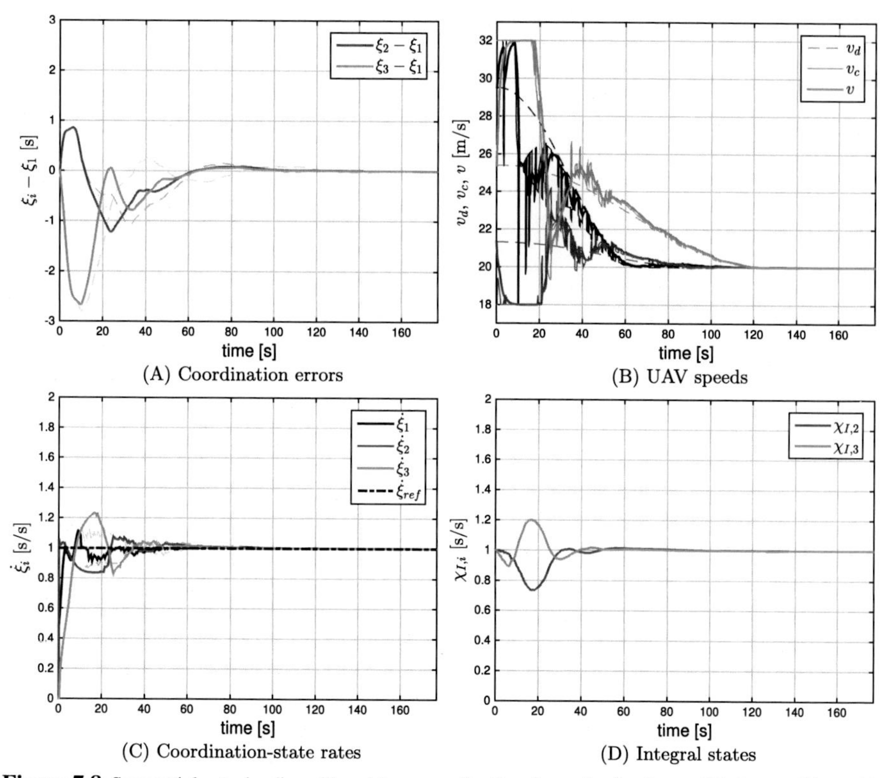

(A) Coordination errors (B) UAV speeds

(C) Coordination-state rates (D) Integral states

Figure 7.8 Sequential auto-landing. Closed-loop coordination dynamics for the modified control law with node estimators and continuous link weights (edge snapping).

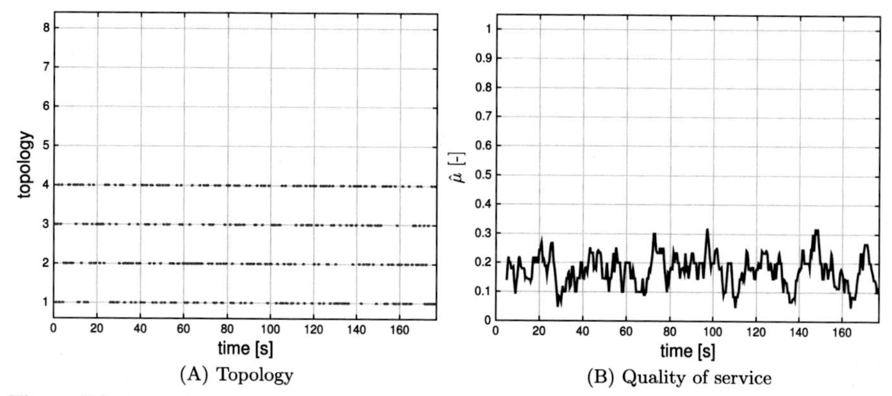

(A) Topology (B) Quality of service

Figure 7.9 Sequential auto-landing. At a given time instant, the information flow is characterized by one of the first four topologies in Fig. 4.2. The resulting graph is only connected in an integral sense, and not pointwise in time.

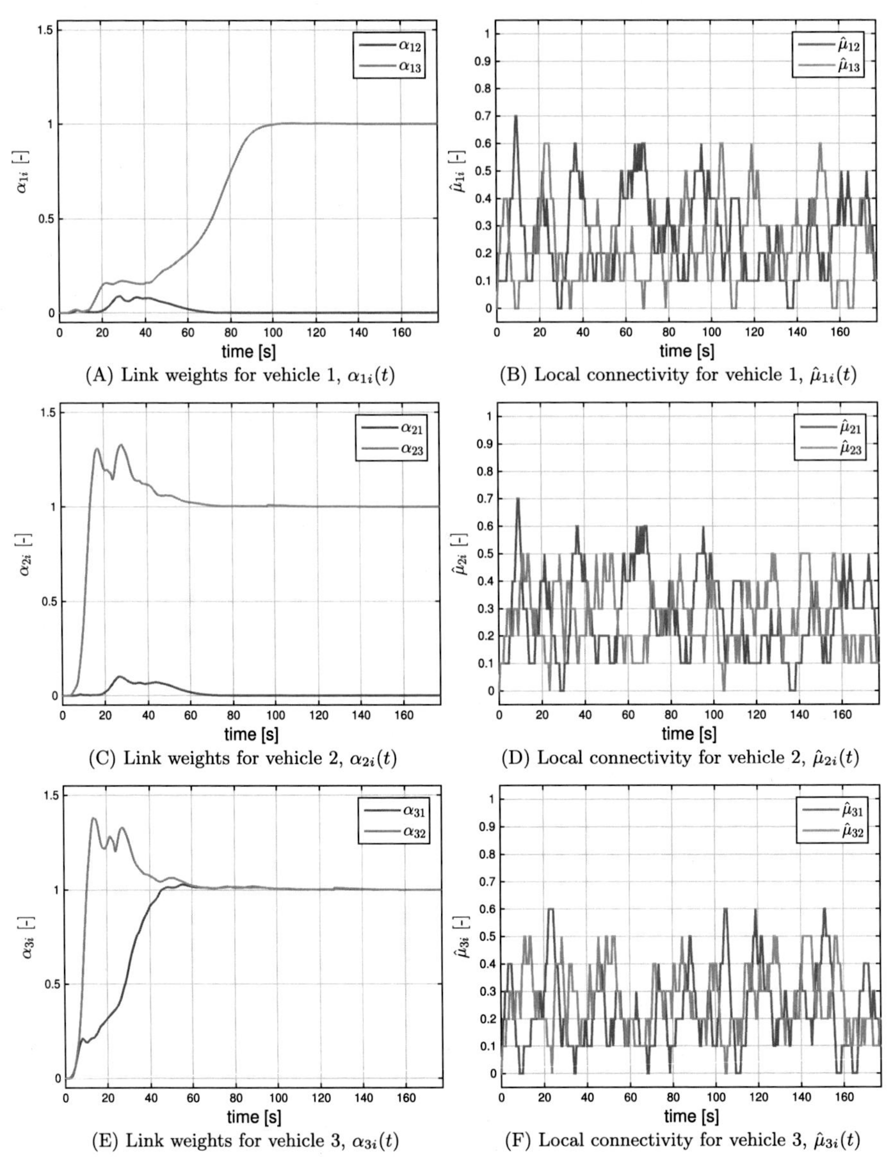

(A) Link weights for vehicle 1, $\alpha_{1i}(t)$

(B) Local connectivity for vehicle 1, $\hat{\mu}_{1i}(t)$

(C) Link weights for vehicle 2, $\alpha_{2i}(t)$

(D) Local connectivity for vehicle 2, $\hat{\mu}_{2i}(t)$

(E) Link weights for vehicle 3, $\alpha_{3i}(t)$

(F) Local connectivity for vehicle 3, $\hat{\mu}_{3i}(t)$

Figure 7.10 Sequential auto-landing. Time-evolution of link weights and local connectivity (edge snapping).

proportional-integral coordination control law described in Chapter 4. The figures also show the evolution of the time-varying network topology along with an estimate of the QoS of the network, computed as in (4.31). This set of simulation results illustrates the benefits of the proposed algorithms in terms of coordination performance, and evidences the different nature of the two topology-control strategies developed in this chapter. While the hybrid approach forces the link weights to continuously switch back and forth from 0 to 1 (Fig. 7.7), the edge-snapping strategy leads to an extended, stable, unweighted network topology with a total of 9 nodes (Fig. 7.10).

References

[1] P. De Lellis, M. di Bernardo, F. Garofalo, M. Porfiri, Evolution of complex networks via edge snapping, IEEE Transactions on Circuits and Systems 57 (2010) 2132–2143.
[2] P. De Lellis, M. di Bernardo, M. Porfiri, Pinning control of complex networks via edge snapping, Chaos 21 (2011) 033119.
[3] S. Kotz, S. Nadarajah, Extreme Value Distributions, Imperial College Press, London, UK, 2000.
[4] Y. Xu, J.P. Hespanha, Communication logics for networked control systems, in: American Control Conference, Boston, MA, June–July 2004, pp. 572–577.
[5] J.K. Yook, D.M. Tilbury, N.R. Soparkar, Trading computation and bandwidth: reducing communication in distributed control systems using state estimators, IEEE Transactions on Control System Technology 10 (2002) 503–518.

Flight Tests: Cooperative Road Search

8

This chapter presents flight-test results for a cooperative road-search mission that show the efficacy of the multi-UAV cooperative framework presented in this book. The flight tests were performed during the quarterly run Tactical Network Topology[1] field experiments conducted through the Field Experimentation Cooperative Program, which is being led by the U.S. Special Operations Command and the Naval Postgraduate School (NPS). The significance of these experiments is twofold. First, the results verify the main stability and convergence properties of the developed cooperative algorithms in a realistic mission scenario, under environmental disturbances, and with the limitations of a real-world communications network. And second, the results demonstrate the validity of the proposed generic theoretical framework in a specific realistic application as well as the feasibility of the onboard implementation of the algorithms.

8.1 Road Search with Multiple Small Autonomous Air Vehicles

8.1.1 *Airborne System Architecture*

The small tactical UAVs employed in this particular mission are two SIG Rascals 110 operated by NPS; see Fig. 8.1. The two UAVs have the same avionics and the same instrumentation onboard, the only difference being the vision sensors. The first UAV has a 1DoF bank-stabilized high-resolution 12-MPx camera, while the second UAV has a full-motion video camera suspended on a 2DoF pan-tilt gimbal. Due to weight and power constraints, each UAV is allowed to carry only one camera at a time and, therefore, the two cameras need to be mounted on different platforms. The rest of the onboard avionics, common to both platforms, includes two PC-104 industrial embedded computers[2] assembled in a stack, a wireless Mobile Ad-hoc Network (MANET) link,[3] and the Piccolo Plus autopilot[4] with its dedicated 900-MHz command and control channel. Details of the complete airborne network-centric architecture are presented in Fig. 8.2.

[1] Information available online at http://www.nps.edu/Academics/Schools/GSOIS/Departments/IS/Research/FX/CBETNT/CBE/TNT.html [Online; accessed 22 January 2017]. See also [5].

[2] Information available online at http://adl-usa.com/products/category/1/embedded-single-board-computers [Online; accessed 22 January 2017].

[3] Information available online at http://www.persistentsystems.com [Online; accessed 22 January 2017].

[4] Information available online at http://www.cloudcaptech.com/piccolo_system.shtm [Online; accessed 22 January 2017].

Time-Critical Cooperative Control of Autonomous Air Vehicles, DOI: 10.1016/B978-0-12-809946-9.00010-5

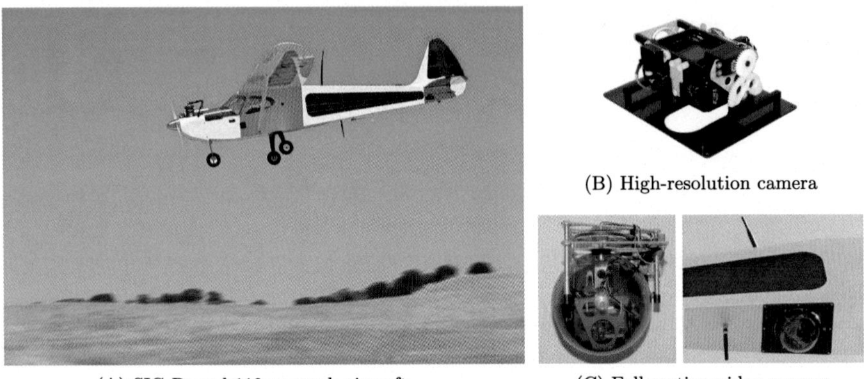

(A) SIG Rascal 110 research aircraft

(B) High-resolution camera

(C) Full-motion video camera

Figure 8.1 SIG Rascal UAV with two different onboard cameras. The SIG Rascal UAVs (A) used for cooperative path-following missions are equipped with complementary vision sensors. The first UAV has a bank-stabilized high-resolution 12-MPx camera (B), while the second UAV has a full-motion video camera suspended on a pan-tilt gimbal (C).

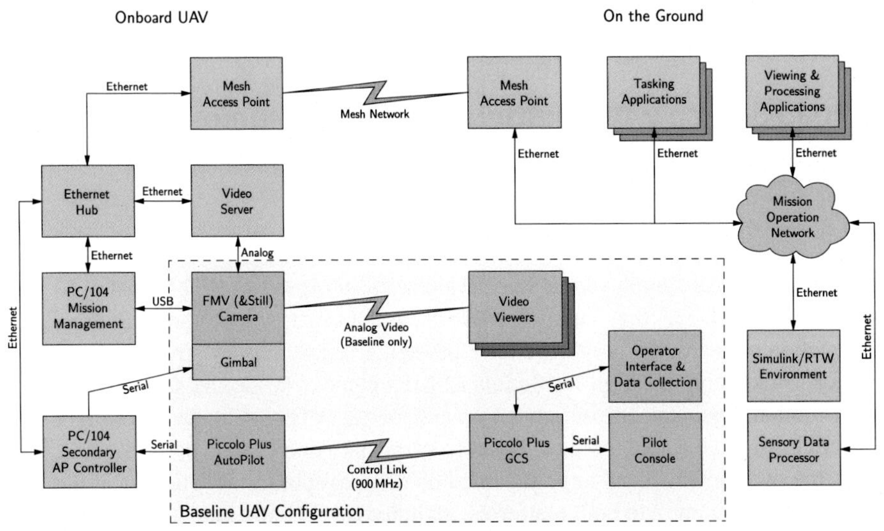

Figure 8.2 Network-centric architecture of the airborne platform. The Rascal UAV avionics include two PC-104 industrial embedded computers assembled in a stack, a wireless Mobile Ad-hoc Network link, and the Piccolo Plus autopilot with its dedicated 900-MHz command and control channel. The PC-104 computers are used to run the cooperative control algorithms in hard real time as well as mission management routines enabling onboard preprocessing and retrieval of sensory data.

The first PC-104 computer runs the cooperative-control algorithms in hard real time at 100 Hz. The computer directly communicates with the Piccolo Plus autopilot at 50 Hz over a dedicated bidirectional serial link. The second PC-104 acts as a mission management computer that implements a set of non-real-time routines enabling onboard preprocessing and retrieval of the sensory data —high-resolution imagery or video— in near real time over the network. Integration of the MANET link allows for robust, transparent inter-vehicle and ground communications, which are needed for both the coordination algorithms and the expedited sensory data delivery to a remote mission operator. In fact, the MANET link provides "any-to-any" connectivity capability, allowing every node —vehicle or ground station— to communicate with every other node. Details on the flight-test architecture and the supporting network infrastructure for coordination control and data dissemination can be found in [1].

8.1.2 Flight-Test Results

The flight-test results for a cooperative road-search mission executed by the two SIG Rascal UAVs are presented next. The objective of the mission is to detect a stationary or moving target along a pre-specified road and, if detection occurs, to collect information about the target. This information is then to be shared over a MANET link so that it can be retrieved by remote mission operators in near real time. Success of the mission relies on the ability to overlap the footprint of the field of views (FoVs) of the two cameras along the road, which increases the probability of target detection [3]. In what follows we provide details about the execution of this coordinated road-search mission, which we divide in four consecutive phases, namely, *initialization*, *transition*, *road search*, and *vision-based target tracking*. The description is supported by one of the flight-test results performed during a Tactical Network Topology field experiment at Camp Roberts, CA. We note that, in these tests, the two UAVs run the coordination control law for relative temporal constraints introduced in Remark 4.6, Chapter 4.

In the *initialization phase*, an operator specifies on a digital map the road of interest. Then, a centralized optimization algorithm generates road-search (sub)optimal paths and desired speed profiles for the two UAVs by taking explicitly into account the vehicle dynamics, collision-avoidance constraints, and mission-specific restrictions such as inter-vehicle and vehicle-to-ground communication limitations, as well as sensory capabilities. In particular, for this mission scenario, the trajectory-generation algorithm is designed to maximize the overlap of the footprints of the FoVs of the high-resolution camera and the full-motion video during the road search, while minimizing at the same time gimbal actuation. In addition to the road-search paths and the corresponding desired speed profiles, the outcome of the trajectory-generation algorithm includes a *sensor trajectory* on the ground to be followed by the vision sensors. The two road-search paths and the sensor path, along with the three corresponding speed profiles, are then transmitted to the UAVs over the MANET link.

In the *transition phase*, the two UAVs fly from their standby starting positions to the initial points of the respective road-search paths. For this purpose, distributed optimization algorithms generate feasible collision-free 3D trajectories to ensure that the two UAVs arrive at the initial points of the road-search paths at the same time. Once these transition trajectories are generated, the two vehicles start operating in cooperative path-following mode. From that moment on, the UAVs follow the transition paths while adjusting their speeds based on coordination information exchanged over the MANET link in order to achieve simultaneous arrival at the starting point of the

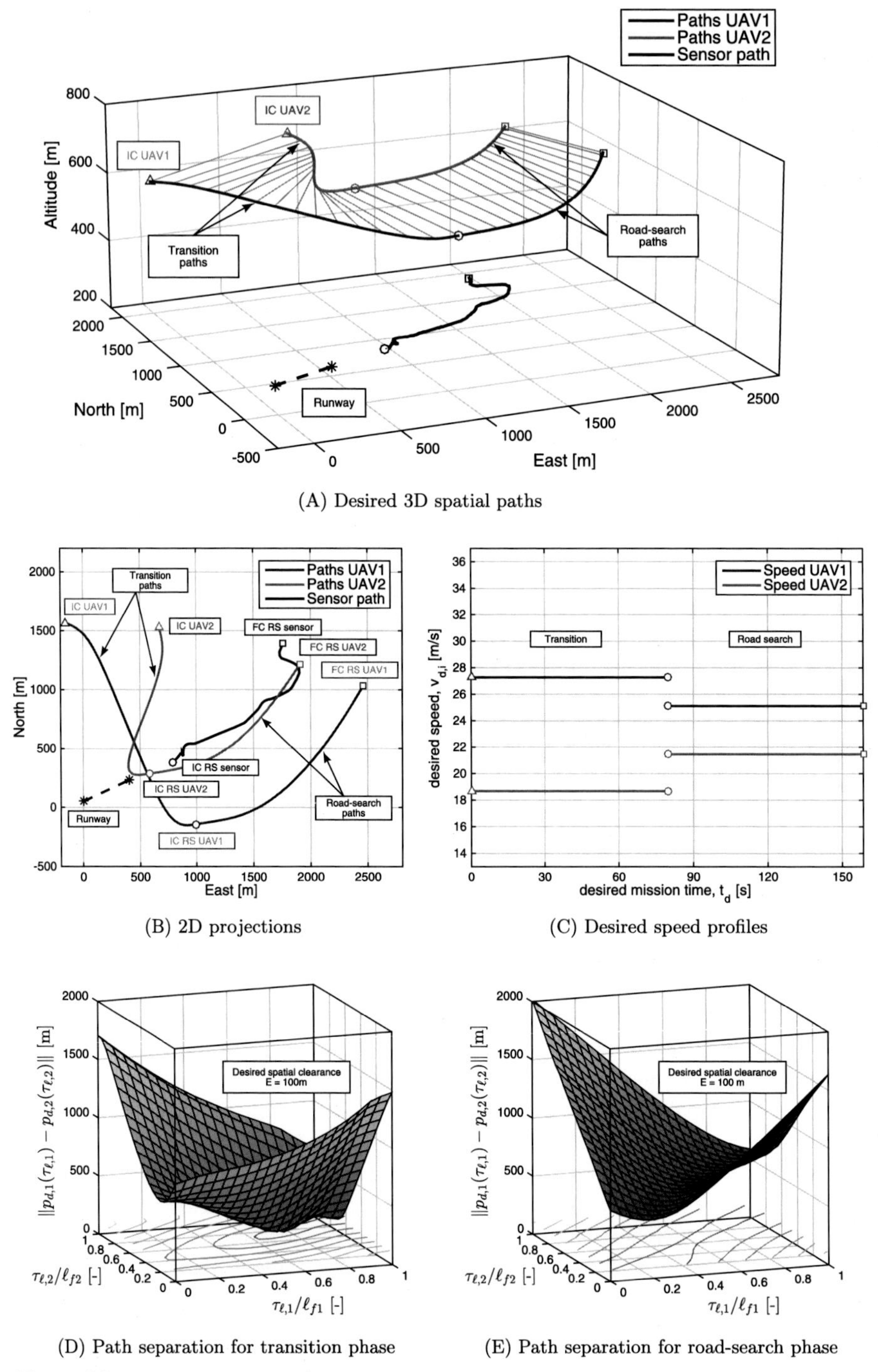

Figure 8.3 Cooperative road-search; trajectory generation.

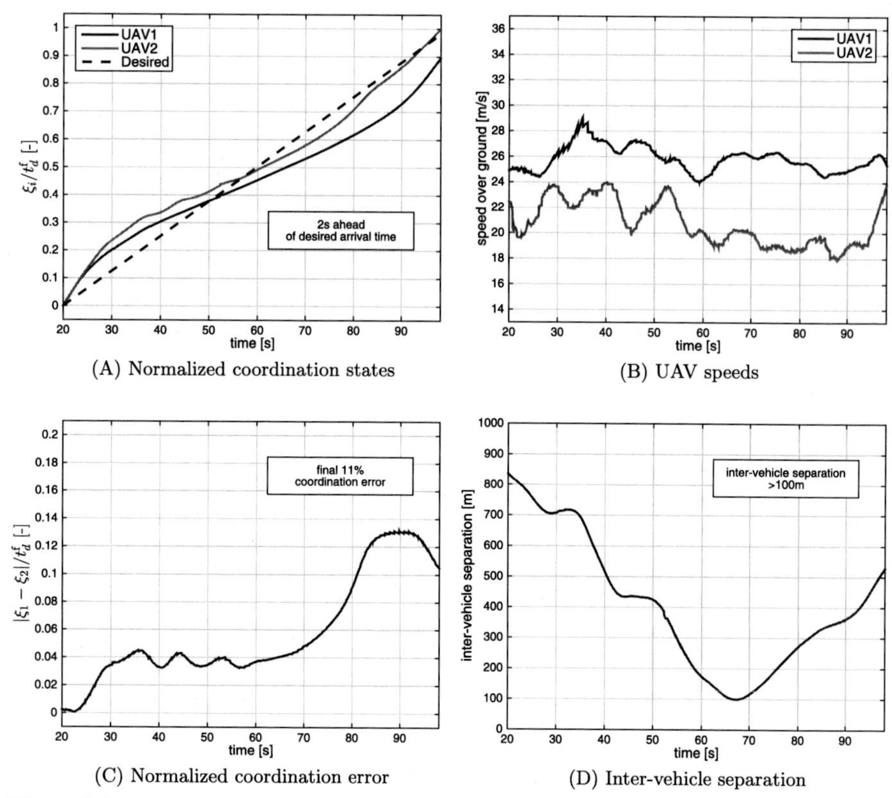

(A) Normalized coordination states (B) UAV speeds

(C) Normalized coordination error (D) Inter-vehicle separation

Figure 8.4 Cooperative road-search; time coordination during the transition phase. The two UAVs arrive at the starting point of the road-search paths with an 11%-error difference.

road-search paths. The transition and road-search paths obtained for this particular mission scenario, together with the corresponding desired speed profiles and the path separations, are shown in Fig. 8.3. Fig. 8.4 illustrates the performance of the coordination control algorithm during the transition phase of the mission. As can be observed, the inter-vehicle separation remains above 100 m and the coordination error remains below 13% during the entire duration of the transition phase, with an 11% error in coordination at the end of this phase.

The third phase addresses the *cooperative road-search mission* itself, in which the two UAVs follow the road-search paths generated in the initialization phase while adjusting their speeds to ensure the required overlap of the FoV footprints of the cameras. In this phase, a virtual vehicle running along the sensor path is implemented on one of the UAVs. For this road-search mission, a natural choice for this sensor path is the road itself, and the virtual vehicle determines thus the spot of the road being observed by the vision sensors mounted onboard the UAVs at a given time. The virtual vehicle is in fact used as a leader in the coordination algorithm, and its speed is also adjusted based on the coordination states of the two UAVs. The coordination state of this virtual vehicle is also transmitted over the tactical network and used in the coor-

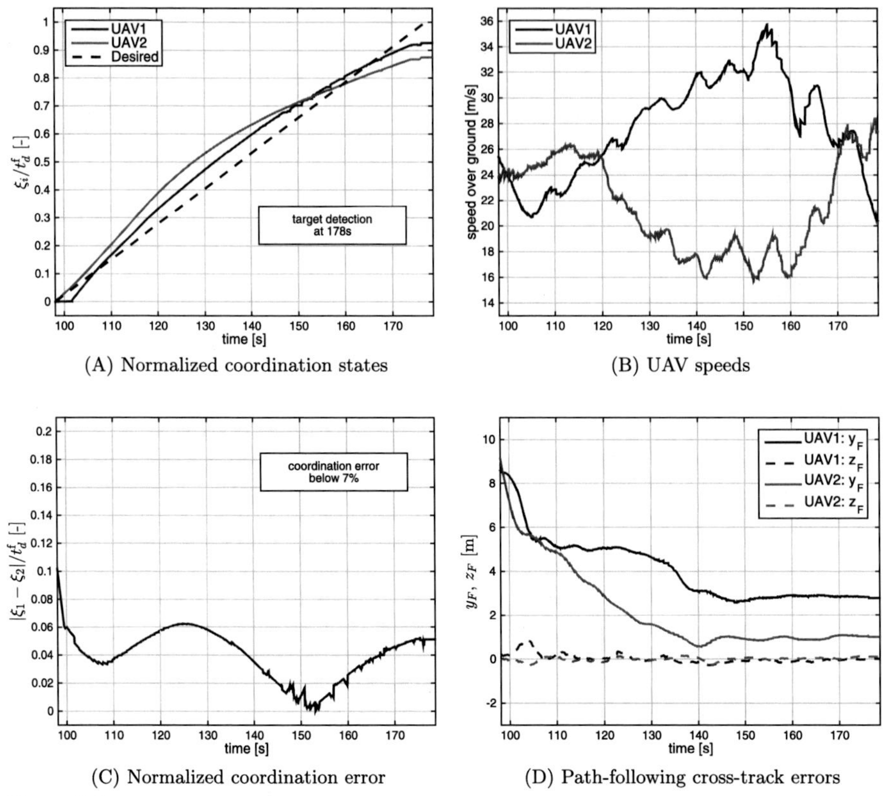

(A) Normalized coordination states (B) UAV speeds

(C) Normalized coordination error (D) Path-following cross-track errors

Figure 8.5 Cooperative road-search; cooperative path-following control during the road-search phase. The coordination errors remain below 7% during the entire duration of the road search, while the path-following cross-track errors converge to a 3-m tube around the desired spatial paths.

dination control laws of the two "real" vehicles. The performance of the cooperative path-following control algorithm is illustrated in Fig. 8.5. For this particular scenario, the path-following cross-track errors converge to a 3-m tube around the desired spatial paths, while the coordination errors remain below 7% during the entire duration of the road search. It is worth noting that significant data dropouts occurred between 145 s and 170 s, especially affecting UAV 1; these data dropouts cause sudden jumps in the normalized coordination states, as can be seen in Figs. 8.5A and 8.5C.

As mentioned above, maintaining a tight coordination along the paths is important to ensure a desired level of FoV overlap with desired image resolution, two key elements for reliable target detection. Fig. 8.6 illustrates the performance of the road-search mission from this perspective. On the one hand, Fig. 8.6A shows a set of estimates of the ground FoV footprints assuming a flat Earth with known ground elevation. These estimates assume a trapezoidal footprint, and are based on experimental data including the inertial position and orientation of the two UAVs, orientation of their cameras, as well as the line-of-sight range to the ground. To provide a quanti-

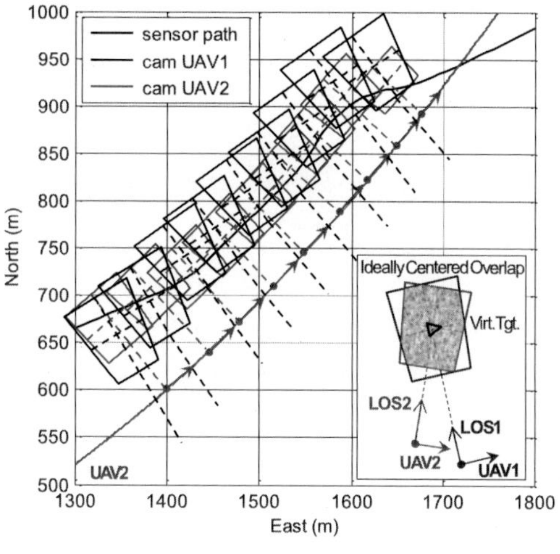

(A) 2D flat-Earth field-of-view footprints

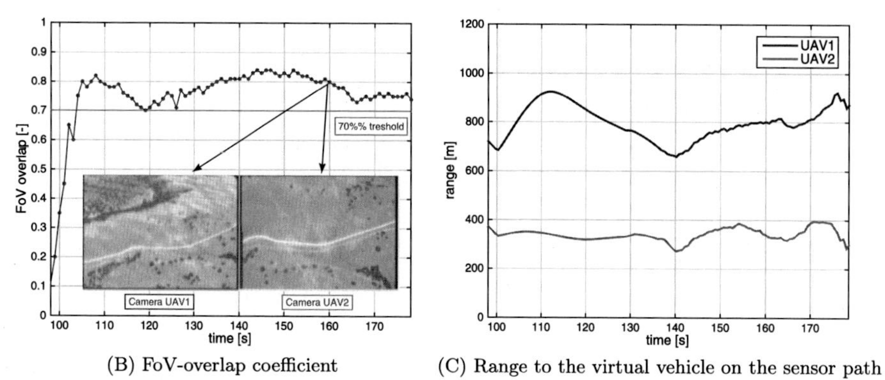

(B) FoV-overlap coefficient (C) Range to the virtual vehicle on the sensor path

Figure 8.6 Cooperative road-search; mission performance. Field-of-view (FoV) overlap and range to the virtual vehicle on the sensor path.

tative measure of the FoV overlap, Fig. 8.6B presents an image-overlap coefficient, sampled at 1 Hz. This coefficient is calculated offline using proprietary technology,[5] and is based on semi-automated alignment and differencing of two synchronous images. As can be seen, except for a 5-s initial transient, the overlap coefficient stays above 0.7 during the cooperative road search. This figure also includes a side-by-side image comparison of the imagery data obtained from the two cameras at approximately 160 s after initiation of the mission; one can easily observe that the two images correspond to the same road segment. On the other hand, Fig. 8.6C shows the range from the two vision sensors to the virtual vehicle on the sensor path; these ranges

[5] Information available online at http://perceptivu.com/TargetTrackingSoftware.html [Online; accessed 22 January 2017].

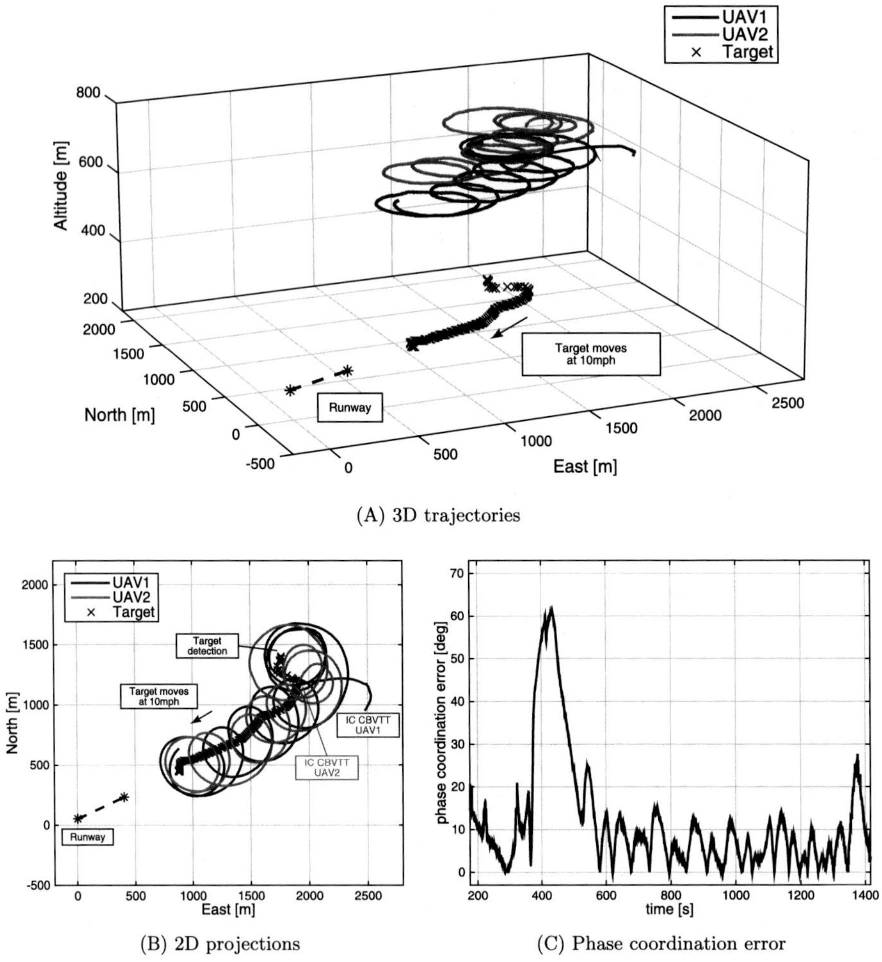

(A) 3D trajectories

(B) 2D projections (C) Phase coordination error

Figure 8.7 Cooperative road-search; coordinated vision-based target tracking (CVBTT). Upon target detection, the two UAVs start tracking the target by means of guidance loops that use visual information for feedback, while simultaneously providing in-situ imagery for precise geo-location of the point of interest. During the target-tracking phase, a coordination algorithm ensures that the two UAVs keep a predefined phase separation of $\frac{\pi}{2}$ rad while "orbiting" around the target.

are always below 1000 m for UAV 1 and 500 m for UAV 2, therefore ensuring desired image resolution for the targets of interest given the characteristics of the two cameras.

Finally, when a target is detected, the two UAVs switch to *cooperative vision-based tracking* mode. In this phase, the UAVs track the target by means of guidance loops that use visual information for feedback, while simultaneously providing in-situ imagery for precise geo-location of the point of interest. During this target-tracking phase, a coordination algorithm ensures that the two UAVs keep a predefined phase separation of $\frac{\pi}{2}$ rad while "orbiting" around the target. This coordination algorithm

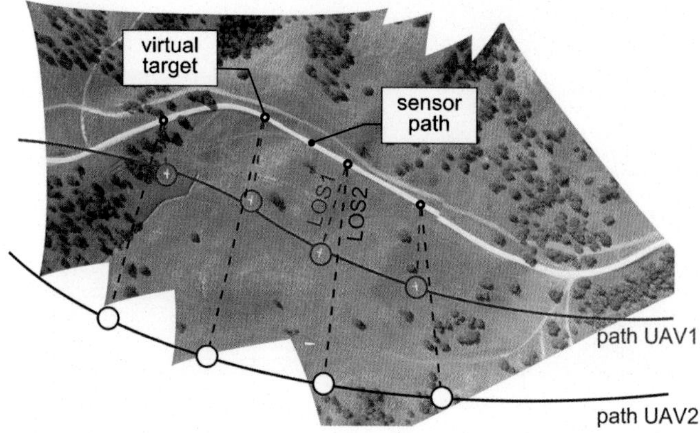

Figure 8.8 Time-critical cooperation in a road-search mission. In this experiment, the road-search paths are intentionally separated by altitude and optimized such that the UAV flying at a lower altitude is continuously present in the field of view of the camera flying at a higher altitude. A mosaic of four consecutive high-resolution images illustrates the progression of the lines of sight (LoSs) connecting the two onboard cameras with the virtual vehicle running along the sensor path.

uses the distributed control law described in Chapter 4 to adjust the speed of the UAVs, with the main difference that *phase on orbit* is now used as a coordination state, rather than the time-variable $\xi_i(t)$. Besides collision avoidance, cooperation through phase-on-orbit coordination allows for several additional benefits, including reduced sensitivity to target escape maneuvers [4]. Performance of the cooperative control algorithm is illustrated in Fig. 8.7, which shows the trajectories of the two UAVs while tracking the target as well as the phase-coordination error. Details about the vision-based guidance loop used in this phase can be found in [2].

8.2 Mission Outcomes

The results presented in the previous section illustrate the benefits of using cooperative control based on the algorithms described in this book when dealing with missions involving multiple vehicles. Such cooperative strategies ensure collision-free maneuvers, and efficiently combine heterogeneous information provided by complementary sensors.

To visually illustrate the effect of time-critical cooperation among the UAVs, Fig. 8.8 presents a mosaic of four consecutive high-resolution images taken during a flight experiment. In this experiment, the road-search paths are intentionally separated by altitude and optimized such that, if the coordination algorithm adequately adjusts the speed of the two UAVs, then the UAV flying at a lower altitude is expected to be continuously present in the FoV of the camera flying at a higher altitude. The figure schematically represents the progression of the lines of sight connecting the two cameras with the virtual vehicle running along the sensor path. Time coordination ensures that cameras observe the same spot on the road and thus maximize the overlap of the footprints of their FoVs, which is critical to provide reliable target discrimination.

(A) Automated 3D terrain extraction from 2D high-resolution imagery and telemetry data obtained using proprietary software from Urban Robotics, Inc (2008).

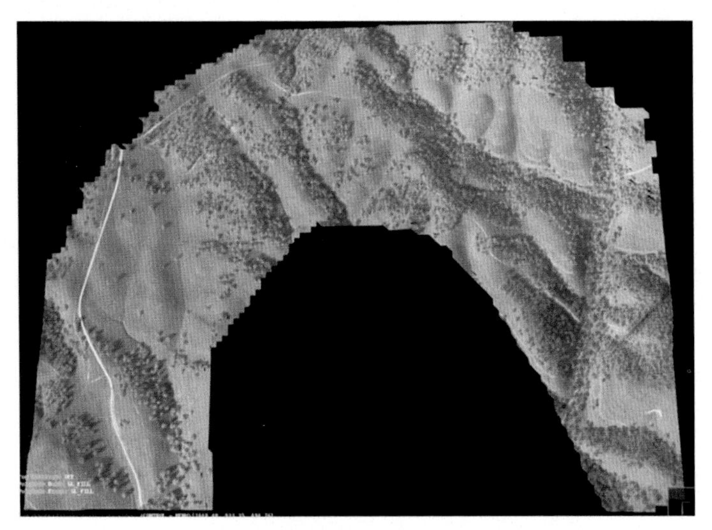

(B) Near-real-time geo-referenced map obtained from high-resolution imagery and telemetry data using proprietary software from 2D3 Sensing, Inc (2008).

Figure 8.9 High-resolution image exploitation. The use of cooperative algorithms in missions involving multiple UAVs can provide valuable mission outcomes, such as (A) 3D geo-referenced models of the operational environment, or (B) geo-referenced maps obtained in near real time from high-resolution imagery.

Also, in order to illustrate possible valuable mission outcomes, Fig. 8.9 presents examples of imagery data exploitation. In Fig. 8.9A, for example, the 3D geo-referenced model of the operational environment is built from 2D high-resolution frames. In Fig. 8.9B, a geo-referenced mosaic is obtained in near real time from high-resolution frames sent by one of the UAVs through the MANET link while in mission.

In summary, the results presented above demonstrate the practical feasibility and efficacy of the systems developed for cooperative control of teams of heterogeneous systems. They also show how the performance of the systems for UAV path following in 3D is superior to that obtained with conventional waypoint guidance methods that are typically implemented on commercial off-the-shelf autopilots. Finally, the results provide a roadmap for further development and onboard implementation of advanced algorithms for cooperative control.

References

[1] M.R. Clement, E. Bourakov, K.D. Jones, V. Dobrokhodov, Exploring network-centric information architectures for unmanned systems control and data dissemination, in: AIAA Infotech@Aerospace, Seattle, WA, April 2009, AIAA 2009-1999.
[2] V. Dobrokhodov, I. Kaminer, K.D. Jones, R. Ghabcheloo, Vision-based tracking and motion estimation for moving targets using small UAVs, Journal of Guidance, Control and Dynamics 31 (2008) 907–917.
[3] M.T. Flynn, R. Juergens, T.L. Cantrell, Employing ISR: SOF best practices, Joint Force Quarterly 50 (2008) 56–61.
[4] S. Morris, E.W. Frew, Cooperative Tracking of Moving Targets by Teams of Autonomous Unmanned Air Vehicles, Tech. Rep. FA9550-04-C-0107, MLB Company and University of Colorado, July 2004.
[5] D. Netzer, A. Bordetsky, TNT testbed for self-organizing tactical networking and collaboration, in: International Command and Control Research and Technology Symposium, Washington, DC, June 2009.

Part Three

Cooperative Control of Multirotor Air Vehicles

3D Path-Following Control of Multirotor Air Vehicles

This chapter describes an outer-loop 3D path-following control algorithm for multirotor UAVs. The control law derived enables a multirotor —equipped with an autopilot capable of tracking angular-rate and thrust reference commands— to converge to and follow a desired three-dimensional path $p_d(\cdot)$ with an arbitrary feasible temporal speed assignment along the path. As will be seen later in Chapter 10, with the present approach the temporal speed assignment along the path can be adjusted online to achieve other objectives such as coordination.

The approach for multirotor 3D path-following proposed in this chapter is similar to the fixed-wing UAV case described in Chapter 3 in that it relies on the use of the Special Orthogonal group $\mathsf{SO}(3)$. At the same time, it also exhibits features that distinguish it from the methodology discussed in Chapter 3. Namely, in the latter the path-following control law was designed so as to align the velocity vector that defined the wind frame of the fixed-wing UAV with the local tangent to the desired path. In addition, it relied on the assumption that the speed of the vehicle had a positive lower bound (see also [4]). However, in the case of multirotor UAVs, which can stop and hover in place, the vehicle-carried wind frame cannot be defined when the velocity of the vehicle is zero. For this reason, we reformulate the path-following problem as follows. We introduce a *body-fixed frame* that is located at the center of mass of the multirotor, with the third axis aligned with the thrust vector, and the first and second axes lying on the plane perpendicular to it. Next, we define a *desired frame* that depends on the position and velocity errors between the vehicle and the desired position along the path. Finally, we reduce the path-following problem to that of driving the attitude error between the body-fixed frame and the desired frame to zero.

The method that we propose employs key concepts and techniques introduced in [7] for trajectory tracking of a quadrotor where the total thrust force and the moment vector generated by the four rotors act as control inputs. Since most multirotors today come equipped with an autopilot capable of tracking total thrust and angular-rate commands, the objective of this chapter is to develop a path-following control algorithm that exploits this feature by using the autopilot commands as control inputs. A rigorous stability analysis is performed to assess the convergence properties of this algorithm for the cases of ideal and non-ideal tracking performance of the autopilot. The main advantage of this approach is the fact that it can be easily used with a variety of commercially available multirotors.

9.1 Problem Formulation

This section formulates the problem of 3D path-following that is the main topic of the present chapter. We first present the equations of motion of a multirotor UAV. This

Time-Critical Cooperative Control of Autonomous Air Vehicles, DOI: 10.1016/B978-0-12-809946-9.00012-9

is followed by the introduction of a virtual target vehicle (that the actual vehicle is required to track) moving along the desired geometric path $\boldsymbol{p}_d(\cdot)$ and formulate a number of assumptions that its speed and acceleration must satisfy. Finally, we define a set of path-following error variables which will be used to formulate the path-following problem.

9.1.1 6-DoF Model for a Multirotor UAV

Let $\{\mathcal{I}\}$ denote an inertial frame $\{\hat{\mathbf{e}}_1, \hat{\mathbf{e}}_2, \hat{\mathbf{e}}_3\}$ and $\{\mathcal{B}\}$ the body frame $\{\hat{\mathbf{b}}_1, \hat{\mathbf{b}}_2, \hat{\mathbf{b}}_3\}$ attached to the center of mass of a multirotor UAV. It is assumed that the vehicle is equipped with an inner-loop autopilot capable of tracking reference commands for angular rates and total thrust [2,5] (see Section 2.2.4), which can thus be viewed as control inputs. We further assume, as in [1,8], that for the types of missions envisioned the atmospheric forces can be neglected. With these assumptions, the six degree of freedom kinematic and dynamic model of the multirotor UAV is given by

$$\begin{aligned} \dot{\boldsymbol{p}} &= \boldsymbol{v}\,, \\ m\dot{\boldsymbol{v}} &= T\hat{\mathbf{b}}_3 - mg\hat{\mathbf{e}}_3\,, \\ \dot{\boldsymbol{R}} &= \boldsymbol{R}(\boldsymbol{\omega})^{\wedge}\,, \end{aligned} \tag{9.1}$$

where $\boldsymbol{p}$ and $\boldsymbol{v}$ are the position and velocity, respectively, of the vehicle's center of mass in the inertial frame with respect to the basis $\{\hat{\mathbf{e}}_1, \hat{\mathbf{e}}_2, \hat{\mathbf{e}}_3\}$, m is the vehicle's mass, T is the total thrust generated by the propellers, $\boldsymbol{R} = \boldsymbol{R}_{\boldsymbol{B}}^{\boldsymbol{I}}$ is the rotation matrix from the body frame to the inertial frame, $\boldsymbol{\omega} = \{\boldsymbol{\omega}_{B/I}\}_B = [p,\, q,\, r]^{\top}$ is the vector of the angular rates of the vehicle with respect to $\{\mathcal{I}\}$, resolved in $\{\mathcal{B}\}$, and $(\cdot)^{\wedge}$ denotes the hat map (see Appendix A.1). The multirotor UAV system is depicted in Fig. 9.1A.

9.1.2 Virtual Target and Virtual Time

Let $\gamma(t)$ be a scalar function that maps actual (clock) time t to the desired mission time t_d defined in Section 2.2.1, i.e.

$$\gamma : \mathbb{R}^{+} \to [0, t_d^{\mathrm{f}}]\,, \tag{9.2}$$

where t_d^{f} is the mission duration. Then $\boldsymbol{p}_d(\gamma(t))$ represents the desired path evaluated at $\gamma(t)$, which denotes the position of the virtual target to be tracked by the vehicle. Note that in this formulation the time-variable $\gamma(t)$ plays a role analogous to that of the free variable $\theta(t)$ introduced in Section 2.2.2, and the parameterizing path length variable $\ell(t)$ introduced in Section 3.1. Differently from Chapter 3, where the dynamics of $\ell(t)$ were directly used to solve the path-following problem, here we use the dynamics of the virtual time (actually its second derivative $\ddot{\gamma}(t)$) as an extra degree of freedom to achieve time coordination. Since coordination is beyond the scope of this chapter, an extended discussion and formal definition of the virtual time will be provided in Chapter 10. Nevertheless, in what follows we derive feasibility limits on the first and second derivative of the virtual time, namely $\dot{\gamma}_{\min}$, $\dot{\gamma}_{\max}$, and $\ddot{\gamma}_{\max}$, which ensure that the dynamics of the virtual target do not violate the dynamic constraints on the speed and acceleration of the vehicle. Namely,

$$v_{\min} \leq \|\dot{\boldsymbol{p}}_{\boldsymbol{d}}(\gamma(t))\| \leq v_{\max}\,, \qquad \|\ddot{\boldsymbol{p}}_{\boldsymbol{d}}(\gamma(t))\| < a_{\max} \leq g\,, \tag{9.3}$$

with $v_{\min} \geq 0$ and $v_{\max}$, $a_{\max} > 0$.

To this end, we start by assuming that the desired path and timing law produced by the cooperative trajectory generation algorithm discussed in Section 2.2.1 satisfy the following constraints [3]:

$$0 \leq v_{d,\min} \leq \|\boldsymbol{p}'_{\boldsymbol{d}}(t_d)\| \leq v_{d,\max}\,, \quad \|\boldsymbol{p}''_{\boldsymbol{d}}(t_d)\| \leq a_{d,\max}\,. \tag{9.4}$$

Then, computing the first derivative of the virtual target position $\boldsymbol{p}_{\boldsymbol{d}}(\gamma(t))$ we obtain

$$\left\|\dot{\boldsymbol{p}}_{\boldsymbol{d}}(\gamma(t))\right\| = \left\|\frac{d\,\boldsymbol{p}_{\boldsymbol{d}}(\gamma(t))}{d\,\gamma(t)}\frac{d\gamma(t)}{dt}\right\| = \left\|\dot{\gamma}(t)\boldsymbol{p}'_{\boldsymbol{d}}(\gamma(t))\right\|\,, \tag{9.5}$$

where $\dot{\boldsymbol{p}}_{\boldsymbol{d}}(\gamma(t)) = d(\boldsymbol{p}_{\boldsymbol{d}}(\gamma(t)))/dt$ denotes the desired speed profile to be tracked by the UAV at time t, while $\boldsymbol{p}'_{\boldsymbol{d}}(\gamma(t)) = d\,\boldsymbol{p}_{\boldsymbol{d}}(\gamma)/d\gamma$ is obtained from the speed profile generated by the trajectory generation algorithm (see Chapter 2). Combining Eqs. (9.4) and (9.5), one can conclude that the speed constraints given in Eq. (9.3) are satisfied if the following bounds hold:

$$\dot{\gamma}_{\min}\, v_{d,\min} \geq v_{\min}\,, \tag{9.6a}$$

$$\dot{\gamma}_{\max}\, v_{d,\max} \leq v_{\max}\,. \tag{9.6b}$$

In other words, Eqs. (9.6a) and (9.6b) relate the limits of the desired speed profile $\dot{\boldsymbol{p}}_{\boldsymbol{d}}(\gamma(t))$ to the limits of $\dot{\gamma}(t)$. Similar limits can be derived for the acceleration profile $\ddot{\boldsymbol{p}}_{\boldsymbol{d}}(\gamma(t))$. In fact, differentiating Eq. (9.5) we get

$$\|\ddot{\boldsymbol{p}}_{\boldsymbol{d}}(\gamma(t))\| = \|\ddot{\gamma}(t)\boldsymbol{p}'_{\boldsymbol{d}}(\gamma(t)) + \dot{\gamma}^2(t)\boldsymbol{p}''_{\boldsymbol{d}}(\gamma(t))\|\,,$$

from which it follows that the constraints given by (9.3) are satisfied if the following inequality holds:

$$|\ddot{\gamma}_{\max}\, v_{d,\max} + \dot{\gamma}^2_{\max}\, a_{d,\max}| \leq a_{\max}\,. \tag{9.7}$$

The equation above relates the limits of the speed and acceleration profiles $\boldsymbol{p}'_{\boldsymbol{d}}(\gamma(t))$ and $\boldsymbol{p}''_{\boldsymbol{d}}(\gamma(t))$ to the limits of $\dot{\gamma}(t)$ and $\ddot{\gamma}(t)$. Thus, when deriving an expression for the desired rate of progression of the UAV along the path (i.e. the dynamics of $\gamma(t)$), the constraints in (9.3) must be taken into consideration. However, because this chapter focuses on the path-following problem, we assume that these bounds are always satisfied, as follows.

Assumption 9.1. *The first and second derivative of the virtual time respect the bounds*

$$\dot{\gamma}_{\min} \leq \dot{\gamma}(t) \leq \dot{\gamma}_{\max}\,, \quad |\ddot{\gamma}(t)| \leq \ddot{\gamma}_{\max}\,,$$

where $\dot{\gamma}_{\min}$, $\dot{\gamma}_{\max}$, and $\ddot{\gamma}_{\max}$ satisfy inequalities (9.6) *and* (9.7). *Thus, the virtual target $\boldsymbol{p_d}(\gamma(t))$ to be tracked by the UAV does not violate the speed and acceleration constraints of the vehicle given by Eq.* (9.3).

9.1.3 Path-Following Error

With the above setup, we now define position error vector $\boldsymbol{e_p} \in \mathbb{R}^3$ as

$$\boldsymbol{e_p} = \boldsymbol{p_d}(\gamma) - \boldsymbol{p}, \tag{9.8}$$

and the corresponding velocity error vector $\boldsymbol{e_v} \in \mathbb{R}^3$ as

$$\boldsymbol{e_v} = \dot{\boldsymbol{p}}_{\boldsymbol{d}}(\gamma) - \dot{\boldsymbol{p}}. \tag{9.9}$$

Following [7] we now introduce an auxiliary frame $\{\mathcal{D}\}$ that is used to (i) shape the approach to the path as a function of the error components $\boldsymbol{e_p}$ and $\boldsymbol{e_v}$ and (ii) impose a desired orientation on the vehicle as it moves along the desired path. Let the rotation matrix from frame $\{\mathcal{D}\}$ to the inertial frame $\{\mathcal{I}\}$ be defined as

$$\boldsymbol{R_c} := \boldsymbol{R}_{\boldsymbol{D}}^{\boldsymbol{I}} = [\hat{\mathbf{b}}_{\mathbf{1D}}\,,\ \hat{\mathbf{b}}_{\mathbf{2D}}\,,\ \hat{\mathbf{b}}_{\mathbf{3D}}]\,,$$

where

$$\hat{\mathbf{b}}_{\mathbf{3D}} = \frac{(k_p + s_p)\boldsymbol{e_p} + (k_v + s_v)\boldsymbol{e_v} + mg\hat{\mathbf{e}}_{\mathbf{3}} + m\ddot{\boldsymbol{p}}_{\boldsymbol{d}}(\gamma)}{\|(k_p + s_p)\boldsymbol{e_p} + (k_v + s_v)\boldsymbol{e_v} + mg\hat{\mathbf{e}}_{\mathbf{3}} + m\ddot{\boldsymbol{p}}_{\boldsymbol{d}}(\gamma)\|}, \tag{9.10}$$

for some $k_p, k_v > 1$, and

$$s_p = \begin{cases} \operatorname{sign}(\boldsymbol{e}_{\boldsymbol{p}}^\top \hat{\mathbf{e}}_{\mathbf{3}}), & \text{if } \|k_p\boldsymbol{e_p} + k_v\boldsymbol{e_v} + mg\hat{\mathbf{e}}_{\mathbf{3}} + m\ddot{\boldsymbol{p}}_{\boldsymbol{d}}(\gamma)\| = 0 \\ 0 & \text{otherwise} \end{cases},$$

$$s_v = \begin{cases} \operatorname{sign}(\boldsymbol{e}_{\boldsymbol{v}}^\top \hat{\mathbf{e}}_{\mathbf{3}}), & \text{if } \|k_p\boldsymbol{e_p} + k_v\boldsymbol{e_v} + mg\hat{\mathbf{e}}_{\mathbf{3}} + m\ddot{\boldsymbol{p}}_{\boldsymbol{d}}(\gamma)\| = 0 \\ 0 & \text{otherwise} \end{cases}.$$

Vector $\hat{\mathbf{b}}_{\mathbf{3D}}$ defines the desired orientation of the $\hat{\mathbf{b}}_{\mathbf{3}}$-axis of the multirotor required in order for it to converge to the desired position $\boldsymbol{p_d}(\gamma(t))$ and track the velocity $\dot{\boldsymbol{p}}_{\boldsymbol{d}}(\gamma(t))$. As an example, Fig. 9.2 illustrates the case where the displacement between the UAV position and the virtual target is along the $\hat{\mathbf{e}}_{\mathbf{1}}$ axis. In this case, the desired orientation of the multirotor's $\hat{\mathbf{b}}_{\mathbf{3}}$-axis is $\hat{\mathbf{b}}_{\mathbf{3D}} = [\hat{\mathbf{b}}_{\mathbf{3D},1}\,,\ 0\,,\ \hat{\mathbf{b}}_{\mathbf{3D},3}]^\top$, i.e. the horizontal component of the $\hat{\mathbf{b}}_{\mathbf{3D}}$ axis along the $\hat{\mathbf{e}}_{\mathbf{2}}$ direction is zero, and the UAV moves solely along the $\hat{\mathbf{e}}_{\mathbf{1}}$ axis, thus reaching the virtual target. Fig. 9.2 also shows how different approaches to the path can be obtained by tuning the control gain k_p (for the sake of simplicity in the illustration the value of the control gain k_v is set to 0). Vector $\hat{\mathbf{b}}_{\mathbf{1D}}$, which describes the desired orientation of the multirotor's $\hat{\mathbf{b}}_{\mathbf{1}}$ axis, can be arbitrarily chosen as long as it is orthonormal to $\hat{\mathbf{b}}_{\mathbf{3D}}$. Consequently, $\hat{\mathbf{b}}_{\mathbf{2D}}$ is chosen to be orthonormal to and form a right hand triad with $\hat{\mathbf{b}}_{\mathbf{1D}}$ and $\hat{\mathbf{b}}_{\mathbf{3D}}$.

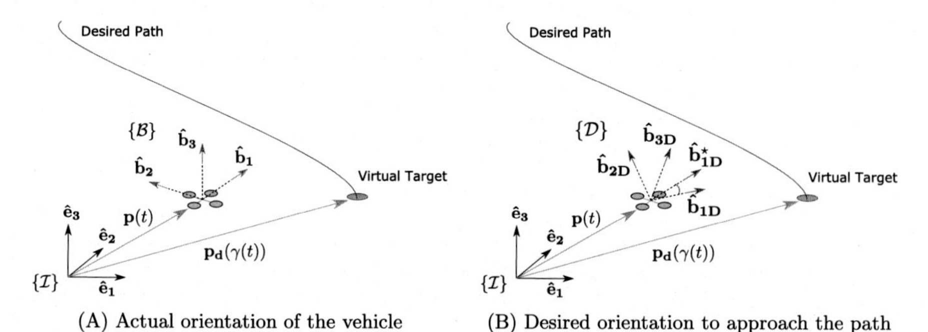

(A) Actual orientation of the vehicle (B) Desired orientation to approach the path

Figure 9.1 Geometry of the path-following problem.

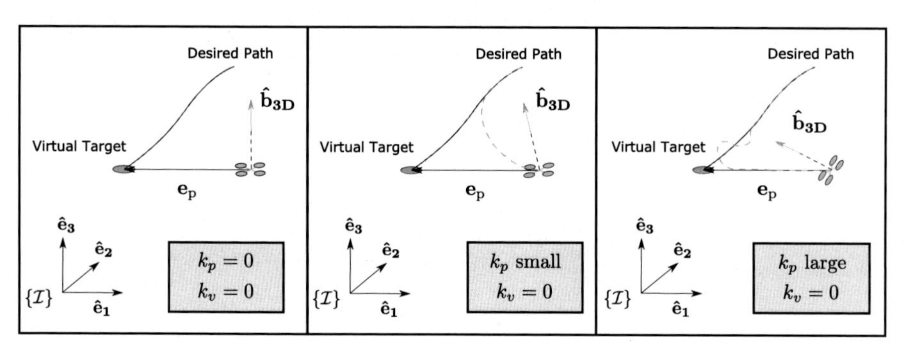

Figure 9.2 Shaping the approach to the path.

Remark 9.1. Given the vector $\hat{\mathbf{b}}_{3D}$ defined in (9.10), let $\hat{\mathbf{b}}^\star_{1D}$ be a vector that describes the desired heading of $\hat{\mathbf{b}}_1$, but that is not orthonormal to $\hat{\mathbf{b}}_{3D}$. Assume that $\hat{\mathbf{b}}^\star_{1D}$ is not parallel to $\hat{\mathbf{b}}_{3D}$ either. Then, $\hat{\mathbf{b}}_{1D}$ and $\hat{\mathbf{b}}_{2D}$ can be formulated as follows [7]:

$$\hat{\mathbf{b}}_{2D} = \frac{\hat{\mathbf{b}}_{3D} \times \hat{\mathbf{b}}^\star_{1D}}{\|\hat{\mathbf{b}}_{3D} \times \hat{\mathbf{b}}^\star_{1D}\|}, \quad \hat{\mathbf{b}}_{1D} = \hat{\mathbf{b}}_{2D} \times \hat{\mathbf{b}}_{3D}\,.$$

Fig. 9.1 depicts the geometry of the problem at hand. △

Lemma 9.1. *From the definition of s_p and s_v, the denominator of $\hat{\mathbf{b}}_{3D}$ in Eq.* (9.10) *is always greater than* 0. *Therefore, vectors $\hat{\mathbf{b}}_{1D}$, $\hat{\mathbf{b}}_{2D}$, and $\hat{\mathbf{b}}_{3D}$ are always well defined.* ◇

Proof. The proof of Lemma 9.1 is given in Appendix B.3.1. □

Next we introduce a set of variables that will be used later to formulate the path-following problem. Let $\tilde{\boldsymbol{R}}$ be the rotation matrix from $\{\mathcal{B}\}$ to $\{\mathcal{D}\}$, that is,

$$\tilde{\boldsymbol{R}} := \boldsymbol{R}^D_B = \boldsymbol{R}_c^\top \boldsymbol{R}\,.$$

Then

$$\dot{\tilde{\boldsymbol{R}}} = \tilde{\boldsymbol{R}}(\tilde{\omega})^\wedge\,,$$

where

$$\tilde{\omega} = \{\omega_{B/D}\}_B = \begin{bmatrix} p \\ q \\ r \end{bmatrix} - \tilde{\boldsymbol{R}}^\top \{\omega_{D/I}\}_D \tag{9.11}$$

and

$$(\{\omega_{D/I}\}_D)^\wedge = \boldsymbol{R}_c^\top \dot{\boldsymbol{R}}_c .$$

Note that if $\tilde{\boldsymbol{R}} = \mathbb{I}_3$, then frame $\{\mathcal{B}\}$ coincides with the desired frame $\{\mathcal{D}\}$. Consider the real-valued function on $\mathsf{SO}(3)$ given by

$$\Psi(\tilde{\boldsymbol{R}}) = \frac{1}{2} tr(\mathbb{I}_3 - \tilde{\boldsymbol{R}}), \tag{9.12}$$

and its time derivative

$$\dot{\Psi}(\tilde{\boldsymbol{R}}) = -\frac{1}{2} tr(\tilde{\boldsymbol{R}}(\tilde{\omega})^\wedge).$$

Finally, let

$$\boldsymbol{e}_{\tilde{R}} = \frac{1}{2}(\tilde{\boldsymbol{R}} - \tilde{\boldsymbol{R}}^\top)^\vee , \tag{9.13}$$

where $(\cdot)^\vee$ denotes the vee map (see Appendix A.1). Using the properties of the $\mathsf{SO}(3)$ group and following derivations similar to those in Chapter 3 it can be shown that

$$\dot{\Psi}(\tilde{\boldsymbol{R}}) = \boldsymbol{e}_{\tilde{R}} \cdot \tilde{\omega} . \tag{9.14}$$

With the above notation, the generalized path-following error vector $\boldsymbol{x}_{pf}(t)$ is now formally defined as

$$\boldsymbol{x}_{pf} := [\boldsymbol{e}_p^\top , \boldsymbol{e}_v^\top , \boldsymbol{e}_{\tilde{R}}^\top]^\top ,$$

with dynamics given by

$$\begin{aligned} \dot{\boldsymbol{e}}_p &= \boldsymbol{e}_v , \\ m\dot{\boldsymbol{e}}_v &= m\ddot{\boldsymbol{p}}_d(\gamma) - T\hat{\mathbf{b}}_3 + mg\hat{\mathbf{e}}_3 , \\ \dot{\Psi}(\tilde{\boldsymbol{R}}) &= \boldsymbol{e}_{\tilde{R}} \cdot \tilde{\omega} . \end{aligned} \tag{9.15}$$

Notice that in the region $\Psi(\tilde{\boldsymbol{R}}) < 1$, if $\boldsymbol{x}_{pf} = 0$, then the path-following position error, the path-following velocity error, and the path-following attitude error are equal to zero, i.e.

$$\boldsymbol{e}_p = \mathbf{0} , \quad \boldsymbol{e}_v = \mathbf{0} , \quad \Psi(\tilde{\boldsymbol{R}}) = 0 .$$

We now define the path-following problem for a multirotor UAV.

Definition 9.1 (Path-Following Problem). For a given multirotor UAV, and for a given virtual target $\boldsymbol{p}_d(\gamma(t))$ moving along the desired path satisfying the bounds given in (9.3), design feedback control laws for the total thrust $T(t)$, roll rate $p(t)$, pitch rate $q(t)$, and yaw rate $r(t)$ such that the generalized path-following error vector $\boldsymbol{x_{pf}} = [\boldsymbol{e}_{\boldsymbol{p}}^\top, \boldsymbol{e}_{\boldsymbol{v}}^\top, \boldsymbol{e}_{\tilde{\boldsymbol{R}}}^\top]^\top$, with the dynamics described in (9.15), converges to a neighborhood of the origin. ♠

Using the setup described, we next derive a path-following control law for a multirotor UAV.

9.2 Path-Following Control Law

Let the total thrust command be governed by

$$T_c = \left((k_p + s_p)\boldsymbol{e}_{\boldsymbol{p}} + (k_v + s_v)\boldsymbol{e}_{\boldsymbol{v}} + mg\hat{\mathbf{e}}_{\mathbf{3}} + m\ddot{\boldsymbol{p}}_{\boldsymbol{d}}(\gamma)\right)^\top \hat{\mathbf{b}}_{\mathbf{3}} . \tag{9.16}$$

In addition, let the angular-rate commands be given by

$$\begin{bmatrix} p_c \\ q_c \\ r_c \end{bmatrix} = \tilde{\boldsymbol{R}}^\top \{\boldsymbol{\omega}_{\boldsymbol{D/I}}\}_D - k_{\tilde{R}} \boldsymbol{e}_{\tilde{\boldsymbol{R}}} , \tag{9.17}$$

for some $k_{\tilde{R}} > 0$.

The following result applies.

Lemma 9.2. *Let the total thrust $T_c(t)$ and the angular-rate commands $p_c(t)$, $q_c(t)$, and $r_c(t)$ of the vehicle be given by* (9.16) *and* (9.17). *Assume ideal performance for the existing inner-loop autopilot (i.e. $T(t) = T_c(t)$ and $[p(t), q(t), r(t)] = [p_c(t), q_c(t), r_c(t)]$ for all $t \geq 0$). Then, there exist $k_p, k_v, k_{\tilde{R}}$ such that the error vector*

$$\boldsymbol{x_{pf}} = [\boldsymbol{e}_{\boldsymbol{p}}^\top, \boldsymbol{e}_{\boldsymbol{v}}^\top, \boldsymbol{e}_{\tilde{\boldsymbol{R}}}^\top]^\top \tag{9.18}$$

converges exponentially to zero with rate of convergence

$$\lambda_{pf} < \frac{c_1(1-c^2)}{m} , \tag{9.19}$$

in the domain of attraction

$$\Omega_{pf} := \left\{ (\boldsymbol{e}_{\boldsymbol{p}}, \tilde{\boldsymbol{R}}) \mid \Psi(\tilde{\boldsymbol{R}}) \leq c^2 , \|\boldsymbol{e}_{\boldsymbol{p}}\| \leq e_{p\,\max} \right\} , \tag{9.20}$$

for some $c_1, e_{p\,\max} > 0$ and $c^2 < 1/2$. In other words, the error vector $\boldsymbol{x_{pf}}$ satisfies

$$\|\boldsymbol{x_{pf}}(t)\| \leq k_{pf} \|\boldsymbol{x_{pf}}(0)\| \mathrm{e}^{-\lambda_{pf} t} ,$$

where

$$k_{pf} := \sqrt{\frac{\lambda_{\max}(\boldsymbol{W_2})}{\lambda_{\min}(\boldsymbol{W_1})}} , \tag{9.21}$$

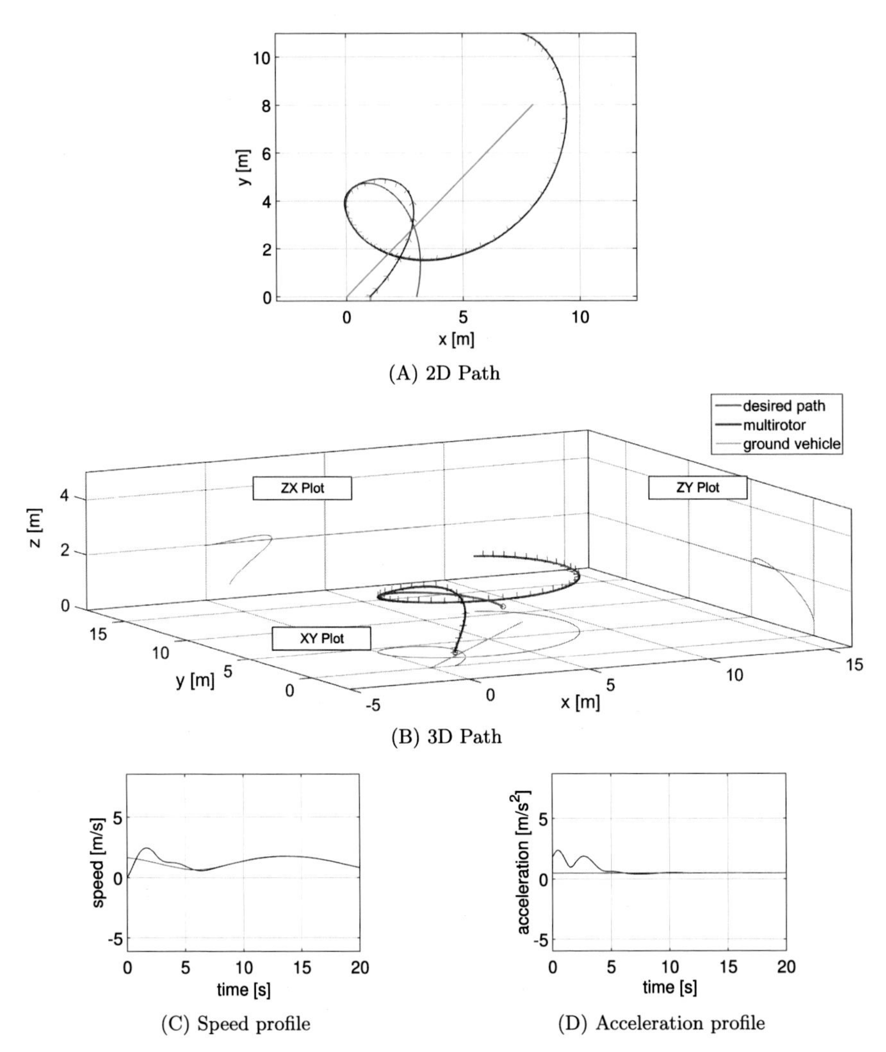

(C) Speed profile (D) Acceleration profile

Figure 9.3 Following a virtual target; simulation scenario.

and

$$W_1 = \begin{bmatrix} \frac{k_p}{2} & -\frac{c_1}{2} & 0 \\ -\frac{c_1}{2} & \frac{m}{2} & 0 \\ 0 & 0 & 1 \end{bmatrix}, \quad W_2 = \begin{bmatrix} \frac{k_p}{2} & -\frac{c_1}{2} & 0 \\ -\frac{c_1}{2} & \frac{m}{2} & 0 \\ 0 & 0 & \frac{1}{1-c^2} \end{bmatrix}. \tag{9.22}$$

◇

Proof. The proof of Lemma 9.2 is given in Appendix B.3.2. □

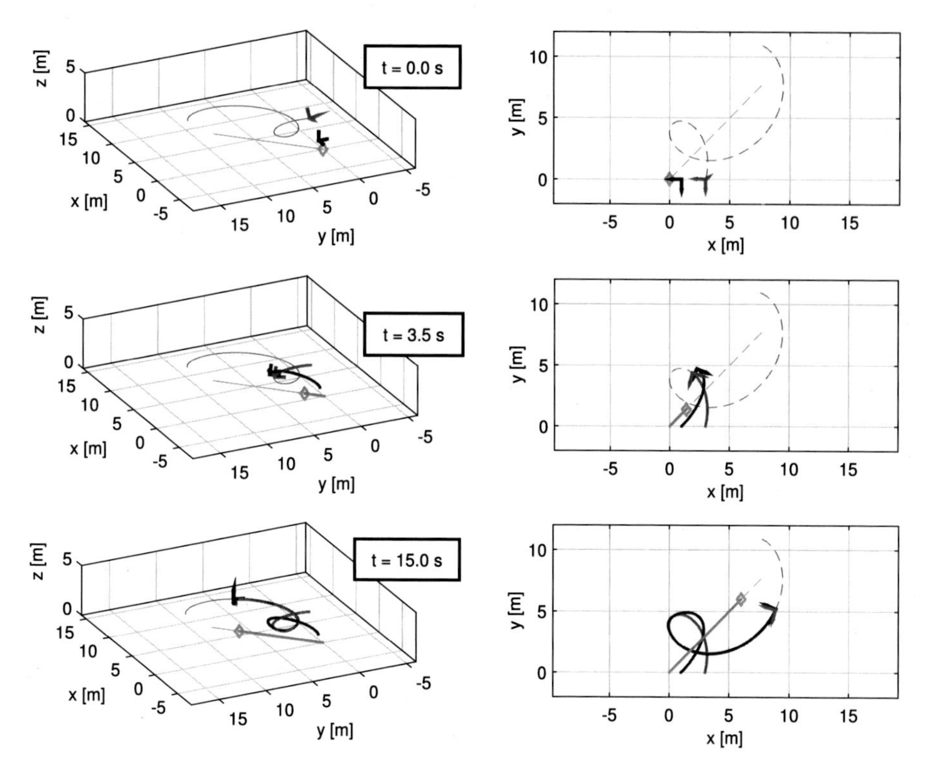

Figure 9.4 Following a virtual target; execution of the mission depicted in Fig. 9.3.

Note that we modified the controller used in [7] in two important ways. First, we redefined it as path-following controller and, second, we modified it to use angular rates and thrust as control inputs to be tracked by an existing inner-loop autopilot. This, in turn, requires that we address the limitations in performance that are introduced by a non-ideal inner-loop controller. For this reason, we now consider the case of non perfect tracking performance of the autopilot, following the approach discussed in Section 2.2.4. This yields the main result of this chapter.

Lemma 9.3. *Let the total thrust $T_c(t)$ and the angular-rate commands $p_c(t)$, $q_c(t)$, and $r_c(t)$ be governed by* (9.16) *and* (9.17). *Let the inner-loop autopilot satisfy the uniform performance bounds*

$$|p_c(t) - p(t)| \leq \gamma_p \,, \quad |q_c(t) - q(t)| \leq \gamma_q \,, \quad |r_c(t) - r(t)| \leq \gamma_r \,,$$
$$|T_c(t) - T(t)| \leq \gamma_T \,,$$

where $T(t)$, $p(t)$, $q(t)$, and $r(t)$ are the actual total thrust and angular rates, and let $\gamma_\omega = \sqrt{\gamma_p^2 + \gamma_q^2 + \gamma_r^2}$. Suppose the performance bounds verify

$$\frac{(c_1/m + 1)\gamma_T + \gamma_\omega}{\lambda_{pf}\lambda_{\min}(\mathbf{W_2})\delta_\lambda} \leq \min(e_{p\,\max}\,,\ (1 - c^2)c^2)\,, \tag{9.23}$$

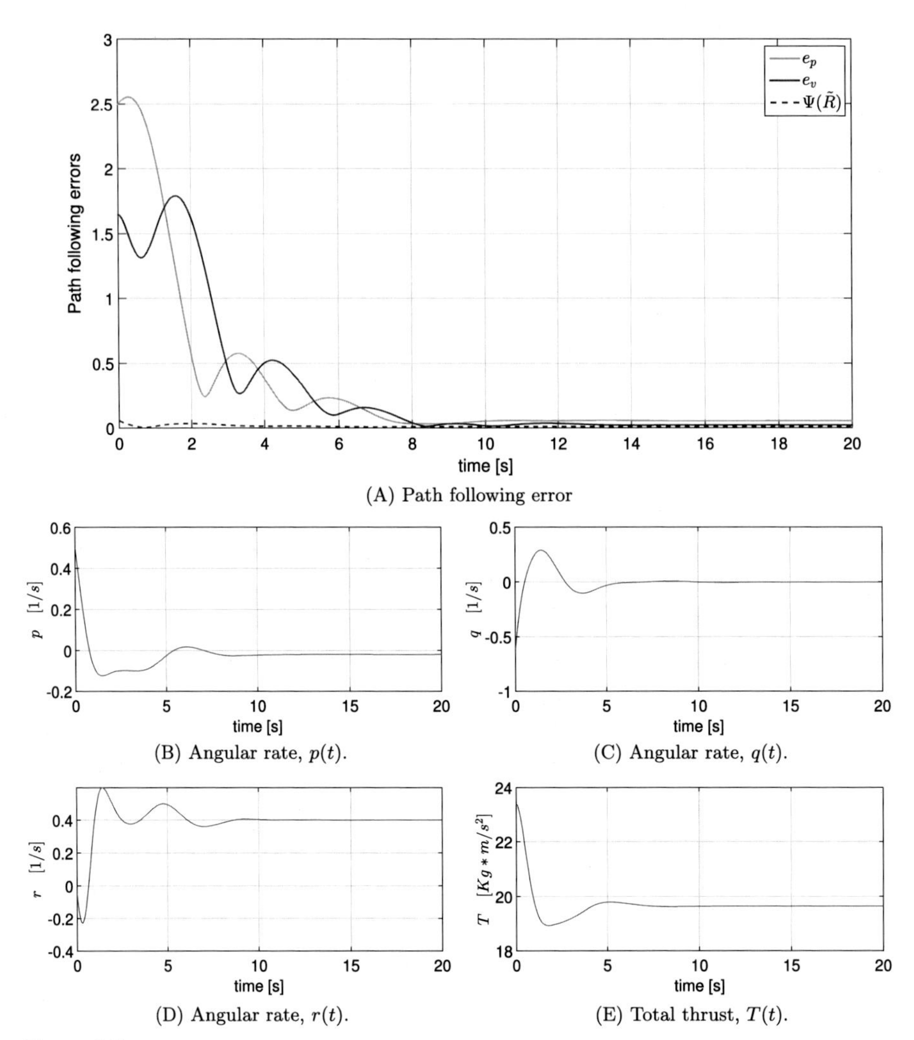

Figure 9.5 Following a virtual target; path-following errors and commands.

where δ_λ satisfies $0 < \delta_\lambda < 1$ and $\mathbf{W_2}$ is defined in (9.22). Then, there exist k_p , k_v , $k_{\tilde{R}}$ such that, for any initial state $\boldsymbol{x_{pf}}(0) \in \Omega_{pf}$, where Ω_{pf} is defined in (9.20), the path-following error is uniformly ultimately bounded. In particular, for any initial state $\boldsymbol{x_{pf}}(0) \in \Omega_{pf}$, there exists a time $T_b \geq 0$ such that the following bounds are satisfied:

$$\|\boldsymbol{x_{pf}}(t)\| \leq k_{pf}\|\boldsymbol{x_{pf}}(0)\|\mathrm{e}^{-\lambda_{pf}(1-\delta_\lambda)t}\,, \qquad \textit{for all } 0 \leq t < T_b\,, \tag{9.24}$$

$$\|\boldsymbol{x_{pf}}(t)\| \leq \rho\,, \qquad \textit{for all } t \geq T_b\,, \tag{9.25}$$

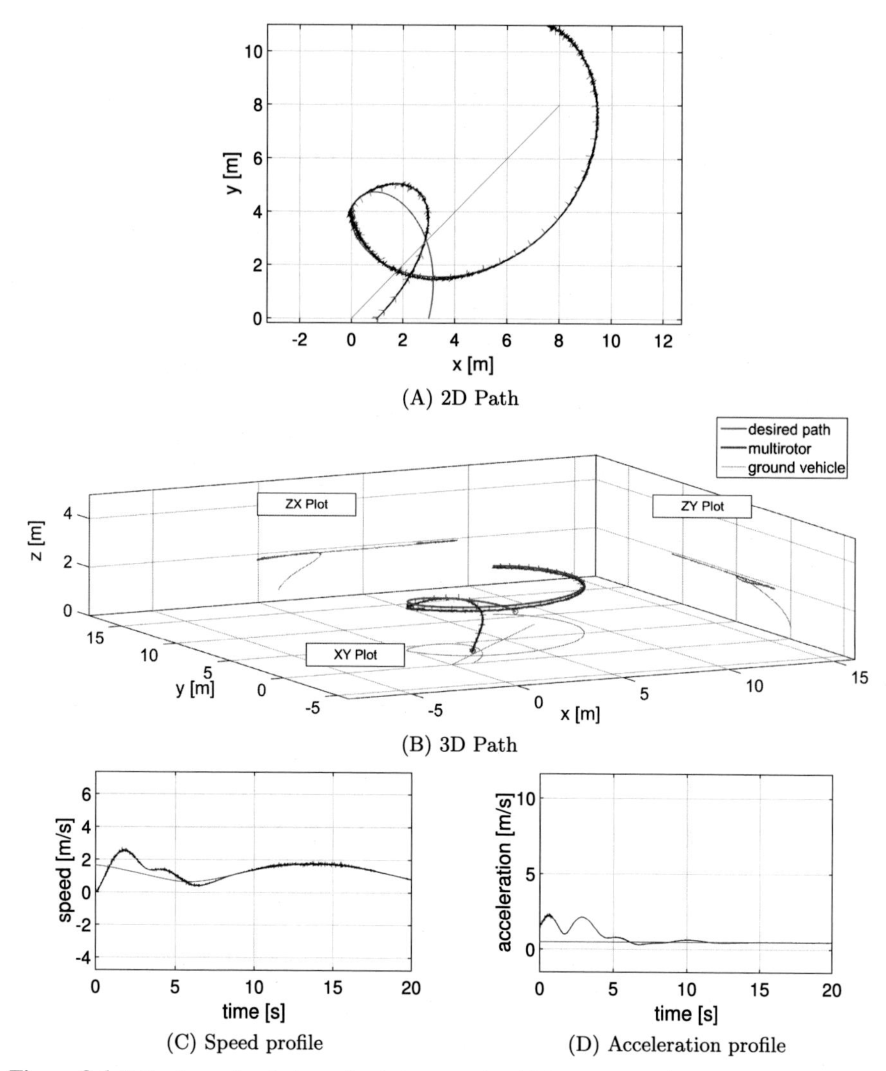

(A) 2D Path

(B) 3D Path

(C) Speed profile

(D) Acceleration profile

Figure 9.6 Following a virtual target; simulation scenario with inner-loop autopilot.

where k_{pf} *is defined in* (9.21) *and*

$$\rho := \sqrt{\frac{\lambda_{\max}(\boldsymbol{W_2})}{\lambda_{\min}(\boldsymbol{W_1})}\left(\frac{(c_1/m+1)\gamma_T+\gamma_\omega}{\lambda_{pf}\lambda_{\min}(\boldsymbol{W_2})\delta_\lambda}\right)}. \tag{9.26}$$

◇

Proof. The proof of Lemma 9.3 is given in Appendix B.3.3. □

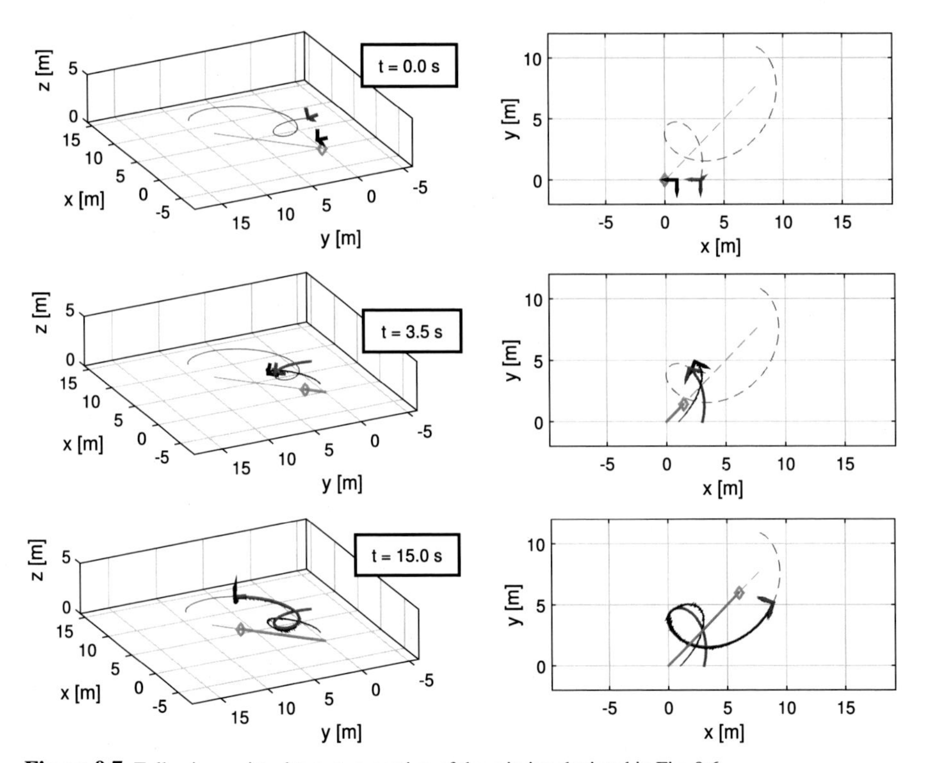

Figure 9.7 Following a virtual target; execution of the mission depicted in Fig. 9.6.

9.3 Simulation Example: Following a Virtual Target

In this section we present simulation results of a scenario in which a multirotor UAV is required to follow a predefined path and track a prescribed speed profile while pointing at a ground vehicle moving at constant speed. We implement the six degree-of-freedom model given by Eq. (9.1). The initial position and velocity of the ground vehicle are $\boldsymbol{p}_t = [0\,,\ 0\,,\ 0]^\top$ m and $\boldsymbol{v}_t = [0.4\,,\ 0.4\,,\ 0]^\top$ m/s, respectively. The UAV is initially positioned at $\boldsymbol{p} = [1\,,\ 0\,,\ 0.5]^\top$ m, with orientation specified by

$$\mathbf{R} = \begin{bmatrix} -1 & 0 & 0 \\ 0 & -1 & 0 \\ 0 & 0 & 1 \end{bmatrix},$$

i.e., the $\hat{\mathbf{b}}_1$ axis is pointing in the negative x-axis direction, while the $\hat{\mathbf{b}}_3$ axis is pointing upwards. The desired orientation for the multirotor UAV's $\hat{\mathbf{b}}_1$-axis is computed according to Remark 9.1 as follows:

$$\hat{\mathbf{b}}^\star_{1\mathrm{D}} = \frac{\boldsymbol{p}_t - \boldsymbol{p}}{\|\boldsymbol{p}_t - \boldsymbol{p}\|}, \quad \hat{\mathbf{b}}_{2\mathrm{D}} = \frac{\hat{\mathbf{b}}_{3\mathrm{D}} \times \hat{\mathbf{b}}^\star_{1\mathrm{D}}}{\|\hat{\mathbf{b}}_{3\mathrm{D}} \times \hat{\mathbf{b}}^\star_{1\mathrm{D}}\|}, \quad \hat{\mathbf{b}}_{1\mathrm{D}} = \hat{\mathbf{b}}_{2\mathrm{D}} \times \hat{\mathbf{b}}_{3\mathrm{D}}\,.$$

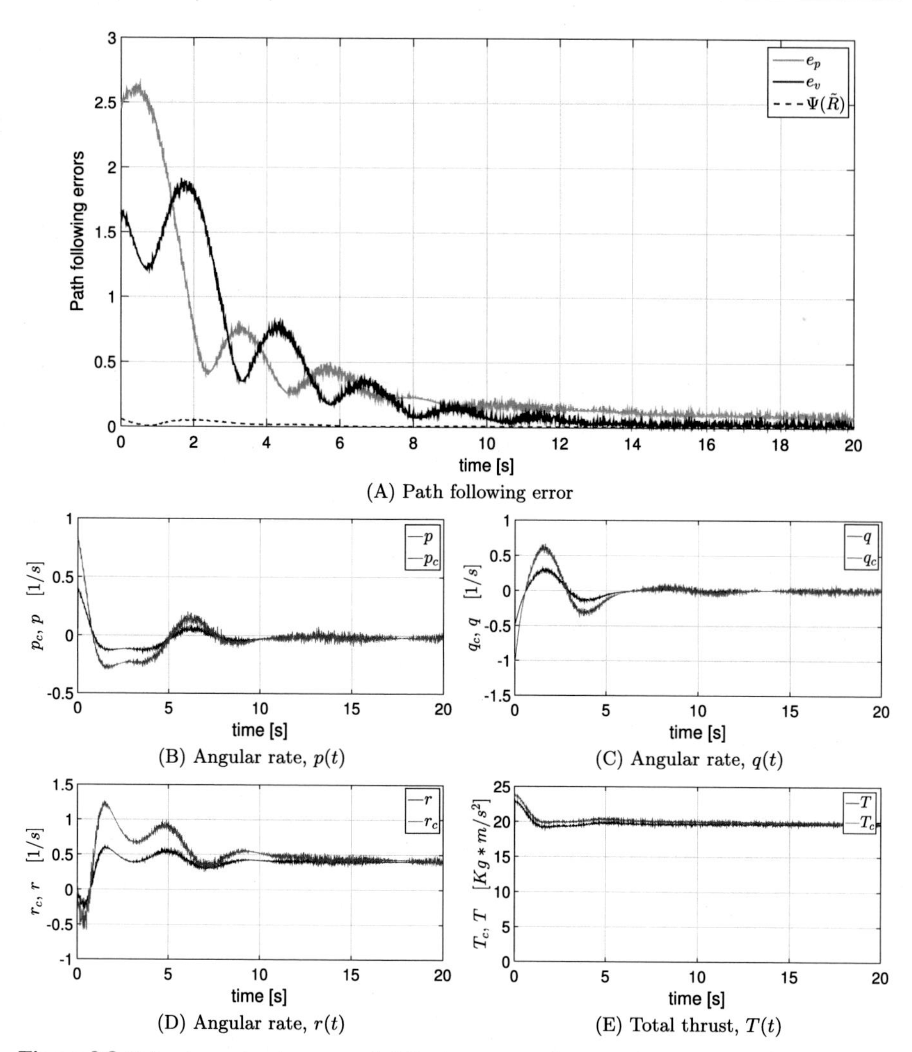

Figure 9.8 Following a virtual target; path-following errors and commands.

Fig. 9.3 includes the simulation results. In particular, Figs. 9.3A and 9.3B illustrate the 2D and 3D plots of the actual and desired trajectories of the UAV, as well as the ground vehicle's trajectory, while Figs. 9.3C and 9.3D show the speed and acceleration profiles, respectively. The execution of the mission at three different instants of time is depicted in Fig. 9.4, which illustrates the desired and actual trajectories of the UAV. The multirotor's body frame $\{\mathcal{B}\}$ (positioned at the vehicle's center of mass) and the desired frame $\{\mathcal{D}\}$ (positioned on the desired path) are also depicted in the same figure, with the $\hat{\mathbf{b}}_1$ and $\hat{\mathbf{b}}_{1\mathbf{D}}$ axes always pointing towards the ground vehicle. Fig. 9.5 highlights the performance of the path-following algorithm. In particular, it is shown that the path-following errors converge to a small neighborhood of zero at $t \approx 8$ s. The

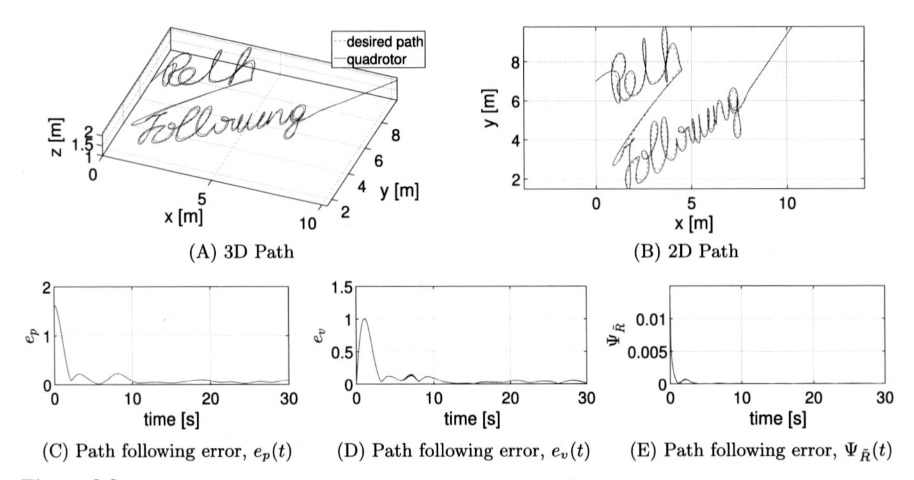

(A) 3D Path (B) 2D Path

(C) Path following error, $e_p(t)$ (D) Path following error, $e_v(t)$ (E) Path following error, $\Psi_{\tilde{R}}(t)$

Figure 9.9 Path following for a multirotor UAV: drone drawing a Bézier curve.

path-following control efforts, namely angular rates and total thrust, are also shown in the same figure.

In order to provide a more realistic simulation scenario, we repeat the above experiment and implement a simple inner-loop autopilot for angular-rate and total thrust command tracking. Additionally, measurement noise and transmission delays have been added. The 2D and 3D plots of the actual and desired trajectories are illustrated in Figs. 9.6A and 9.6B, respectively, while Figs. 9.6C and 9.6D show the actual and desired speed and acceleration profiles. Fig. 9.7 illustrates the performance of the mission at three different instants of time. The performance of the path-following controller is illustrated in Fig. 9.8. Fig. 9.8A shows the path-following error, converging to a neighborhood of zero at a rate that is slower when compared with the previous case (see Fig. 9.5A). Finally, the path-following commands and actual angular rates and total thrust of the vehicle are shown in Figs. 9.8B–9.8E.

Finally, Fig. 9.9 presents an additional simulation scenario in which the multirotor UAV is asked to perform more aggressive maneuvers by "drawing" the phrase *path following* in the air. The overall mission is executed in 260 s. However, for the sake of clarity, Figs. 9.9C, 9.9D, and 9.9E show the path-following error in the first 20 s of the mission. The desired path is produced by connecting four Bézier curves, which are generated through Bernstein polynomial approximation using a total of 400 points. For further details on Bézier curves and Bernstein polynomial approximation the reader is referred to [6].

In the simulation scenarios described in this section, the control gains have been selected as follows:

$$k_p = 3\,, \quad k_v = 3\,, \quad k_{\tilde{R}} = 5\,.$$

References

[1] P. Castillo, R. Lozano, A. Dzul, Stabilization of a mini rotorcraft with four rotors, IEEE Control Systems Magazine 25 (2005) 45–55.
[2] V. Cichella, R. Choe, S.B. Mehdi, E. Xargay, N. Hovakimyan, I. Kaminer, V. Dobrokhodov, A 3D path-following approach for a multirotor UAV on $\mathsf{SO}(3)$, IFAC Proceedings Volumes 46 (2013) 13–18.
[3] V. Cichella, R. Choe, S.B. Mehdi, E. Xargay, N. Hovakimyan, A.C. Trujillo, I. Kaminer, Trajectory generation and collision avoidance for safe operation of cooperating UAVs, in: AIAA Guidance, Navigation, and Control Conference (GNC), 2014, AIAA 2014-0972.
[4] V. Cichella, I. Kaminer, V. Dobrokhodov, E. Xargay, N. Hovakimyan, A. Pascoal, Geometric 3D path-following control for a fixed-wing UAV on $\mathsf{SO}(3)$, in: AIAA Guidance, Navigation and Control Conference, Portland, OR, August 2011, AIAA 2011-6415.
[5] V. Cichella, I. Kaminer, E. Xargay, V. Dobrokhodov, N. Hovakimyan, A.P. Aguiar, A.M. Pascoal, A Lyapunov-based approach for time-coordinated 3D path-following of multiple quadrotors, in: 51st IEEE Conference on Decision and Control (CDC), IEEE, 2012, pp. 1776–1781.
[6] R.T. Farouki, The Bernstein polynomial basis: a centennial retrospective, Computer Aided Geometric Design 29 (2012) 379–419.
[7] T. Lee, M. Leok, N.H. McClamroch, Control of complex maneuvers for a quadrotor UAV using geometric methods on $\mathsf{SE}(3)$. Available online arXiv:1003.2005v4, 2011.
[8] A. Tayebi, S. McGilvray, Attitude stabilization of a VTOL quadrotor aircraft, IEEE Transactions on Control Systems Technology 14 (2006) 562–571.

Time Coordination of Multirotor Air Vehicles

10

This chapter addresses the problem of coordinating a fleet of multirotor UAVs. As described in previous chapters, the cooperative missions considered require that each vehicle follow a feasible collision-free path, and that all vehicles arrive at their respective final destinations at the same time, or at different times but meeting a desired inter-vehicle schedule. Two simple examples of coordination between multirotor UAVs are shown in Fig. 10.1. Fig. 10.1A illustrates the case where two vehicles are required to follow two paths of different lengths while coordinating along the x-axis[1], while Fig. 10.1B captures a scenario in which eight vehicles must exchange positions with each other while coordinating their motions, in order to avoid collisions.

We offer a solution to the problem of cooperative motion control defined above that departs considerably from previous work reported in the literature, of which [1,2, 5–10] and the references therein are representative examples. We tackle the problem of decentralized cooperative control with time-varying communications networks using a Lyapunov-based approach and derive performance bounds as a function of the quality of service of the communications network. Furthermore, we study the degradation in performance that arises when the inner-loop vehicle autopilots responsible for tracking reference commands in thrust and angular rates exhibit bounded tracking errors. In particular, we show that all time-coordination errors remain bounded and that the vehicles remain close to their desired positions.

The solution described in this chapter, initially proposed in [3,4], is motivated by the results presented in Chapters 3 and 4; see also [11,12]. A key step in the approach proposed in the above chapter is the derivation of a path-following algorithm that reduces significantly the complexity of the coordination problem at hand by simplifying the coordination dynamics to n integrators, where n is the number of UAVs. However, as discussed in the previous chapter, the solution described in Chapter 4 for cooperative control of fixed-wing UAVs cannot be used directly for UAVs with hovering capability, such as multirotors. This drastic limitation prompted us to reformulate the coordination problem defined before and to derive a solution that can be used with a more general class of UAVs.

In the approach proposed in this chapter, the time-coordination problem is solved by controlling a suitably defined coordination variable (i.e. the virtual time $\gamma(t)$ introduced in Chapter 9). We also demonstrate that the solution proposed yields guaranteed performance in the presence of time-varying communications networks that may arise due to temporary loss of communication links and switching communication topologies.

[1] Fig. 10.1A is the screen shot of the video of an actual flight test that can be downloaded from https://www.youtube.com/watch?v=izXmgetsBYw. See Chapter 11 for a description of the test and the results obtained.

Time-Critical Cooperative Control of Autonomous Air Vehicles, DOI: 10.1016/B978-0-12-809946-9.00013-0

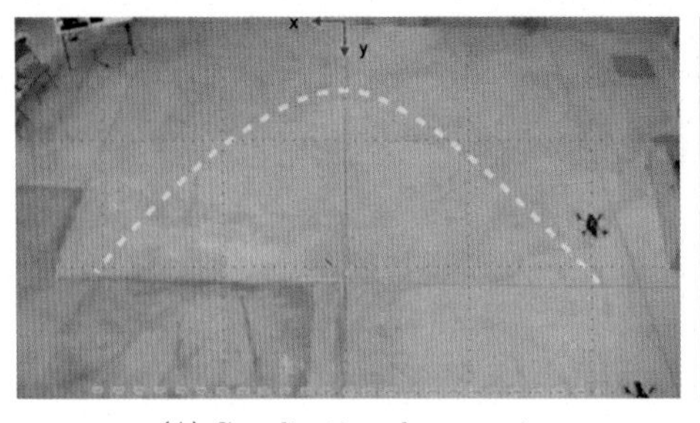

(A) Coordination along x-axis.

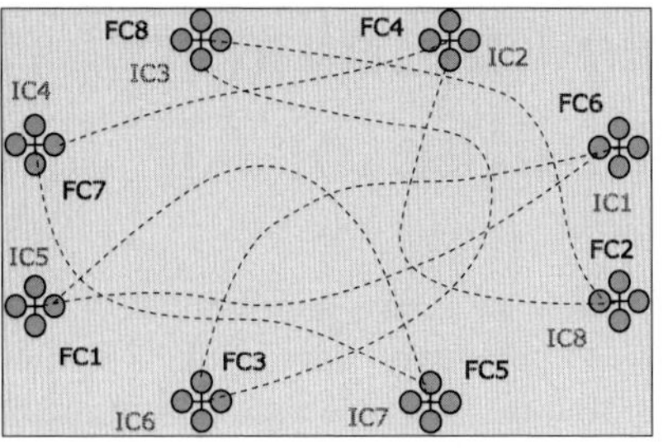

(B) Coordinated exchange of position.

Figure 10.1 Examples of coordinated maneuvers.

10.1 Coordination States and Maps

This section provides the mathematical formulation for the multirotor coordination problem, whereby the UAVs are required to follow predefined geometric paths while satisfying prescribed temporal requirements. We start by defining a set of coordination error variables which will be used to formulate the problem at hand.

Recall from Section 9.1 that the desired position assigned to the ith vehicle at time t is given by $\boldsymbol{p}_{d,i}(\gamma_i(t))$, where $\boldsymbol{p}_{d,i}(\cdot)$ is a geometric path produced by a trajectory generation algorithm, and the path parameter $\gamma_i(t)$ is the virtual time defined in Eq. (9.2) as

$$\gamma_i : \mathbb{R}^+ \to [0, t_d^{\mathrm{f}}], \tag{10.1}$$

where t_d^{f} is the nominal total mission duration. We notice that, if $\dot{\gamma}_i(t) = 1$, the desired speed profile at which the ith vehicle is required to fly is equal to the speed profile produced by the trajectory generation algorithm (i.e. $\dot{\gamma}_i(t) = 1$ implies that the mission is executed at the desired pace set during the trajectory generation phase.). On the other hand, $\dot{\gamma}_i(t) > 1$ ($\dot{\gamma}_i(t) < 1$) implies a faster (slower) execution of the mission. As will become clear, the virtual time and its first time derivative play a crucial role in the time-coordination problem. In fact, from the definition of γ_i in Eq. (10.1), and given that the paths produced by the trajectory generation algorithm satisfy the simultaneous time-of-arrival requirement introduced in Section 2.2.1, we say that the vehicles are coordinated at time t if

$$\gamma_i(t) - \gamma_j(t) = 0\,, \qquad \forall\, i, j \in \{1, \ldots, n\}\,, \quad i \neq j\,. \tag{10.2}$$

Moreover, as discussed earlier, if

$$\dot{\gamma}_i - 1 = 0\,, \qquad \forall\, i \in \{1, \ldots, n\}\,, \tag{10.3}$$

then the desired speed at which the vehicles are required to converge is equal to the desired speed profile established at the trajectory-generation level. Thus, Eqs. (10.2) and (10.3) capture the time-coordination objective, and require that a control law for $\ddot{\gamma}_i$ be formulated to ensure that the objective will be achieved.

Let $\dot{\boldsymbol{\gamma}}(t) := [\dot{\gamma}_1(t), \ldots, \dot{\gamma}_n(t)]^\top$ and define the coordination error vectors as

$$\boldsymbol{\zeta}_1(t) := \boldsymbol{Q}\boldsymbol{\gamma}(t) \qquad \in \mathbb{R}^{n-1}\,, \tag{10.4}$$

$$\boldsymbol{\zeta}_2(t) := \dot{\boldsymbol{\gamma}}(t) - \mathbf{1}_n \qquad \in \mathbb{R}^{n}\,, \tag{10.5}$$

where the $(n-1) \times n$ matrix $\boldsymbol{Q}$ introduced in Section 2.2.3, Eq. (2.8), satisfies $\boldsymbol{Q}\mathbf{1}_n = \mathbf{0}$. Notice that if $\boldsymbol{\zeta}_1(t) = \mathbf{0}$, then $\gamma_i - \gamma_j = 0, \forall i, j \in \{1, \ldots, n\}$. Furthermore, convergence of $\boldsymbol{\zeta}_2(t)$ to zero implies that the individual parameterizing variables $\gamma_i(t)$ evolve at the desired normalized rate of 1.

Using the above notation and the time-coordination problem definition in Section 2.2.3, we now formulate the time-coordination problem for multirotor UAVs.

Definition 10.1 (Time-Coordination Problem)**.** Consider a set of n multirotor UAVs. Let each UAV be assigned a desired path and a timing law that satisfy the constraints given by (9.4). Assume that the vehicles are equipped with path-following controllers that solve the path-following problem defined in Section 9.1, Definition 9.1. Then, the objective of coordination is to design feedback control laws for $\ddot{\gamma}_i(t)$ for all the vehicles such that the time-coordination generalized error vector defined as

$$\boldsymbol{x}_{cd} := [\boldsymbol{\zeta}_1^\top, \boldsymbol{\zeta}_2^\top]^\top,$$

converges to a neighborhood of zero and Assumption 9.1 in Section 9.1.2 is verified. ♠

Before introducing the control laws and stating the main result of this chapter, we conclude the present section with the following remark.

Remark 10.1. In mission scenarios where the vehicles are requested to follow their respective paths while maintaining prespecified time intervals, the coordination states in (10.4) and (10.5) need to be redefined following a procedure similar to that adopted in Section 4.5.2. △

10.2 Coordination Control Law

To solve the time-coordination problem, we let the evolution of $\gamma_i(t)$ be governed by the equations

$$\ddot{\gamma}_i = -b(\dot{\gamma}_i - 1) - a \sum_{j \in \mathcal{N}_i} (\gamma_i - \gamma_j) - \bar{\alpha}_i(\boldsymbol{x}_{pf,i}),$$
$$\gamma_i(0) = 0, \quad \dot{\gamma}_i(0) = 1,$$

where a and b are positive coordination control gains, while $\bar{\alpha}_i(\boldsymbol{x}_{pf,i})$ is defined as

$$\bar{\alpha}_i(\boldsymbol{x}_{pf,i}) := \frac{\dot{\boldsymbol{p}}_{d,i}(\gamma_i)^\top \boldsymbol{e}_{p,i}}{\|\dot{\boldsymbol{p}}_{d,i}(\gamma_i)\| + \delta},$$

where δ is a positive design parameter and $\boldsymbol{e}_{p,i}$ is the path-following position error of vehicle i defined in Eq. (9.8). The dynamics of $\boldsymbol{\gamma}(t)$ can be written in compact form as

$$\ddot{\boldsymbol{\gamma}} = -b\boldsymbol{\zeta}_2 - a\boldsymbol{L}\boldsymbol{\gamma} - \bar{\boldsymbol{\alpha}}(\boldsymbol{x}_{pf}), \quad \boldsymbol{\gamma}(0) = \mathbf{0}_n, \ \dot{\boldsymbol{\gamma}}(0) = \mathbf{1}_n, \tag{10.6}$$

where

$$\begin{aligned} \boldsymbol{x}_{pf} &:= [\boldsymbol{x}_{pf,1}^\top, \ldots, \boldsymbol{x}_{pf,n}^\top]^\top && \in \mathbb{R}^{9n}, \\ \bar{\boldsymbol{\alpha}}(\boldsymbol{x}_{pf}) &:= [\bar{\alpha}_1(\boldsymbol{x}_{pf,1}), \ldots, \bar{\alpha}_n(\boldsymbol{x}_{pf,n})]^\top && \in \mathbb{R}^n. \end{aligned}$$

The following result is obtained.

Theorem 10.1. *Consider a group of n multirotor UAVs. Assume that the Laplacian of the graph that models the inter-vehicle communication topology satisfies the PE-like condition* (2.8) *for some parameters $\mu, T > 0$. Let the vehicles be equipped with inner-loop autopilots and path-following controllers that ensure uniform ultimate boundedness of the path-following error (recall the result in Lemma 9.3). Let the time-coordination error vector $\boldsymbol{x}_{cd}(t)$ at time $t = 0$ and the path-following performance bound ρ introduced in Lemma 9.3, Eq. (9.26), satisfy*

$$\max\left(\|\boldsymbol{x}_{cd}(0)\|, \rho\right) \\ \leq \min\left(\frac{1 - \frac{v_{i,\min}}{v_{d,\min}}}{(\kappa_1 + \kappa_2)}, \frac{\frac{v_{i,\max}}{v_{d,\max}} - 1}{(\kappa_1 + \kappa_2)}, \frac{\sqrt{\frac{a_{i,\max}}{a_{d,\max}}} - 1}{(\kappa_1 + \kappa_2)}, \frac{a_{i,\max} - \dot{\gamma}_{i,\max}^2 a_{d,\max}}{v_{d,\max}(b\kappa_1 + b\kappa_2 + 1)}\right), \tag{10.7}$$

for some $\kappa_1, \kappa_2 > 0$, where $v_{i,\min}$, $v_{i,\max}$ and $a_{i,\max}$ are the speed and acceleration limits of vehicle i, and $v_{d,\min}$, $v_{d,\max}$ and $a_{d,\max}$ are the velocity and acceleration constraints imposed by the trajectory generation algorithm. Finally, let $\ddot{\boldsymbol{\gamma}}$ be governed by the control law given by Eq. (10.6)*. Then, there exist control gains a, b, and δ such that the time-coordination error is uniformly bounded. In particular, the time-coordination error satisfies*

$$\|\boldsymbol{x}_{cd}(t)\| \leq \kappa_1 \|\boldsymbol{x}_{cd}(0)\| e^{-\lambda_{cd} t} + \kappa_2 \sup_{t \geq 0}(\|\boldsymbol{x}_{pf}(t)\|), \tag{10.8}$$

with guaranteed rate of convergence

$$\lambda_{cd} := \frac{a}{b} \frac{n\mu\delta_{\bar{\lambda}}}{T(1 + \frac{a}{b} nT)^2}, \quad 0 < \delta_{\bar{\lambda}} < 1. \tag{10.9}$$

◇

Proof. The proof of Theorem 10.1 is given in Appendix B.3.4. □

Remark 10.2. Notice that the maximum convergence rate λ_{cd} is obtained when the control gains a and b satisfy

$$\frac{a}{b} = \frac{1}{nT}. \tag{10.10}$$

Substituting (10.10) in (10.9) yields

$$\max_{a,b>0}(\lambda_{cd}) = \frac{\mu\delta_{\bar{\lambda}}}{4T^2},$$

that is, the convergence rate depends on the quality of service of the network only. △

Corollary 10.1. *Suppose that each UAV is equipped with an ideal inner-loop autopilot (i.e. the path-following algorithm designed in Chapter 9 is exponentially stable, see Lemma 9.2). Then the time-coordination error converges exponentially fast to zero, that is,*

$$\|\boldsymbol{x}_{cd}(t)\| \leq \bar{\kappa}_1 \|\boldsymbol{x}_{cd}(0)\| e^{-\lambda_{cd} t} + \bar{\kappa}_2 \|\boldsymbol{x}_{pf}(0)\| e^{-\frac{\lambda_{pf}+\lambda_{cd}}{2} t}, \tag{10.11}$$

for some $\bar{\kappa}_1, \bar{\kappa}_2 > 0$. ◇

Proof. The proof of Corollary 10.1 is given in Appendix B.3.5. □

Remark 10.3. Notice that the time-coordination control law introduced in (10.6) depends also on the path-following error. By virtue of the path-following dependent term (i.e. $\bar{\boldsymbol{\alpha}}(\boldsymbol{x}_{pf})$), if for example one vehicle is away from the desired position ($\|\boldsymbol{e}_p\| \neq 0$), its assigned virtual target speeds up or slows down in order to reduce the path-following error; then, as a direct consequence, the other vehicles involved in the cooperative mission will also adjust their speeds to maintain coordination. This point will become clear in the next section where the simulation results are presented. △

Remark 10.4. The formulation of the time-critical cooperative path-following problem described above assumes only relative temporal constraints in the execution of a given mission. Absolute temporal constraints, such as specifications on the desired final time of the mission, are not considered. As shown in Chapter 5, such constraints can be easily incorporated in the problem formulation, and enforced by judiciously modifying the coordination control laws presented in this chapter. △

10.3 Simulation Results

In this section we present simulation results for the scenario introduced in Fig. 10.1B, where eight quadrotor UAVs, initially positioned along the perimeter of a 40 m × 40 m square area, must exchange their positions and arrive at their final destinations at the same time. Before the mission starts, a set of 2D trajectories are generated which ensure temporal deconfliction ($E = 1$ m) of the UAVs, i.e.

$$\min_{\substack{j,k=1,\dots,8 \\ j \neq k}} \|\boldsymbol{p}_{d,j}(t_d) - \boldsymbol{p}_{d,k}(t_d)\|^2 \geq E^2, \qquad \text{for all } t_d \in [0, t_d^{\mathrm{f}}].$$

The corresponding paths are shown in Fig. 10.2.

In the remainder of this section, we analyze the performance of the algorithm developed for cooperative motion control via three numerical simulations. In the first simulation we consider the case of ideal all-to-all communication between the vehicles, and assume ideal performance of the path-following algorithm, i.e. $\|\boldsymbol{x}_{pf}(t)\| = 0$ for all $t \geq 0$. In the second simulation, a similar experiment is performed, but consid-

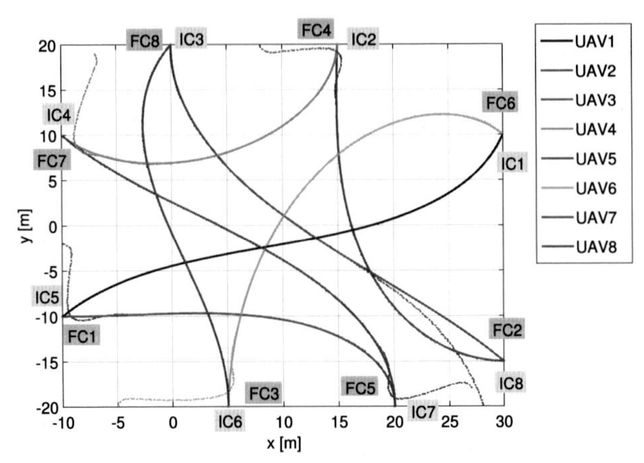

Figure 10.2 Coordinated exchange of positions; simulation results with eight quadrotor UAVs. The solid lines indicate the UAV trajectories in the case of ideal path following; the dashed lines depict the UAVs trajectories when a displacement between the initial desired and actual positions is introduced.

ering non-ideal communications. In the third simulation, we introduce path-following errors by implementing the path-following controller introduced in Chapter 9 with a simple inner-loop autopilot for angular-rate and total thrust command tracking. In all the experiments, the control gains are chosen to be $a = 1.5$, $b = 3.6$, and $\delta = 3$. To illustrate the convergence properties of the solution, the virtual times are initialized as follows:

$$\gamma_1(0) = 2\,, \quad \gamma_4(0) = 3\,, \quad \gamma_6(0) = 1\,, \quad \gamma_8(0) = 1.5\,,$$
$$\gamma_2(0) = \gamma_3(0) = \gamma_5(0) = \gamma_7(0) = 0\,.$$

Ideal Communications — Ideal Path Following

In this simulation, all the vehicles communicate with each other all the time, i.e. $L_{ii}(t) = 7\,, L_{ij}(t) = -1\,, \forall t \geq 0\,, \forall i, j \in \{1\,, \ldots\,, 8\}\,, i \neq j$, where the $L_{ij}(t)$'s are the entries of the Laplacian matrix $\boldsymbol{L}(t)$. Moreover, we let $\|\boldsymbol{x}_{pf}(t)\| = 0\,, \forall t \geq 0$, i.e. we assume the path-following algorithm performs ideally.

At time $t = 0$ the vehicles start navigating the room and follow the predefined trajectories until they reach their final destinations, at time $t \approx 8.8$ s. In Fig. 10.2, the solid lines indicate the trajectories of each UAV, while ICi and FCi indicate the initial and final positions of ith UAV, respectively.

Fig. 10.3 shows the coordination variables. At the beginning of the mission, vehicles 1, 4, 6 and 8 speed up, while vehicles 2, 3, 5 and 7 slow down (see Figs. 10.3B and 10.3C) until, at time $t \approx 2$ s, coordination is achieved and maintained thereafter. Fig. 10.3A is a plot of the evolution of the virtual times γ_i. Notice that after a transient phase they become identical and evolve at the same rate.

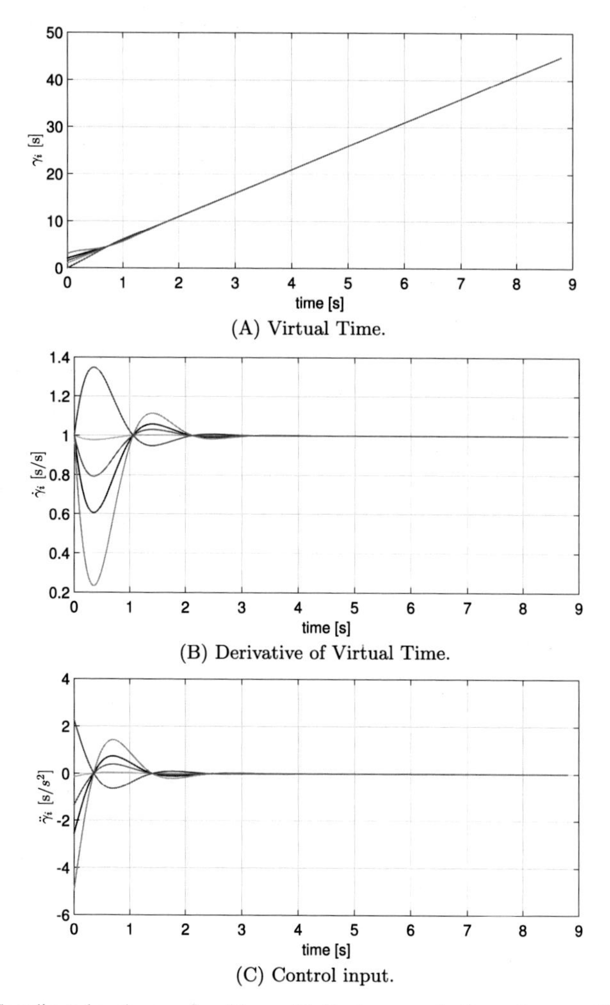

(A) Virtual Time.

(B) Derivative of Virtual Time.

(C) Control input.

Figure 10.3 Coordinated exchange of positions with ideal communications; time coordination.

Range-Based Communications — Ideal Path Following

The previous simulation is repeated, but in this case we let UAV i and UAV j communicate with each other at time $t \geq 0$ only if $\|\boldsymbol{p}_i(t) - \boldsymbol{p}_j(t)\| \leq 20$ m.

Fig. 10.4 shows the estimate of the quality of service of the network computed as (see Chapter 4)

$$\hat{\mu}(t) := \lambda_{\min}\left(\frac{1}{n}\frac{1}{T}\int_{t-T}^{t} \bar{L}(\tau)d\tau\right), \qquad t \geq T ,$$

with $n = 8$ and $T = 1$ s. As seen in the figure, the estimate of the quality of service is maximum at $t \approx 4$–5 s, when the vehicles are positioned near the center of the area,

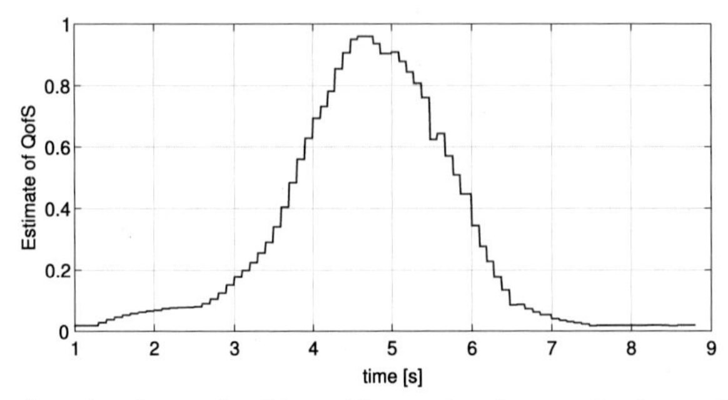

Figure 10.4 Coordinated exchange of positions with range-based communications; estimate of the quality of service (QoS).

and therefore closer to each other. On the other hand, the quality of service is smaller at the beginning and the end of the mission, when the vehicles communicate with only a few neighbors. Fig. 10.5 illustrates the performance of the time-coordination algorithm. Notice that the time-coordination variables converge to the desired values at time $t \approx 4$ s, that is, slower than in the case with ideal communications.

Range-Based Communications — Non-Ideal Path Following

In this last simulation, to illustrate the impact of non-zero path-following error, we implemented the path-following control law described in Chapter 9 together with a simple inner-loop autopilot to track angular-rate and total thrust commands. According to the result obtained in Lemma 9.3, the path-following error is ultimately bounded, and the time-coordination error satisfies (10.8). The communication topology is the one used in the previous simulation.

The vehicles start, at $t = 0$, with an initial displacement from the desired positions, and track the desired paths. In Fig. 10.2 the dashed lines indicate the actual trajectories of the UAVs. Fig. 10.6 shows the time history of the time-coordination variables. Fig. 10.7 captures the time history of the norm of the time-coordination error state, $\|\boldsymbol{x}_{cd}(t)\|$, and compares it with the two cases described above. As expected, the coordination error converges to a neighborhood of the origin, and remains bounded. Finally, Fig. 10.8 shows the evolution of the minimum distance between the vehicles throughout the mission, defined as

$$\min_{i,j}\{\|\boldsymbol{p}_i(t) - \boldsymbol{p}_j(t)\|\}, \tag{10.12}$$

in three different cases: (*i*) ideal path following; (*ii*) non-ideal path following (that is, path following is executed with errors) and time-coordination law given by (10.6); (*iii*) non-ideal path following and coordination law given by (10.6) but without the third term $\bar{\boldsymbol{\alpha}}(\boldsymbol{x}_{pf})$ (that is, without the dependence on the path-following error). While in case (*i*) temporal separation is guaranteed at the trajectory generation level, when the UAVs are away from the desired position, the time-coordination algorithm must take into account the path-following error in order to ensure that the actual UAV posi-

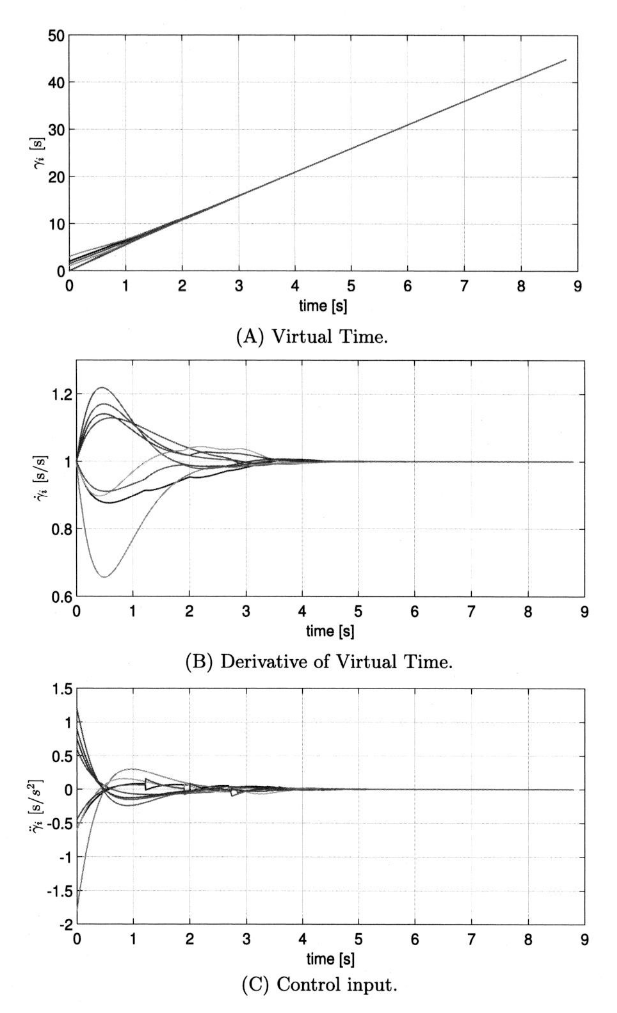

(A) Virtual Time.

(B) Derivative of Virtual Time.

(C) Control input.

Figure 10.5 Coordinated exchange of positions with range-based communications; time coordination.

tions are separated. As pointed out in Remark 10.3, the third term in Eq. (10.6) enables the UAVs to maintain coordination even in the presence of path-following errors, which in turn implies that a minimum separation between the vehicles is guaranteed. Notice in Figs. 10.6B and 10.6C that because UAV 8 is initially at a considerable distance from its desired position, when the mission starts the virtual time γ_8 associated with it decelerates significantly ($\dot{\gamma}_8 < 1$ and $\ddot{\gamma}_8 < 0$) by virtue of $\bar{\boldsymbol{\alpha}}(\boldsymbol{x}_{pf})$. As a consequence, γ_1 decelerates also to coordinate with γ_8, thus allowing the actual vehicles to synchronize with each other along the paths and maintain a desired separation. In the absence of the term $\bar{\boldsymbol{\alpha}}(\boldsymbol{x}_{pf})$, the virtual times associated with the vehicles would maintain coordination among them, without taking into account the actual position of the UAVs, thus leading to potential collisions (see Fig. 10.8).

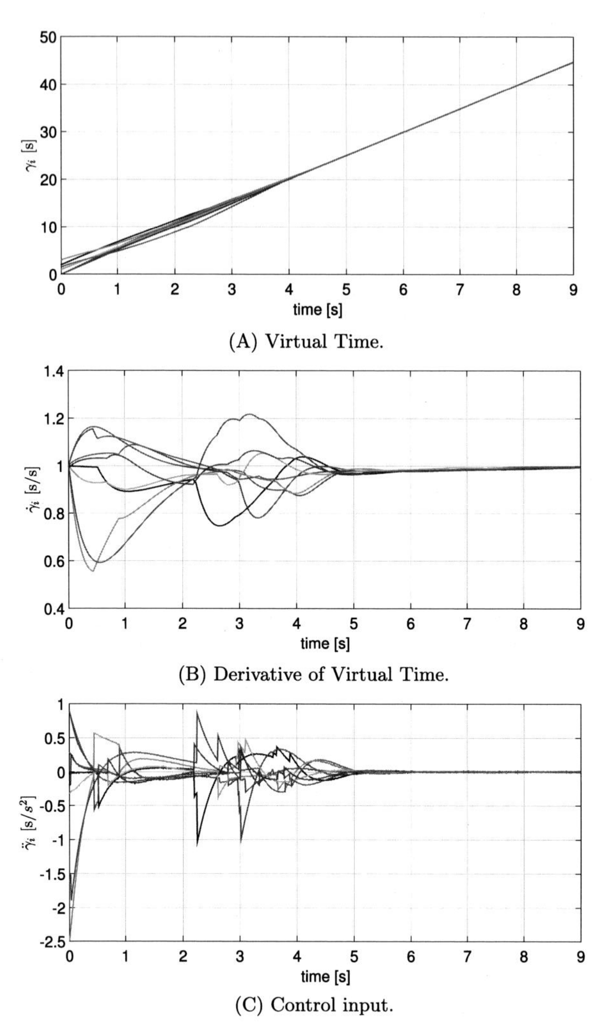

(A) Virtual Time.

(B) Derivative of Virtual Time.

(C) Control input.

Figure 10.6 Coordinated exchange of positions with range-based communications and inner-loop autopilots; time coordination.

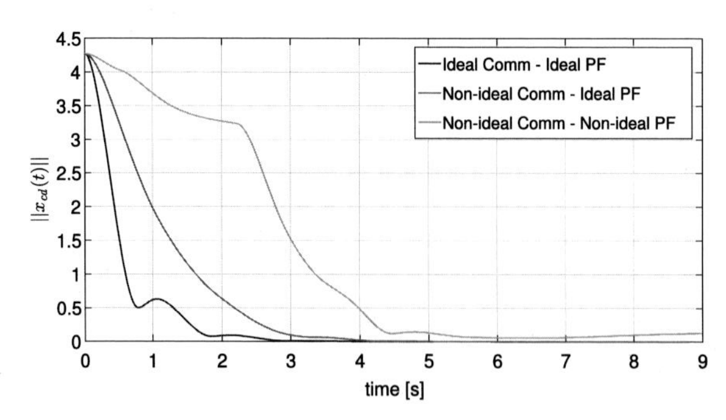

Figure 10.7 Effect of network connectivity on time-coordination; norm of the coordination error vector.

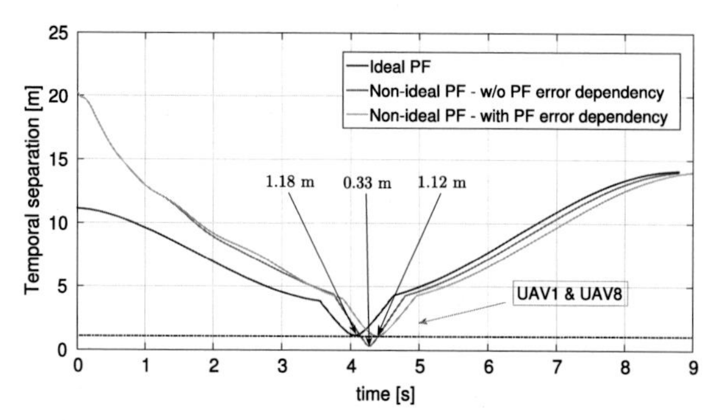

Figure 10.8 Effect of path-following correction terms $\bar{\alpha}_i(\cdot)$ on inter-vehicle separation.

References

[1] J.J. Acevedo, B.C. Arrue, J.M. Diaz-Bañez, I. Ventura, I. Maza, A. Ollero, One-to-one coordination algorithm for decentralized area partition in surveillance missions with a team of aerial robots, Journal of Intelligent & Robotic Systems 74 (2014) 269–285.

[2] J. Capitan, M.T. Spaan, L. Merino, A. Ollero, Decentralized multi-robot cooperation with auctioned POMDPs, The International Journal of Robotics Research 32 (2013) 650–671.

[3] V. Cichella, I. Kaminer, V. Dobrokhodov, E. Xargay, R. Choe, N. Hovakimyan, A.P. Aguiar, A.M. Pascoal, Cooperative path following of multiple multirotors over time-varying networks, IEEE Transactions on Automation Science and Engineering 12 (2015) 945–957.

[4] V. Cichella, I. Kaminer, E. Xargay, V. Dobrokhodov, N. Hovakimyan, A.P. Aguiar, A.M. Pascoal, A Lyapunov-based approach for time-coordinated 3D path-following of multiple quadrotors, in: IEEE Conference on Decision and Control, 2012, pp. 1776–1781.

[5] U. Gurcuoglu, G.A. Puerto-Souza, F. Morbidi, G.L. Mariottini, Hierarchical control of a team of quadrotors for cooperative active target tracking, in: IEEE/RSJ International Conference on Intelligent Robots and Systems, 2013, pp. 5730–5735.

[6] A. Kushleyev, D. Mellinger, C. Powers, V. Kumar, Towards a swarm of agile micro quadrotors, Autonomous Robots 35 (2013) 287–300.

[7] R. Ritz, R. D'Andrea, Carrying a flexible payload with multiple flying vehicles, in: IEEE/RSJ International Conference on Intelligent Robots and Systems, 2013, pp. 3465–3471.

[8] M. Saska, T. Krajník, V. Vonásek, Z. Kasl, V. Spurný, L. Přeučil, Fault-tolerant formation driving mechanism designed for heterogeneous MAVs-UGVs groups, Journal of Intelligent & Robotic Systems 73 (2014) 603–622.

[9] M. Saska, V. Vonásek, T. Krajník, L. Přeučil, Coordination and navigation of heterogeneous MAV–UGV formations localized by a 'hawk-eye'-like approach under a model predictive control scheme, The International Journal of Robotics Research 33 (2014) 1393–1412.

[10] M. Turpin, N. Michael, V. Kumar, Concurrent assignment and planning of trajectories for large teams of interchangeable robots, in: IEEE International Conference on Robotics and Automation, 2013, pp. 842–848.

[11] E. Xargay, Time-Critical Cooperative Path-Following Control of Multiple Unmanned Aerial Vehicles, PhD thesis, University of Illinois at Urbana-Champaign, Urbana, IL, United States, May 2013.

[12] E. Xargay, I. Kaminer, A.M. Pascoal, N. Hovakimyan, V. Dobrokhodov, V. Cichella, A.P. Aguiar, R. Ghabcheloo, Time-critical cooperative path following of multiple unmanned aerial vehicles over time-varying networks, Journal of Guidance, Control and Dynamics 36 (2013) 499–516.

Flight Tests of Multirotor UAVs

11

This chapter presents the results of flight tests with two quadrotor UAVs aimed at verifying experimentally the stability and convergence properties associated with the time-critical cooperative control framework presented in Chapter 10. We consider two operational scenarios in which the quadrotors are required to execute simple cooperative tasks. Namely, *phase on orbit coordination* and *spatial coordination along one axis*. The reader is referred to [1], where videos of additional experiments can be found. In what follows, we first describe the system architecture and the indoor facility used to conduct the experiments, after which we discuss the flight-test results in detail.

11.1 System Architecture and Indoor Facility

The flight tests were performed at the Center for Autonomous Vehicle Research (CAVR), Naval Postgraduate School, Monterey, CA [5] (see Fig. 11.1). The facility is equipped with eight VICON T-160 cameras [4] connected into one network to provide precise synthetic geopositioning with resolution on the order of 1 mm in the volume of $30 \times 30 \times 20$ ft. The position, velocity, and attitude data of the two vehicles are transmitted to MATLAB/Simulink, running on a Linux OS (ground station), at the frequency of up to 200 Hz. The quadrotor UAVs employed in these flight tests are the Parrot AR.Drone [2] depicted in Fig. 11.2. This platform, commercially available to the general public, is equipped with a sonar height sensor, an Inertial Measurement Unit (IMU) capable of acquiring and processing data on the linear acceleration, angular velocity and orientation of the UAV, two cameras (one pointing forward, and the other one facing downwards), and an on-board computer running proprietary software. The software includes an inner-loop stabilizer which uses the sonar sensor and IMU data in order to track roll, pitch, yaw-rate, and vertical-speed commands. For this reason, the path-following problem discussed in Chapter 9 was reformulated and a new path-following algorithm that uses the above control inputs was derived; see [3] for complete details. The path-following controller and the time-coordination algorithm were implemented on the host machine (i.e. the corresponding MATLAB/Simulink code were implemented on the ground station). Thus, for simplicity, even though the cooperative path-following control scheme adopted is suitable for distributed systems, the control algorithms were implemented in a centralized fashion. The host machine sends the path-following commands to the Parrot AR.Drone vehicles via a wireless

Time-Critical Cooperative Control of Autonomous Air Vehicles, DOI: 10.1016/B978-0-12-809946-9.00014-2

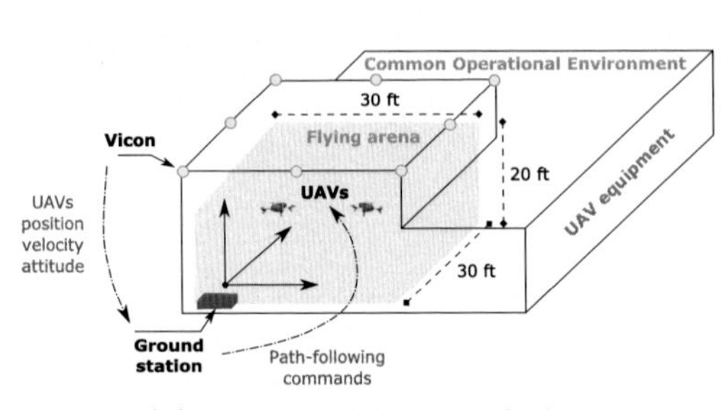

(A) Schematic of the indoor facility.

(B) CAVR, Naval Postgraduate School.

Figure 11.1 Indoor facility at the Naval Postgraduate School.

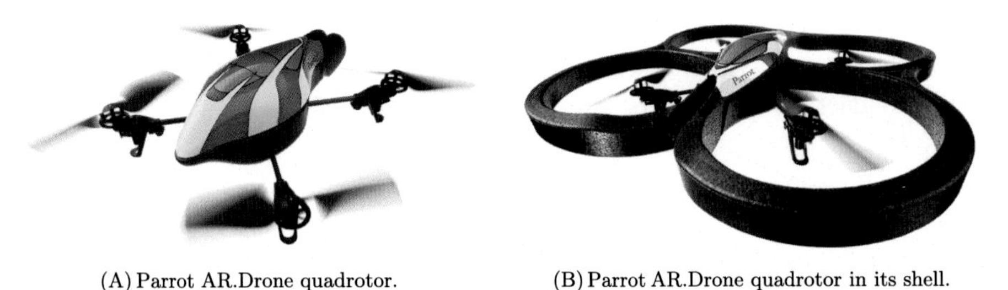

(A) Parrot AR.Drone quadrotor. (B) Parrot AR.Drone quadrotor in its shell.

Figure 11.2 Parrot AR.Drone [2]; quadrotor employed in the flight-test experiments.

ad-hoc connection with an update rate of up to 50 Hz. The coordination variables are exchanged among the UAVs at a data transfer rate of 100 Hz (imposed via Simulink). A schematic of the indoor facility and the system architecture is shown in Fig. 11.1A.

11.2 Flight-Test Results

This section describes the results of two flight-test experiments with two quadrotor UAVs. In the experiments, the control law in Eq. (10.6) was implemented with the control gains set to

$$a = 3\,, \quad b = 5\,, \quad \delta = 5\,.$$

In the first scenario, called *phase on orbit coordination*, two Parrot AR.Drone quadrotors are required to follow accurately a circular planar path while maintaining a desired time-varying phase shift (angular displacement between the two vehicles on the circle). In the second scenario, which we refer to as *spatial coordination along one axis*, the UAVs are prescribed to follow planar paths of different lengths while adjusting their speeds in order to maintain coordination along one direction. We now discuss the experimental results in detail.

11.2.1 Phase on Orbit Coordination

In this scenario two quadrotor UAVs are required to follow a circular path of radius 2 m. Fig. 11.3 shows the desired orbit and the actual trajectories of the two quadrotors. Since the UAVs are tasked to follow the same orbit, a phase-on-orbit separation is required between the vehicles to avoid collision. This separation is specified interactively from the ground station, and it varies according to operator commands. The UAVs are initially required to keep a 180-deg phase separation (face-to-face); at approximately $t = 94$ s, the required phase separation switches to 90 deg; the two quadrotors keep this configuration for about 14 s, when the required phase separation switches back to 180 deg; finally, in the last part of the experiment, the UAVs are required to keep a phase separation of 270 deg. The scenario is illustrated in Fig. 11.4 at six different instants of time. A video of the flight test can be downloaded from http://naira.mechse.illinois.edu/quadrotor-uavs/. The desired phase-on-orbit separation, along with the actual phase separation between the two UAVs, are shown in Fig. 11.5. Finally, Fig. 11.6 shows the convergence of $\dot{\gamma}_1$ and $\dot{\gamma}_2$ to the desired rate 1 s/s, as well as the convergence of the difference between the virtual times to a neighborhood of zero. From these figures it can be noticed that when the desired phase shift changes, the coordination errors diverge slightly from zero, and then converge again after a short transient by virtue of the time-coordination controller.

11.2.2 Spatial Coordination Along One Axis

In this scenario, the Parrot AR.Drone quadrotors are tasked to follow two paths of different lengths, and to coordinate along one direction. In particular, UAV_1 is required to follow a straight line of length 5 m along the x-axis, while UAV_2 has to follow an arc of a circle of radius 2.5 m. Fig. 11.7 depicts the desired and actual paths of UAV_1 and UAV_2. During the mission, the UAVs adjust their speed profiles in order to coordinate along the x-axis, therefore facing each other. Fig. 11.8 shows snapshots of the mission at four different times, illustrating the performance of the coordination algorithm. The speed profiles are shown in Fig. 11.9. Notice how the speed of UAV_2 is always greater than the one of UAV_1, because the desired path assigned to the second quadrotor is longer than the one assigned to the first vehicle and the two vehicles must complete their maneuvers at the same time. Finally, Fig. 11.10 presents the time history of the coordination error $(\gamma_1(t) - \gamma_2(t))$ and the coordination-state rates $\dot{\gamma}_1(t)$ and $\dot{\gamma}_2(t)$. As can be seen in the figure, $(\gamma_1(t) - \gamma_2(t))$ converges to a neighborhood of zero, while $\dot{\gamma}_1(t)$ and $\dot{\gamma}_2(t)$ converge to a neighborhood of the desired rate 1 s/s.

11.2.3 Additional Flight Tests

For additional flight-test experiments, we refer the reader to the authors' web page: http://naira.mechse.illinois.edu/quadrotor-uavs/. By navigating the above website, two

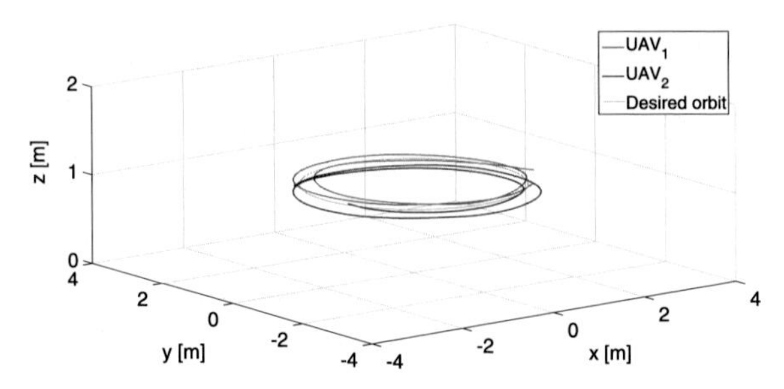

Figure 11.3 Phase on orbit coordination; desired and actual vehicle paths in 3D.

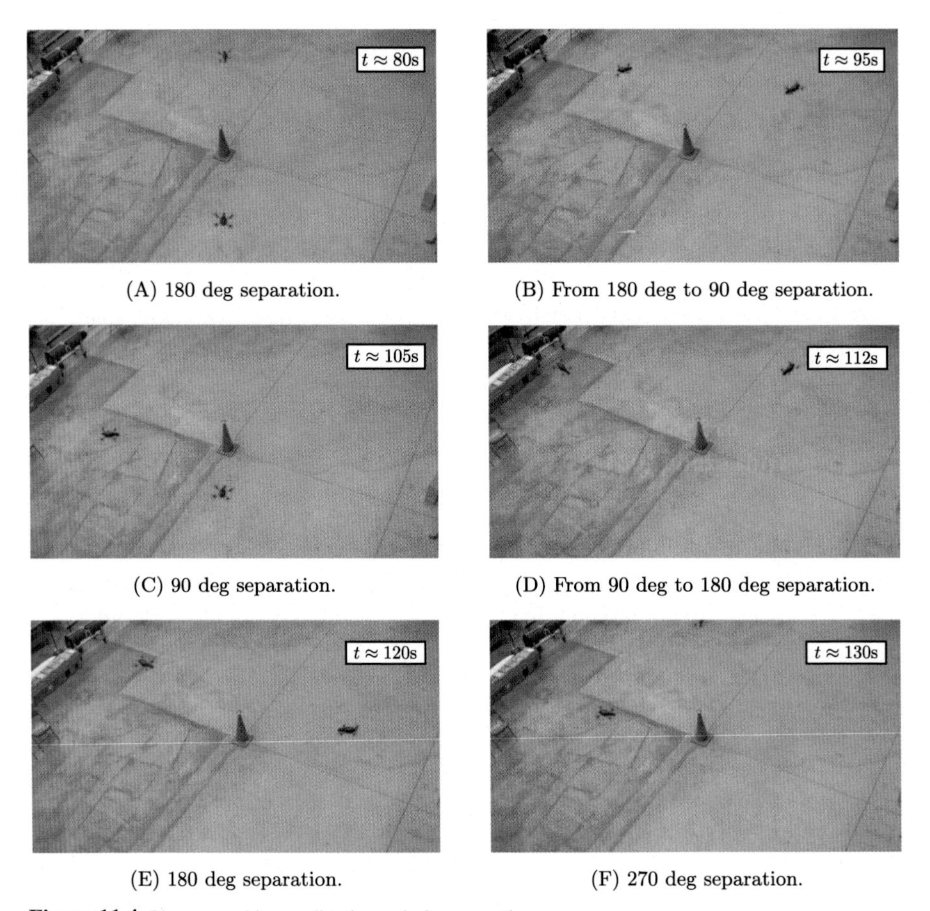

(A) 180 deg separation.

(B) From 180 deg to 90 deg separation.

(C) 90 deg separation.

(D) From 90 deg to 180 deg separation.

(E) 180 deg separation.

(F) 270 deg separation.

Figure 11.4 Phase on orbit coordination; mission execution.

videos can be found. The first one (also available at https://www.youtube.com/watch?v=izXmgetsBYw) shows two quadrotors undertaking a set of cooperative tasks (such as coordinated exchange of positions, face-to-face phase-on-orbit coordination, and phase shift on vertical orbit), including the ones described above. The second

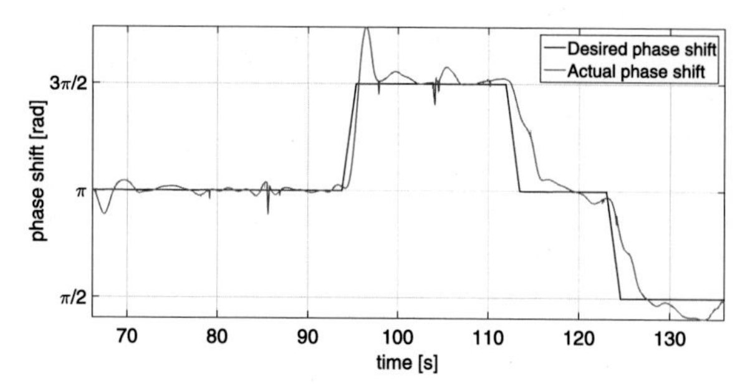

Figure 11.5 Phase on orbit coordination; phase shift.

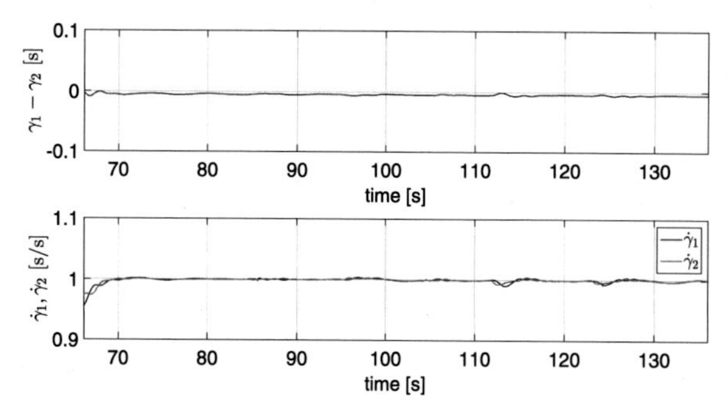

Figure 11.6 Phase on orbit coordination; coordination error and coordination-state rates.

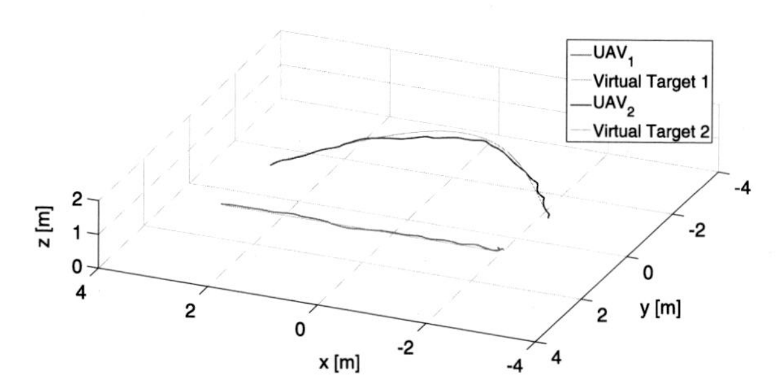

Figure 11.7 Spatial coordination along one axis; desired and actual vehicle paths in 3D.

video (also available at http://www.youtube.com/watch?v=OBtLCf1Lfiw) shows two quadrotors coordinating along with a tango tune. In the latter example, the sound wave of the song plays the role of a *virtual vehicle* with which the quadrotors are required to coordinate; this demonstrates that, similarly to the fixed-wing UAVs coordination discussed in Part II, the solution proposed in Chapter 10 can be easily extended to the case where absolute temporal constraints are enforced.

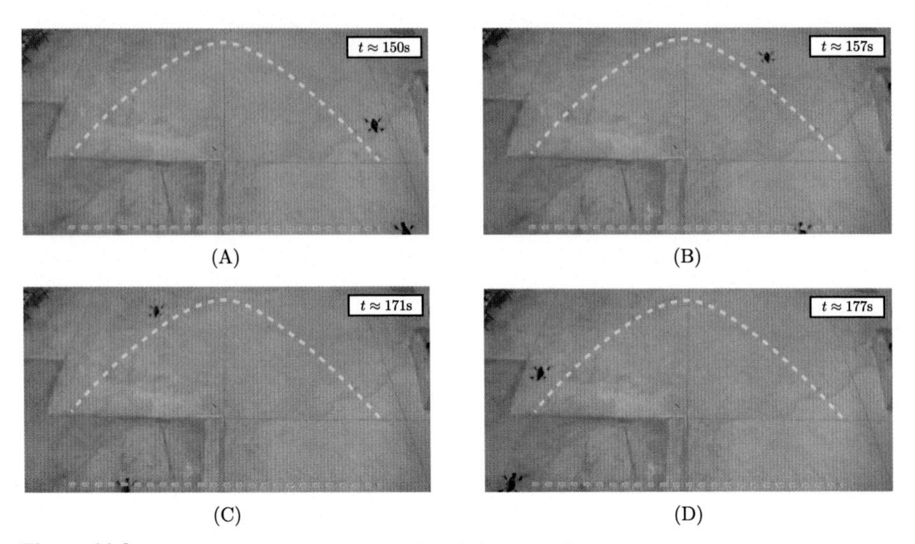

Figure 11.8 Spatial coordination along one axis; mission execution.

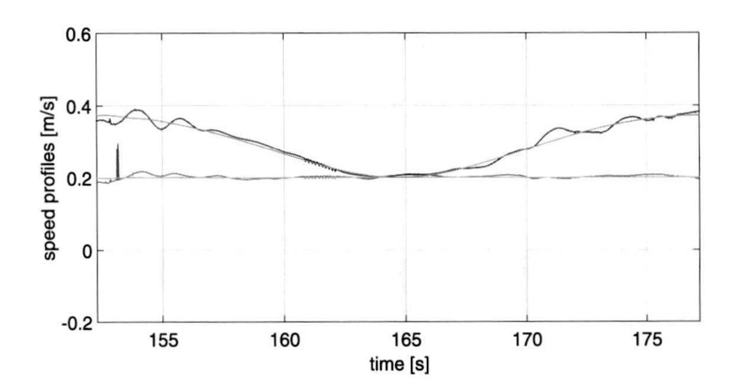

Figure 11.9 Spatial coordination along one axis; speed profiles.

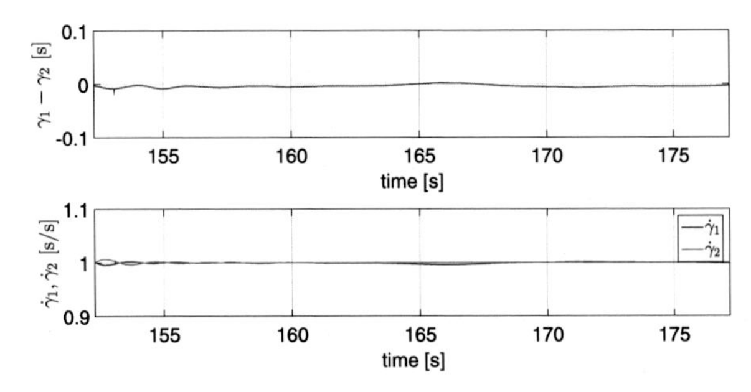

Figure 11.10 Spatial coordination along one axis; coordination error and coordination-state rates.

References

[1] Advanced Controls Research Laboratory at the University of Illinois at Urbana Champaign, Quadrotor UAVs, http://naira.mechse.illinois.edu/quadrotor-uavs/ [Online; accessed 25 July 2016].

[2] Parrot AR.Drone, http://www.parrot.com [Online; accessed 25 July 2016].

[3] V. Cichella, I. Kaminer, E. Xargay, V. Dobrokhodov, N. Hovakimyan, A.P. Aguiar, A.M. Pascoal, A Lyapunov-based approach for time-coordinated 3D path-following of multiple quadrotors, in: IEEE Conference on Decision and Control, 2012, pp. 1776–1781.

[4] Motion Capture System from Vicon, https://www.vicon.com/ [Online; accessed 25 July 2016].

[5] Naval Postgraduate School, Monterey, CA, Center for Autonomous Vehicle Research, http://my.nps.edu/web/cavr [Online; accessed 25 July 2016].

Part Four

Final Considerations

Summary and Concluding Remarks

12

12.1 Summary

This book addressed the problem of steering a group of unmanned aerial vehicles (UAVs) along desired 3D spatial paths while meeting relative and absolute temporal constraints. The methodology adopted to solve this problem unfolds in three basic steps. First, using a multiple vehicle motion planning algorithm, each vehicle is assigned a feasible path and a prescribed speed profile along it. Together, the paths and speed assignments satisfy mission requirements, respect vehicle dynamic constraints, and ensure collision-free maneuvers. Then, path-following algorithms ensure that each vehicle will follow the path assigned to it, regardless of the temporal assignments of the mission. Finally, the vehicles coordinate their positions along the paths by exchanging coordination information over the supporting communications network. With this setup, the vehicles have the capability to react to off-nominal situations —such as unpredictable disturbances and temporary loss of performance by isolated vehicles— by negotiating their speeds along the paths so as to try and meet mission objectives. This capability is key to the success of the methodology proposed for time-critical cooperative control. The above three steps are accomplished by decoupling space and time in the formulation of the path-following and coordination problems and by relying on existing inner-loop onboard controllers (autopilots). The latter are capable of tracking reference commands generated by the outer-loop controllers that implement selected path-following and time-coordination algorithms. The methodology derived integrates various concepts and tools from a broad spectrum of disciplines and yields a streamlined design procedure for time-critical cooperative path-following control of autonomous vehicles. The approach presented applies to teams of heterogeneous systems and departs considerably from well known algorithms used to obtain swarming behavior, which is not appropriate for a large number of practical mission scenarios.

The networked cooperative control methodology that is the focal point of this book is sufficiently general to be applicable to a number of fields that require the use of autonomous air and marine vehicles. In the present work, however, the exposition was focused on two types of autonomous platforms: fixed-wing and multirotor UAVs. Starting with the *fixed-wing UAVs* case, the book put forward a new singularity-free, path-following control law on $\mathsf{SO}(3)$ and introduced a set of coordination states for cooperative motion control to deal explicitly with path-dependent speed profiles. Using results from nonlinear systems, differential inclusions, and algebraic graph theory, conditions were derived under which the algorithms proposed solve the cooperative path-following control problem in the presence of switching communication topologies and the exchange of quantized information. Lower bounds were also derived on the convergence rate of the network-related error dynamics as a function of the number of (possibly virtual) leaders included in the coordination control law and the quality

Time-Critical Cooperative Control of Autonomous Air Vehicles, DOI: 10.1016/B978-0-12-809946-9.00016-6

of service (QoS) of the supporting communications network. The latter, in the context of this book, is a measure of the level of connectivity of the inter-vehicle communications graph. It was shown that the guaranteed rate of convergence of the coordination control loop is limited by the QoS of the communications network, which implies that in communication-limited environments long times may be required in order for the vehicles to reach agreement and coordinate their positions along the paths. To address this issue and improve the convergence rate of the coordination dynamics in low-connectivity scenarios, a coordination algorithm was proposed that integrates network estimators with topology-control strategies. The proposed approach leads to an evolving extended network, the topology of which depends on the local exchange of information among its nodes.

The cooperative path-following methodology was then extended to deal with *multirotor UAVs* which, due to their dynamics, pose specific design challenges. A modified $\mathsf{SO}(3)$ path-following control law was presented that is applicable to vehicles that may have a zero velocity vector over finite, non-empty intervals of time. The multiple vehicle coordination problem was also re-formulated in order to accommodate this class of vehicles. Similarly to the fixed-wing UAVs case, cooperative control algorithms were derived that exhibit robust performance in spite of temporary communication failures and switching network topologies. The expected performance of the cooperative control framework was also assessed in terms of the QoS of the supporting communications network.

In order to bridge the gap between theory and practice, this book included and analyzed the results of extensive flight-tests with multiple small fixed-wing and quadrotor UAVs executing cooperative missions. The results illustrated the efficacy of the algorithms proposed. They also demonstrated the validity of the general theoretical framework adopted for cooperative motion control in realistic applications as well as the feasibility of the onboard implementation of the algorithms derived. It is relevant to emphasize that, even at the path-following control level, the new control algorithms proposed for accurate UAV path following in 3D outperform conventional waypoint navigation methods that are typically implemented on commercial off-the-shelf autopilots.

12.2 Open Problems

The methodologies proposed for time-critical cooperative UAV motion control hold promise for the development and operation of advanced systems that must meet strict mission constraints. There is however ample room for improvement and intensive research efforts are warranted in a number of key topics, of which the ones listed below are but representative examples.

Cooperative Trajectory Generation

A key element in the execution of multiple vehicle, time-critical cooperative missions akin to those described in this book, is the availability of efficient trajectory-generation algorithms. The latter must be capable of generating trajectories that do not violate the dynamic constraints of each vehicle and ensure that the vehicles maintain a predefined spatial clearance among them, account for communication constraints, and verify desired temporal and mission-dependent spatial requirements. Furthermore, the algorithms must lend themselves to seamless implementation onboard heterogeneous autonomous vehicles in a decentralized fashion and run in real-time. Their complexity should also scale with the size of the vehicle fleet. In the context of time-critical missions, the work in [11] proposes a novel framework for cooperative trajectory generation that addresses these challenges. Future developments must also include explicitly the effect of the wind in the cooperative motion problem formulation, which can significantly reduce the amount of replanning required for the successful execution of a cooperative mission. The use of inverse simulation techniques for trajectory generation is also a topic of special interest, as such methods can help evaluate the feasibility of the generated trajectories, estimate mission effectiveness, and provide corrections to turn unfeasible maneuvers into feasible ones.

Coordination Under Communication Constraints

Future research should also explore the development of coordination algorithms for effective execution of cooperative missions in communication-limited and communication-denied environments. In particular, it would be interesting to investigate in further detail the stability and convergence properties of the coordination control law proposed in Chapter 7 of this book for low-connectivity mission scenarios. The derivation of design constraints under which this control law may improve the convergence rate of the closed-loop coordination dynamics appears to be rather challenging, as the proposed approach leads to an evolving network with time-varying link weights and unidirectional communications.

In some cooperative missions, the QoS of the supporting communications network can be adjusted by means of motion-control algorithms for connectivity maintenance. Future efforts will investigate the integration of such algorithms into the framework for cooperative path following described in this book. Of particular interest are the motion strategies that arise from a differential game-theoretic formulation of the connectivity-maintenance problem; see, for example, [8,9] and references therein. These strategies could be designed, for instance, to ensure that the connectivity condition (2.8) be satisfied during the entire execution of the mission, rather than assuming it a priori. Moreover, because in such a setup the vehicles would be allowed to deviate from the corresponding paths in order to maintain connectivity, the resulting solution would require the integration of algorithms for (online) collision avoidance.

Another topic that warrants future research is the impact of latency in the communications, channel noise, and random link failures on the achievable level of performance that can be expected in the types of cooperative missions considered in this book. An exciting and challenging subject of research is the development of a unifying framework that would capture these communication constraints, along with quantization, while accounting at the same time for the random nature and inherent probabilistic properties of data transmission. Such a framework could be valuable both for analysis purposes and as an effective tool for the design of new coordination control laws that are less sensitive to link failures and channel constraints. Another avenue of research that is worth exploring is the use of techniques for multiple agent systems synchronization using self-triggered and event-triggered broadcast as a means to reduce the amount of information transmitted among the different vehicles. The key objective is to broadcast information in a communications network only when deemed necessary, in accordance with appropriately defined triggering conditions; see for example [4,5, 7,12,13,21,22] and the references therein.

Autonomy

Increasingly complex multiple UAV mission scenarios pose several new challenges to the design and integration of autonomous systems, especially in terms of autonomy, cooperation, endurance, and resilience. In fact, it is anticipated that future operations will require teams of autonomous systems working in cooperation to achieve common objectives, while being able to operate safely in highly uncertain, remote areas for periods of time that might range from hours to years. It is therefore important to develop energy-harvesting cooperative solutions that provide guaranteed levels of performance in the presence of faulty communications networks, limited sensing capabilities, and partial vehicle failures that might span months. To this end, new breakthroughs will be required in a vast number of areas that include multi-vehicle cooperative motion planning, guidance, and control, resilient control, battery and solar-panel efficient systems, energy management, big data processing, integration and fusion, wireless communications, numerical weather prediction and analysis, meteorological data assimilation, real-time computation, and aerodynamics. The execution of future envisioned missions will also pose formidable challenges from a human factor perspective. In fact, considerable research effort is required to yield new concepts and tools for the design of systems that will enable effective human-machine interaction. Particular emphasis is to be placed on situation awareness interfaces to ensure that automation, whose behavior is too often opaque, is made transparent to system operators.

12.3 Cooperative Control in Future Airspace Scenarios

Fast paced developments in miniaturized sensors and actuators, embedded powerful computers, airframes, and advanced algorithms for networked navigation and motion control are steadily paving the way for the emergence of a new breed of aerial robots capable of performing challenging cooperative missions in the presence of harsh environmental constraints. Examples of representative missions include cooperative environment mapping and adaptive sampling, inspection of critical infrastructures, surveillance and protection of civilian and military assets, and transportation of goods. Envisioned developments target not only the operation of multiple autonomous vehicles, but also manned systems with control automation capabilities onboard and the combination of manned and unmanned vehicles with humans and machines working side by side in a symbiotic relationship.

Manned aviation will also benefit from future developments in cooperative aerial vehicles with a view to improving the capacity, speed, and safety of air transportation in congested airspace. In fact, it is expected that critical cooperative control solutions proposed for manned aviation will be technologically similar to those developed for unmanned vehicles, but targeting significantly higher levels of safety and reliability. Elements of the envisioned future technologies have already been implemented over the last 10–15 years since the inception of NASA's NextGen [19] project. NextGen is a multi-billion-dollar technology development and modernization effort that will make air travel safer, more flexible, and efficient. Interestingly enough, the keystone of the Trajectory Based Operation (TBO) paradigm that is part of the NextGen project bears affinity with the time and space separation strategy for cooperative motion control adopted in this book. Information sharing, as an enabler of cooperative control, is another core concept in NextGen, called Net-Centric Operations, that is being designed to provide secure information access by means of integrating modernized radar-based Air Traffic Control (ATC) centers with onboard Automatic Dependent Surveillance-Broadcast (ADS-B) [14] systems and satellite-based communication links. The Position, Navigation, and Timing (PNT) services of NextGen are the guidance and control elements of the cooperative path following that enable sequencing of aircraft along the path to satisfy safety separation on route or sequential arrival in the terminal airspace of any airport.

In a more distant future, the most exciting developments will most likely occur in application areas that mandate the combined operation of manned and unmanned cooperative vehicles. In fact, the new solutions that must be sought to guarantee the safety of mutual operations are expected to yield complementary manned and unmanned systems with unprecedented capabilities. This trend is clearly visible in the NASA effort to integrate UAVs in the national airspace [17].

Aerospace applications are a seemingly narrow niche where time-critical cooperative control strategies can yield better performance and lead to increased operational safety. Nevertheless, the methods developed in this book are general enough to be applicable to other types of systems as the operational environment considered in the task formulation has been specifically made "unforgiving of human and engineering errors". Therefore, the solutions proposed hold considerable promise to be extended to other application areas. Representative examples include autonomous driving cars [15], highway traffic control [6], distributed power generation [10] or other applications that require optimal resource allocation. The most vivid recognition of potential of this technology is the recent announcement [20] of the U.S. Department of transportation ruling to mandate vehicle-to-vehicle (V2V) communications on light vehicles, allowing cars to "talk" to each other to avoid a multitude of crashes. There is also tremendous potential to extend the solutions proposed to the operation of cooperative marine vehicles or even cooperative air and marine vehicles in a multitude of tasks that involve marine habitat mapping, geotechnical surveying, and the detection and assessment of the spatial extend of pollutant spills [1–3,16,18,23].

References

[1] P. Abreu, H. Morishita, A. Pascoal, J. Ribeiro, H. Silva, Marine vehicles with streamers for geotechnical surveys: modeling, positioning, and control, IFAC-PapersOnLine 49 (2016) 458–464.
[2] P.C. Abreu, A.M. Pascoal, Formation control in the scope of the MORPH project. Part I: Theoretical foundations, IFAC-PapersOnLine 48 (2015) 244–249.
[3] P.C. Abreu, M. Bayat, A.M. Pascoal, J. Botelho, P. Góis, J. Ribeiro, M. Ribeiro, M. Rufino, L. Sebastião, H. Silva, Formation control in the scope of the MORPH project. Part II: Implementation and results, IFAC-PapersOnLine 48 (2015) 250–255.
[4] J. Almeida, C. Silvestre, A.M. Pascoal, Event-triggered output synchronization of heterogeneous multi-agent systems, International Journal of Robust and Nonlinear Control 27 (2017) 1302–1338.
[5] J. Almeida, C. Silvestre, A.M. Pascoal, Synchronization of multi-agent systems using event-triggered and self-triggered broadcasts, IEEE Transactions on Automatic Control (2017), in press.
[6] L. Alvarez, R. Horowitz, P. Li, Traffic flow control in automated highway systems, Control Engineering Practice 7 (1999) 1071–1078.
[7] A. Anta, P. Tabuada, To sample or not to sample: self-triggered control for nonlinear systems, IEEE Transactions on Automatic Control 55 (2010) 2030–2042.
[8] S. Bhattacharya, T. Başar, Graph-theoretic approach for connectivity maintenance in mobile networks in the presence of a jammer, in: IEEE Conference on Decision and Control, Atlanta, GA, December 2010, pp. 3560–3565.
[9] S. Bhattacharya, T. Başar, N. Hovakimyan, Singular surfaces in multi-agent connectivity maintenance games, in: IEEE Conference on Decision and Control, Orlando, FL, December 2011, pp. 261–266.
[10] F. Blaabjerg, R. Teodorescu, M. Liserre, A.V. Timbus, Overview of control and grid synchronization for distributed power generation systems, IEEE Transactions on Industrial Electronics 53 (2006) 1398–1409.
[11] R. Choe, Distributed Cooperative Trajectory Generation for Multiple Autonomous Vehicles Using Pythagorean Hodograph Bézier Curves, PhD thesis, University of Illinois at Urbana-Champaign, Urbana, IL, United States, April 2017.
[12] C. De Persis, R. Sailer, F. Wirth, On a small-gain approach to distributed event-triggered control, IFAC Proceedings Volumes 44 (2011) 2401–2406.
[13] D.V. Dimarogonas, E. Frazzoli, K.H. Johansson, Distributed event-triggered control for multi-agent systems, IEEE Transactions on Automatic Control 57 (2012) 1291–1297.
[14] FAA, NextGen, automatic dependent surveillance-broadcast (ADS-B), https://www.faa.gov/nextgen/programs/adsb/, 12 2016.
[15] A. Hars, Top misconceptions of autonomous cars and self-driving vehicles, http://www.driverless-future.com, 12 2016.
[16] G. Indiveri, G. Antonelli, F. Arrichiello, A. Caffaz, A. Caiti, G. Casalino, N.C. Volpi, I.B. de Jong, D. De Palma, H. Duarte, et al., Overview and first year progress of the widely scalable mobile underwater sonar technology H2020 project, IFAC-PapersOnLine 49 (2016) 430–433.
[17] D. Konstantinos, V. Kimon, P. Les, On Integrating Unmanned Aircraft Systems into the National Airspace System: Issues, Challenges, Operational Restrictions, Certification, and Recommendations, vol. 54, second ed., Springer Science & Business Media, 2011.
[18] F. Maurelli, M. Carreras, K. Rajan, D. Lane, Guest editorial: Special issue on long-term autonomy in marine robotics, Autonomous Robots 40 (2016) 1111–1112.

[19] NASA, Concept of Operations for the Next Generation Air Transportation System, version 3.2, Technical Report, Joint Planning and Development Office, 1500 K St NW Ste 500, Washington, DC, 20005, 2011.
[20] U.S.DOT, U.S. DOT advances deployment of Connected Vehicle Technology to prevent hundreds of thousands of crashes, NHTSA 34-16, 12 2016.
[21] X. Wang, M.D. Lemmon, Event-triggering in distributed networked control systems, IEEE Transactions on Automatic Control 56 (2011) 586–601.
[22] X. Wang, Y. Sun, N. Hovakimyan, Asynchronous task execution in networked control systems using decentralized event-triggering, Systems & Control Letters 61 (2012) 936–944.
[23] R.B. Wynn, V.A.I. Huvenne, T.P. Le Bas, B.J. Murton, D.P. Connelly, B.J. Bett, H.A. Ruhl, K.J. Morris, J. Peakall, D.R. Parsons, E.J. Sumner, S.E. Darby, R.M. Dorrell, J.E. Hunt, Autonomous Underwater Vehicles (AUVs): Their past, present and future contributions to the advancement of marine geoscience, Marine Geology 352 (2012) 451–468.

Mathematical Background

This appendix summarizes key concepts and results from linear algebra, nonlinear systems, and graph theory that are used throughout the book.

A.1 The Hat and Vee Maps

Let $\mathbb{R}^3$ denote the three-dimensional Euclidean space equipped with the usual Euclidean norm and $\mathfrak{so}(3)$ the set of 3×3 skew-symmetric matrices defined over $\mathbb{R}$. The *hat map* $(\cdot)^\wedge : \mathbb{R}^3 \to \mathfrak{so}(3)$ is defined as

$$(\boldsymbol{x})^\wedge = \begin{bmatrix} 0 & -x_3 & x_2 \\ x_3 & 0 & -x_1 \\ -x_2 & x_1 & 0 \end{bmatrix}$$

for $\boldsymbol{x} = [x_1,\ x_2,\ x_3]^\top \in \mathbb{R}^3$. The inverse of the hat map is referred to as the *vee map* $(\cdot)^\vee : \mathfrak{so}(3) \to \mathbb{R}^3$. A property of the hat and vee maps used in this book is given in terms of the equality

$$\operatorname{tr}\left[\boldsymbol{M}(\boldsymbol{x})^\wedge\right] = -\boldsymbol{x} \cdot \left(\boldsymbol{M} - \boldsymbol{M}^\top\right)^\vee , \tag{A.1}$$

which holds for any $\boldsymbol{x} \in \mathbb{R}^3$ and $\boldsymbol{M} \in \mathbb{R}^{3\times 3}$, with $\cdot$ being the inner product operation. We refer the reader to [5] for further details on the hat and vee maps.

A.2 Nonlinear Stability Theory

In this section, for the sake of completeness, we recall important concepts and results on the stability of equilibrium points of autonomous and nonautonomous nonlinear systems. The definitions and the proofs of the theorems can be found in [4,6,7]. The exposition follows closely the seminal reference [4].

A.2.1 Lipschitz Functions, Existence and Uniqueness of Solutions

To ensure some fundamental properties such as existence and uniqueness of the solutions of the differential equation (initial value problem)

$$\dot{x} = f(t, x), \qquad x(t_0) = x_0 ,$$

we assume that the function $f(t, x)$ satisfies the Lipschitz condition

$$\|f(t, x) - f(t, y)\| \leq L\|x - y\| \tag{A.2}$$

Time-Critical Cooperative Control of Autonomous Air Vehicles, DOI: 10.1016/B978-0-12-809946-9.00027-0

for all (t, x) and (t, y) in some neighborhood of (t_0, x_0). A function satisfying inequality (A.2) f is said to be *Lipschitz* in x, and the positive constant L is called a *Lipschitz constant*. The following existence and uniqueness theorem holds.

Theorem A.1 (Local Existence and Uniqueness). *Let $f(t, x)$ be piecewise continuous in t and satisfy the Lipschitz condition $\|f(t, x) - f(t, y)\| \leq L\|x - y\|$, for all $x, y \in \mathcal{B}_r$ and for all $t \in [t_0, t_1]$, where $\mathcal{B}_r = \{x \in \mathbb{R}^n \,|\, \|x - x_0\| \leq r\}$ for some positive r. Then there exists some $\delta > 0$ such that the differential equation $\dot{x} = f(t, x)$ with $x(t_0) = x_0$ has a unique solution over $[t_0, t_0 + \delta]$.* ◇

Often, it is important to indicate explicitly the domain over which the Lipschitz condition holds. This leads naturally to the definitions below, where D and W are a domain (open and connected set) and a general set in $\mathbb{R}^n$, respectively.

Definition A.1. A function $f(t, x)$ is

- locally Lipschitz in x on $[a, b] \times D \subset \mathbb{R} \times \mathbb{R}^n$ if each point $x \in D$ has a neighborhood D_0 such that f satisfies (A.2) on $[a, b] \times D_0$ with some Lipschitz constant L_0;
- Lipschitz in x on $[a, b] \times W \subset \mathbb{R} \times \mathbb{R}^n$ if f satisfies (A.2) for all $t \in [a, b]$ and all points in W, with the same Lipschitz constant L;
- globally Lipschitz in x if it is Lipschitz in x on $[a, b] \times \mathbb{R}^n \subset \mathbb{R} \times \mathbb{R}^n$. ♠

The Lipschitz property of a function is stronger than continuity and, as stated in the following lemmas, weaker than continuous differentiability.

Lemma A.1. *If $f(t, x)$ and $[\partial f/\partial x](t, x)$ are continuous on $[a, b] \times D$, for some domain $D \subset \mathbb{R}^n$, then f is locally Lipschitz in x on $[a, b] \times D$.* ◇

Lemma A.2. *If $f(t, x)$ and $[\partial f/\partial x](t, x)$ are continuous on $[a, b] \times \mathbb{R}^n$, then f is globally Lipschitz in x, on $[a, b] \times \mathbb{R}^n$ if and only if $[\partial f/\partial x]$ is uniformly bounded on $[a, b] \times \mathbb{R}^n$.* ◇

We are now in a position to state a global existence and uniqueness result.

Theorem A.2 (Global Existence and Uniqueness). *Suppose that $f(t, x)$ is piecewise continuous in t and satisfies the Lipschitz condition $\|f(t, x) - f(t, y)\| \leq L\|x - y\|$, for all $x, y \in \mathbb{R}^n$ and for all $t \in [t_0, t_1]$. Then, the differential equation $\dot{x} = f(t, x)$, with $x(t_0) = x_0$, has a unique solution over $[t_0, t_1]$.* ◇

If the solution of the differential equation is known to lie in a compact set, it is possible to derive a global existence and uniqueness theorem that requires the function f to be only locally Lipschitz.

Theorem A.3. *Assume that $f(t, x)$ is piecewise continuous in t and locally Lipschitz in x for all $t \geq t_0$ and all x in a domain $D \subset \mathbb{R}^n$. Let W be a compact subset of D, $x_0 \in W$, and suppose it is known that every solution of the differential equation $\dot{x} = f(t, x)$, with $x(t_0) = x_0$, lies entirely in W. Then, there is a unique solution that is defined for all $t \geq t_0$.* ◇

The next result allows for the computation of bounds on the solution of a differential equation without computing the solution explicitly.

Lemma A.3 (Comparison Lemma). *Consider the scalar differential equation*

$$\dot{x} = f(x,t)\,, \qquad x(t_0) = x_0\,,$$

where f is continuous in t and locally Lipschitz in x for all $t \geq 0$ and $x \in M \subset \mathbb{R}$. Let $[t_0, T)$ be the maximal interval of existence of the solution $x(t)$ (where T could be infinity) and suppose $x(t) \in M$ for all $t \in [t_0, T)$. Let $v(t)$ be a C^1 function that satisfies the differential inequality

$$\dot{v} \leq f(v,t)\,, \qquad v(t_0) \leq x_0\,,$$

with $v \in M$ for all $t \in [t_0, T)$. Then, $v(t) \leq x(t)$ for all $t \in [t_0, T)$. ◇

A.2.2 Autonomous Systems

Suppose the autonomous system[1]

$$\dot{x} = f(x)\,, \qquad x(0) = x_0 \tag{A.3}$$

has an equilibrium point, which is assumed to be at the origin of $\mathbb{R}^n$, that is, $f(0) = 0$. There is no loss of generality in doing so because any equilibrium point can be shifted to the origin via a change of variables.

Definition A.2. The equilibrium point $x = 0$ of (A.3) is

- stable if, for each $\epsilon > 0$, there is $\delta = \delta(\epsilon) > 0$ such that

$$\|x(0)\| < \delta \Rightarrow \|x(t)\| < \epsilon\,, \qquad \forall\, t \geq 0\,;$$

- unstable if it is not stable;
- asymptotically stable if it is stable and δ can be chosen such that

$$\|x(0)\| < \delta \Rightarrow \lim_{t\to\infty} x(t) = 0\,.$$

♠

In order to demonstrate that the origin is a stable equilibrium point, for each selected value of ϵ one must produce a value of δ, possibly dependent on ϵ, such that a trajectory starting in the δ neighborhood of the origin will never leave its ϵ neighborhood. It is possible to determine stability by examining the derivatives of some particular functions, without having to know explicitly the solution of (A.3).

Theorem A.4 (Lyapunov's stability theorem). *Let $x = 0$ be an equilibrium point for* (A.3) *and $D \subset \mathbb{R}^n$ be a domain containing $x = 0$. Let $V : D \to \mathbb{R}$ be a continuously*

[1] A system in which the function f does not depend explicitly on time.

differentiable function such that

$$V(0) = 0 \quad \textit{and} \quad V(x) > 0 \quad \textit{in } D \setminus \{0\}, \tag{A.4}$$

$$\dot{V}(x) \leq 0 \quad \textit{in } D. \tag{A.5}$$

Then, $x = 0$ is stable. Moreover, if

$$\dot{V}(x) < 0 \quad \textit{in } D \setminus \{0\},$$

then $x = 0$ is asymptotically stable. ◇

A function $V(x)$ satisfying condition (A.4) is said to be *positive definite*. If it satisfies the weaker condition $V(x) \geq 0$ for $x \neq 0$ it is said to be *positive semidefinite*. A function $V(x)$ is said to be *negative definite* or *negative semidefinite* if $-V(x)$ is *positive definite* or *positive semidefinite*, respectively. A continuously differentiable function $V(x)$ satisfying (A.4) and (A.5) is called a *Lyapunov function*, after the Russian mathematician A. M. Lyapunov who pioneered stability theory. The following result extends stability to a global setting.

Theorem A.5. *Let $x = 0$ be an equilibrium point for* (A.3). *Let $V : \mathbb{R}^n \to \mathbb{R}$ be a continuously differentiable function such that*

$$V(0) = 0 \quad \textit{and} \quad V(x) > 0, \qquad \forall x \neq 0, \tag{A.6}$$

$$\|x\| \to \infty \Rightarrow V(x) \to \infty, \tag{A.7}$$

$$\dot{V}(x) < 0, \qquad \forall x \neq 0. \tag{A.8}$$

Then $x = 0$ is globally asymptotically stable. ◇

A function satisfying condition (A.7) is said to be *radially unbounded*.

A.2.3 The Invariance Principle

The asymptotic stability theorems of Section A.2.2 require finding a Lyapunov function whose time derivative is negative definite. If in a domain about the origin, however, a Lyapunov function can be found whose derivative along the trajectories of the system is only negative semidefinite, asymptotic stability of the origin might still be proven, provided that no trajectory can remain entirely in the region consisting of those points where $\dot{V}(x) = 0$, except at the origin. The key results in this area are due to Barbashin, Krasovskii, and later LaSalle, whose name is associated with the celebrated *Invariance Principle*. For our purposes, the following result suffices.

Theorem A.6 (Barbashin-Krasovskii theorem)**.** *Let $x = 0$ be an equilibrium point of* (A.3). *Let $V : \mathbb{R}^n \to \mathbb{R}$ be a continuously differentiable, radially unbounded, positive definite function such that $\dot{V}(x) \leq 0$ for all $x \in \mathbb{R}^n$. Let $S = \{x \in \mathbb{R}^n | \dot{V}(x) = 0\}$ and suppose that no solution can stay identically in S, other than the trivial solution $x(t) \equiv 0$. Then, the origin is globally asymptotically stable.* ◇

A.2.4 Nonautonomous Systems

The notions of stability and asymptotic stability of equilibrium points of nonautonomous systems are similar to those introduced in Definition A.2 for autonomous systems. The difference lies in the fact that while the solutions of an autonomous system depend only on $(t - t_0)$, with t_0 being the initial time, the solutions of the nonautonomous system

$$\dot{x} = f(t, x), \qquad x(t_0) = x_0 \tag{A.9}$$

depend on both t and t_0. For this reason, stability and asymptotic stability need to be redefined as uniform properties with respect to the initial time.

The origin is an equilibrium point of (A.9) at $t = 0$ if

$$f(t, 0) = 0, \qquad \forall\, t \geq 0.$$

Again, there is no loss of generality in studying the stability of the origin, since an equilibrium point at zero could be a translation of a nonzero equilibrium point or even a nonzero solution of the system.

Definition A.3. The equilibrium point $x = 0$ of (A.9) is

- stable if, for each $\epsilon > 0$ and any $t_0 \geq 0$, there is $\delta = \delta(\epsilon, t_0) > 0$ such that

$$\|x(t_0)\| < \delta \Rightarrow \|x(t)\| < \epsilon, \qquad \forall\, t \geq t_0 \geq 0; \tag{A.10}$$

- uniformly stable if, for each $\epsilon > 0$, there is $\delta = \delta(\epsilon) > 0$, independent of t_0, such that (A.10) is satisfied;
- unstable if it is not stable;
- asymptotically stable if it is stable and there is a positive constant $c = c(t_0)$ such that $x(t) \to 0$ as $t \to \infty$, for all $\|x(t_0)\| < c$;
- uniformly asymptotically stable if it is uniformly stable and there is a positive constant c, independent of t_0, such that for all $\|x(t_0)\| < c$, $x(t) \to 0$ as $t \to \infty$ uniformly in t_0; that is, for each $\eta > 0$, there is $T = T(\eta) > 0$ such that

$$\|x(t)\| < \eta, \qquad \forall\, t \geq t_0 + T(\eta), \ \forall\, \|x(t_0)\| < c;$$

- globally uniformly asymptotically stable if it is uniformly stable, $\delta(\epsilon)$ can be chosen to satisfy $\lim_{\epsilon \to \infty} \delta(\epsilon) = \infty$, and, for each pair of positive numbers η and c, there is $T = T(\eta, c) > 0$ such that

$$\|x(t)\| < \eta, \qquad \forall\, t \geq t_0 + T(\eta, c), \ \forall\, \|x(t_0)\| < c.$$ ♠

Equivalent definitions, more convenient for the approach followed in this book to deal with the different control problems, can be given using two comparison functions, known as class $\mathcal{K}$ and $\mathcal{KL}$ functions.

Definition A.4. A continuous function $\alpha : [0, a) \to [0, \infty)$ is said to belong to class $\mathcal{K}$ if it is strictly increasing and $\alpha(0) = 0$. It is said to belong to class $\mathcal{K}_\infty$ if $a = \infty$ and $\alpha(r) \to \infty$ as $r \to \infty$. ♠

Definition A.5. A continuous function $\beta : [0, a) \times [0, \infty) \to [0, \infty)$ is said to belong to class $\mathcal{KL}$ if, for each fixed s, the mapping $\beta(r, s)$ belongs to class $\mathcal{K}$ with respect to r and, for each fixed r, the mapping $\beta(r, s)$ is decreasing with respect to s and $\beta(r, s) \to 0$ as $s \to \infty$. ♠

The next lemma redefines uniform stability and uniform asymptotic stability using class $\mathcal{K}$ and class $\mathcal{KL}$ functions.

Lemma A.4. *The equilibrium point* $x = 0$ *of* (A.9) *is*

- *uniformly stable if and only if there exist a class* $\mathcal{K}$ *function* α *and a positive constant* c*, independent of* t_0*, such that*

$$\|x(t)\| < \alpha(\|x(t_0)\|)\,, \qquad \forall\, t \geq t_0 \geq 0\,,\ \forall\, \|x(t_0)\| < c\,;$$

- *uniformly asymptotically stable if and only if there exist a class* $\mathcal{KL}$ *function* β *and a positive constant* c*, independent of* t_0*, such that*

$$\|x(t)\| < \beta(\|x(t_0)\|, t - t_0)\,, \qquad \forall\, t \geq t_0 \geq 0\,,\ \forall\, \|x(t_0)\| < c\,; \tag{A.11}$$

- *globally uniformly asymptotically stable if and only if inequality* (A.11) *is satisfied for any initial state* $x(t_0)$. ◇

When the class $\mathcal{KL}$ function β in (A.11) takes the form $\beta(r, s) = kre^{-\lambda s}$, a special case of uniform asymptotic stability, called *exponential stability*, is obtained. See the next definition.

Definition A.6. The equilibrium point $x = 0$ of (A.9) is exponentially stable if there exist positive constants c, k, and λ such that

$$\|x(t)\| \leq k\|x(t_0)\|e^{-\lambda(t-t_0)}\,, \qquad \forall\, \|x(t_0)\| < c\,, \tag{A.12}$$

and globally uniformly exponentially stable (GUES) if (A.12) is satisfied for any initial state $x(t_0)$. ♠

Lyapunov theory for autonomous systems admits an elegant extension to nonautonomous systems. The following theorems summarize key results on uniform stability, uniform asymptotic stability, and exponential stability of nonautonomous systems.

Theorem A.7. *Let* $x = 0$ *be an equilibrium point for* (A.9) *and* $D \subset \mathbb{R}^n$ *a domain containing* $x = 0$*. Let* $V : [0, \infty) \times D \to \mathbb{R}$ *be a continuously differentiable function such that*

$$W_1(x) \leq V(t, x) \leq W_2(x) \tag{A.13}$$

$$\frac{\partial V}{\partial t} + \frac{\partial V}{\partial x} f(t, x) \leq 0 \tag{A.14}$$

for all $t \geq 0$ *and for all* $x \in D$*, where* $W_1(x)$ *and* $W_2(x)$ *are continuous positive definite functions on* D*. Then,* $x = 0$ *is uniformly stable.* ◇

Theorem A.8. *Suppose the assumptions of Theorem* A.7 *are satisfied with inequality* (A.14) *strengthened to*

$$\frac{\partial V}{\partial t} + \frac{\partial V}{\partial x} f(t,x) \le -W_3(x)$$

for all $t \ge 0$ *and for all* $x \in D$, *where* $W_3(x)$ *is a continuous positive definite function defined on* D. *Then,* $x = 0$ *is uniformly asymptotically stable. Moreover, if positive constants* r *and* c *are chosen such that* $\mathcal{B}_r = \{x \in \mathbb{R}^n \,|\, \|x\| \le r\} \subset D$ *and* $c < \min_{\|x\|=r} W_1(x)$, *then every trajectory starting in* $\{x \in \mathcal{B}_r \mid W_2(x) \le c\}$ *satisfies*

$$\|x(t)\| \le \beta(\|x(t_0)\|, t - t_0), \qquad \forall\, t \ge t_0 \ge 0$$

for some class $\mathcal{KL}$ *function* β. *Finally, if* $D = \mathbb{R}^n$ *and* $W_1(x)$ *is radially unbounded, then* $x = 0$ *is globally uniformly asymptotically stable.* ◇

Theorem A.9. *Let* $x = 0$ *be an equilibrium point for* (A.9) *and* $D \subset \mathbb{R}^n$ *be a domain containing* $x = 0$. *Let* $V : [0, \infty) \times D \to \mathbb{R}$ *be a continuously differentiable function that satisfies*

$$k_1\|x\|^a \le V(t,x) \le k_2\|x\|^a$$

$$\frac{\partial V}{\partial t} + \frac{\partial V}{\partial x} f(t,x) \le -k_3\|x\|^a$$

for all $t \ge 0$ *and for all* $x \in D$, *where* k_1, k_2, k_3, *and* a *are positive constants. Then,* $x = 0$ *is exponentially stable. If the assumptions hold globally, then* $x = 0$ *is globally exponentially stable.* ◇

A.2.5 Boundedness

Even when a system has no equilibrium points, Lyapunov analysis can be used to show boundedness of its solutions, as per the definition below. Consider

$$\dot{x} = f(t,x)\,, \qquad x(t_0) = x_0\,. \tag{A.15}$$

Definition A.7. The solutions of (A.15) are

- uniformly bounded if there exists a positive constant c, independent of $t_0 \ge 0$, and for every $a \in (0, c)$, there is $\beta = \beta(a) > 0$, independent of t_0, such that

$$\|x(t_0)\| \le a \Rightarrow \|x(t)\| \le \beta\,, \qquad \forall\, t \ge t_0\,; \tag{A.16}$$

- globally uniformly bounded if (A.16) holds for arbitrarily large a;
- uniformly ultimately bounded with ultimate bound b if there exist positive constants b and c, independent of $t_0 \ge 0$, and for every $a \in (0, c)$, there is $T = T(a, b) \ge 0$, independent of t_0, such that

$$\|x(t_0)\| \le a \Rightarrow \|x(t)\| \le b\,, \qquad \forall\, t \ge t_0 + T\,; \tag{A.17}$$

- globally uniformly ultimately bounded if (A.17) holds for arbitrarily large a. ♠

For autonomous systems, the word "uniformly" may be omitted since the solution depends only on $t - t_0$.

The next Lyapunov-like theorem can be used to prove uniform boundedness and ultimate boundedness of the solutions of (A.15).

Theorem A.10. *Let $D \subset \mathbb{R}^n$ be a domain that contains the origin and $V : [0, \infty) \times D \to \mathbb{R}$ be a continuously differentiable function that satisfies*

$$\alpha_1(\|x\|) \leq V(t, x) \leq \alpha_2(\|x\|)$$
$$\frac{\partial V}{\partial t} + \frac{\partial V}{\partial x} f(t, x) \leq -W_3(x), \qquad \forall\, \|x\| \geq \mu > 0$$

for all $t \geq 0$ and for all $x \in D$, where α_1 and α_2 are class $\mathcal{K}$ functions and $W_3(x)$ is a continuous positive definite function. Take $r > 0$ such that $\mathcal{B}_r \subset D$ and suppose that

$$\mu < \alpha_2^{-1}(\alpha_1(r)) .$$

Then, there exists a class $\mathcal{KL}$ function β and for every initial state $x(t_0)$, satisfying $\|x(t_0)\| \leq \alpha_2^{-1}(\alpha_1(r))$, there is $T \geq 0$ (dependent on $x(t_0)$ and μ) such that the solution of (A.15) *satisfies*

$$\|x(t)\| \leq \beta(\|x(t_0)\|, t - t_0), \qquad \forall\, t_0 \leq t \leq t_0 + T$$
$$\|x(t)\| \leq \alpha_1^{-1}(\alpha_2(\mu)), \qquad \forall\, t \geq t_0 + T .$$

◇

A.2.6 Input-to-State Stability

Consider now the forced system

$$\dot{x} = f(t, x, u), \qquad x(t_0) = x_0, \tag{A.18}$$

where $f : [0, \infty) \times \mathbb{R}^n \times \mathbb{R}^m \to \mathbb{R}^n$ is piecewise continuous in t and locally Lipschitz in x and u, and the input $u(t)$ is a piecewise continuous function of t, for all $t \geq 0$. Suppose the unforced system

$$\dot{x} = f(t, x, 0), \qquad x(t_0) = x_0, \tag{A.19}$$

has a globally uniformly asymptotically stable equilibrium point at the origin $x = 0$. It is possible to view the system (A.18) as a perturbation of the unforced system (A.19). Under certain conditions, if the input $u(t)$ is bounded, that is, its supremum norm $\|u_{[t_0,\infty)}\|$ is finite, then the state $x(t)$ is also bounded. This motivates the following definition of *input-to-state stability*, which captures the influence of both the initial condition $x(t_0)$ and the input u on the evolution of the state x.

Definition A.8. The system (A.18) is said to be input-to-state stable if there exist a class $\mathcal{KL}$ function β and a class $\mathcal{K}$ function γ such that for any initial state $x(t_0)$ and any bounded input $u(t)$, the solution $x(t)$ exists for all $t \geq t_0$ and satisfies

$$\|x(t)\| \leq \beta(\|x(t_0)\|, t - t_0) + \gamma\left(\|u_{[t_0,t]}\|\right) . \tag{A.20}$$

♠

Inequality (A.20) guarantees that for any bounded input $u(t)$ the state $x(t)$ will in fact be bounded. Furthermore, if $u(t)$ converges to zero as $t \to \infty$, so does $x(t)$. The following theorem gives sufficient conditions for input-to-state stability of system (A.18).

Theorem A.11. *Let $V : [0, \infty) \times \mathbb{R}^n \to \mathbb{R}$ be a continuously differentiable function such that*

$$\alpha_1(\|x\|) \leq V(t, x) \leq \alpha_2(\|x\|) \tag{A.21}$$

$$\frac{\partial V}{\partial t} + \frac{\partial V}{\partial x} f(t, x, u) \leq -W_3(x), \qquad \forall\ \|x\| \geq \rho(\|u\|) > 0 \tag{A.22}$$

for all $(t, x, u) \in [0, \infty) \times \mathbb{R}^n \times \mathbb{R}^m$, where α_1 and α_2 are class $\mathcal{K}_\infty$ functions, ρ is a class $\mathcal{K}$ function, and $W_3(x)$ is a continuous positive definite function on $\mathbb{R}^n$. Then, the system (A.18) *is input-to-state stable with $\gamma = {\alpha_1}^{-1} \circ \alpha_2 \circ \rho$.* ◇

For autonomous systems, conditions (A.21) and (A.22) are also necessary for input-to-state stability. It is common in the literature to simply abbreviate input-to-state stability as ISS and to refer to the function V of Theorem A.11 as an ISS-Lyapunov function. The following theorem establishes an important connection between global exponential stability and ISS.

Theorem A.12. *Suppose $f(t, x, u)$ is continuously differentiable and globally Lipschitz in (x, u), uniformly in t. If the unforced system* (A.19) *has a globally exponentially stable equilibrium point at $x = 0$, then the system* (A.18) *is input-to-state stable.*

A.3 Graph Theory

This section summarizes key concepts and results in graph theory that play an important role in the study of the communications networks that allow for cooperation among multiple vehicles. The definitions presented borrow from the expositions in [3] and [8] for undirected and directed graphs, respectively.

A.3.1 Basic Definitions

A *digraph* (or *directed graph*) is a pair $\Gamma = (\mathcal{V}, \mathcal{E})$, where $\mathcal{V}$ is a finite nonempty set of *nodes*, and $\mathcal{E} \subseteq \mathcal{V} \times \mathcal{V}$ is a set of ordered pairs of nodes, called *edges*. If the pairs of nodes are unordered, then Γ is an *undirected graph*. The number of nodes of a graph Γ defines its *order*, written as $|\Gamma|$; the *size* of a graph Γ is defined as its number of edges, and denoted as $\|\Gamma\|$.

Undirected graphs. A node $v \in \mathcal{V}$ is *incident* with an edge $e \in \mathcal{E}$ if $v \in e$; then e is an edge at v. The two nodes v, w that are incident with an edge e are called *ends* of e, and one writes $e = vw$. In a graph, the ends of an edge are always different, and no two edges have the same ends. For this reason, graphs are sometimes referred to as *simple graphs*, to distinguish them from *multigraphs*, which can have loops and multiple edges. The *degree* of a node v is the cardinality of the set of all its incident edges. A node is *isolated* if it has degree 0. Two nodes $v, w \in \mathcal{V}$ are *adjacent*, or *neighbors*, if $vw \in \mathcal{E}$, and two edges $e, f \in \mathcal{E}$ are called *adjacent* if they have an end in common. In the main text we use $\mathcal{N}_v$ to define the set of all neighbors of node v. A graph of order n whose nodes are all pairwise adjacent is called *complete*, and is denoted by $\mathcal{K}^n$. The complement of an undirected graph $\Gamma = (\mathcal{V}, \mathcal{E})$ is the undirected graph with the same vertex set $\mathcal{V}$ in which two vertices are adjacent if and only if they are not adjacent in Γ.

Directed graphs. In digraphs, if the edge e is directed from v to w, then w is called the *head end* while v is called the *tail end*, and we write $e = vw$. The number of head ends at a node v is called the *in-degree* of v, while the number of tail ends is its *out-degree*. A digraph is called *balanced* if, for all nodes, the in-degree is equal to the out-degree.

A.3.2 Connectivity

Consider a graph $\Gamma = (\mathcal{V}, \mathcal{E})$. Node $v_k \in \mathcal{V}$ is said to be *reachable* from node $v_0 \in \mathcal{V}$ if there exists a sequence of vertices $v_0, v_1, \ldots, v_k$, $v_i \in \mathcal{V}$, such that the edges $v_{i-1}v_i$ are in $\mathcal{E}$, for all $1 \leq i \leq k$. The non-empty graph $\mathcal{P}$ with vertex set $\mathcal{V}_p = \{v_0, v_1, \ldots, v_k\}$ and edge set $\mathcal{E}_p = \{v_0v_1, v_1v_2, \ldots, v_{k-1}v_k\}$ is called a *directed path* from v_0 *to* v_k. The vertices $v_1, \ldots, v_{k-1}$ are the *inner* vertices of $\mathcal{P}$. If a node is reachable from any other node then it is *globally reachable*.

In an undirected graph Γ, two nodes v and w are *connected* if Γ contains a path from v to w. Otherwise, they are *disconnected*. The undirected graph Γ is said to be connected if every pair of nodes is connected.

A digraph Γ is said to be *weakly connected* if its underlying undirected graph (that is, the undirected graph obtained by replacing all of the directed edges with undirected ones) is connected. The digraph Γ is *connected* if, for every pair of nodes v and w, it contains a directed path either from v to w or from w to v. It is called *strongly connected* if every node is globally reachable. Finally, the digraph is *disconnected* if its underlying undirected graph is disconnected.

A.3.3 Algebraic Graph Theory

If an arbitrary enumeration is assigned to the vertices and edges of a graph Γ, it is possible to represent the graph in terms of matrices. Algebraic graph theory studies the relationships between the structure and properties of a graph and its matricial representations.

The *adjacency matrix* $\boldsymbol{A}(\Gamma)$ of an undirected graph is the symmetric square matrix of size $|\Gamma|$, defined by $\boldsymbol{A}_{ij} = 1$ if $v_iv_j \in \mathcal{E}$, and $\boldsymbol{A}_{ij} = 0$ otherwise. The adjacency matrix uniquely defines a graph, although for a given graph, $\boldsymbol{A}$ is not itself unique, as it depends on the enumeration of the nodes. The *degree matrix* of an undirected graph is the square matrix $\boldsymbol{D}(\Gamma)$ in which the elements of the diagonal are the degrees (out-degrees) of the corresponding nodes. Then, the *Laplacian* of an undirected graph is defined as

$$\boldsymbol{L}(\Gamma) = \boldsymbol{D}(\Gamma) - \boldsymbol{A}(\Gamma)\,. \tag{A.23}$$

In the case of digraphs, one can define the *incidence matrix* $\boldsymbol{M}(\Gamma)$ as the matrix with rows and columns indexed by the vertices and edges of Γ, with elements

$$m_{ij} = \begin{cases} +1, & \text{if } v_i \text{ is the head of edge } e_j \\ -1, & \text{if } v_i \text{ is the tail of edge } e_j \\ 0, & \text{otherwise} \end{cases}$$

In this case, the Laplacian is defined as

$$\boldsymbol{L}(\Gamma) = \boldsymbol{M}(\Gamma)\boldsymbol{M}(\Gamma)^\top\,.$$

By construction, the Laplacian $\boldsymbol{L}$ of a graph is positive semi-definite.

Next we state some results for undirected graphs that are used throughout the book. The interested reader is referred to [1,2] for further details on spectral graph theory. In what follows, we let Γ be an undirected graph of order n, and we write $\lambda_1 \le \lambda_2 \le \cdots \le \lambda_n$ for its eigenvalues.

Theorem A.13. *All eigenvalues of* $\boldsymbol{L}(\Gamma)$ *are real nonnegative.* ◇

Theorem A.14. *The all-ones vector* $\mathbf{1}_n$ *is a right eigenvector of* $\boldsymbol{L}(\Gamma)$ *with eigenvalue* $\lambda_1 = 0$. ◇

Theorem A.15. $\lambda_2 = 0$ *if and only if the graph* Γ *is disconnected.* ◇

Theorem A.16. $\lambda_n \le n$, *with equality only if the complement of* Γ *is disconnected.* ◇

References

[1] N. Biggs, Algebraic Graph Theory, Cambridge University Press, New York, NY, 1993.
[2] D.M. Cvetković, M. Doob, H. Sachs, Spectra of Graphs, Theory and Application, Academic Press, New York, NY, 1980.
[3] R. Diestel, Graph Theory, 3rd ed., Springer-Verlag, Heidelberg–New York, 2005.
[4] H.K. Khalil, Nonlinear Systems, Prentice Hall, Englewood Cliffs, NJ, 2002.
[5] T. Lee, M. Leok, N.H. McClamroch, Control of complex maneuvers for a quadrotor UAV using geometric methods on $\mathsf{SE}(3)$. Available online arXiv:1003.2005v4, 2011.
[6] E.D. Sontag, Input to state stability: basic concepts and results, in: P. Nistri, G. Stefani (Eds.), Nonlinear and Optimal Control Theory, Springer-Verlag, Berlin, 2006.
[7] E.D. Sontag, Y. Wang, On characterizations of the input-to-state stability property, Systems and Control Letters 5 (1995) 351–359.
[8] R.J. Wilson, Introduction to Graph Theory, Fourth Edition, Addison Wesley, Harlow, UK, 1996.

Proofs and Derivations

B.1 Proofs and Derivations in Part I

B.1.1 The Coordination Projection Matrix

The equality $\boldsymbol{Q}^\top \boldsymbol{Q} = \boldsymbol{\Pi}_\xi$ is used several times throughout this book and plays an important role in the derivation of the results described in the text. Next, we provide a proof for this equality.

Lemma B.1. *Let $\boldsymbol{Q}$ be an $(n-1) \times n$ matrix such that $\boldsymbol{Q}\mathbf{1}_n = \mathbf{0}$ and $\boldsymbol{Q}\boldsymbol{Q}^\top = \mathbb{I}_{n-1}$. Then, the following equality holds:*

$$\boldsymbol{Q}^\top \boldsymbol{Q} = \boldsymbol{\Pi}_\xi := \mathbb{I}_n - \frac{\mathbf{1}_n \mathbf{1}_n^\top}{n} . \qquad \diamond$$

Proof. We start by partitioning matrix $\boldsymbol{Q}$ as

$$\boldsymbol{Q} = \begin{bmatrix} \tilde{\boldsymbol{q}}_1^\top \\ \vdots \\ \tilde{\boldsymbol{q}}_{n-1}^\top \end{bmatrix}, \qquad \tilde{\boldsymbol{q}}_i \in \mathbb{R}^n .$$

From $\boldsymbol{Q}\boldsymbol{Q}^\top = \mathbb{I}_{n-1}$, it follows that

$$\begin{aligned} \tilde{\boldsymbol{q}}_i^\top \tilde{\boldsymbol{q}}_i &= 1 , & i &= 1, \ldots, n-1 , \\ \tilde{\boldsymbol{q}}_i^\top \tilde{\boldsymbol{q}}_j &= 0 , & i, j &= 1, \ldots, n-1 , \quad i \neq j , \end{aligned}$$

which implies that $\{\tilde{\boldsymbol{q}}_1, \ldots, \tilde{\boldsymbol{q}}_{n-1}\}$ is a set of $(n-1)$ orthonormal vectors in $\mathbb{R}^n$. From $\boldsymbol{Q}\mathbf{1}_n = \mathbf{0}$, it also follows that

$$\tilde{\boldsymbol{q}}_i^\top \mathbf{1}_n = 0 , \qquad i = 1, \ldots, n-1 ,$$

and therefore $\{\tilde{\boldsymbol{q}}_1, \ldots, \tilde{\boldsymbol{q}}_{n-1}, \mathbf{1}_n\}$ is an orthogonal basis of $\mathbb{R}^n$.

Next, we prove that the vector $\mathbf{1}_n$ spans the null space of matrix $\boldsymbol{Q}$. We prove this result by contradiction. Assume there exists a vector $\boldsymbol{v} \in \mathbb{R}^n$, $\boldsymbol{v} \neq \gamma \mathbf{1}_n$, $\gamma \in \mathbb{R}$, such that $\boldsymbol{Q}\boldsymbol{v} = \mathbf{0}$. Since $\{\tilde{\boldsymbol{q}}_1, \ldots, \tilde{\boldsymbol{q}}_{n-1}, \mathbf{1}_n\}$ forms a basis of $\mathbb{R}^n$, vector $\boldsymbol{v}$ can be expressed as a linear combination of these basis vectors, that is,

$$\boldsymbol{v} = \alpha_0 \mathbf{1}_n + \sum_{k=1}^{n-1} \alpha_i \tilde{\boldsymbol{q}}_k , \qquad \alpha_i \in \mathbb{R} .$$

From this equation it can be concluded that

$$\mathbf{0} \;=\; \boldsymbol{Q}\boldsymbol{v} \;=\; \alpha_0 \boldsymbol{Q}\mathbf{1}_n + \sum_{k=1}^{n-1} \alpha_i \boldsymbol{Q}\tilde{\boldsymbol{q}}_k \;=\; \mathbf{0} + \begin{bmatrix} \alpha_1 \\ \vdots \\ \alpha_{n-1} \end{bmatrix},$$

Time-Critical Cooperative Control of Autonomous Air Vehicles, DOI: 10.1016/B978-0-12-809946-9.00028-2

which implies that $\alpha_i = 0$, $i = 1, \ldots, n-1$. Therefore, $\boldsymbol{v} = \alpha_0 \mathbf{1}_n$, which contradicts the assumption that $\boldsymbol{v} \neq \gamma \mathbf{1}_n$, $\gamma \in \mathbb{R}$.

We now note that by multiplying $\boldsymbol{Q}\boldsymbol{Q}^\top = \mathbb{I}_{n-1}$ on the right by $\boldsymbol{Q}$ one obtains $\boldsymbol{Q}\boldsymbol{Q}^\top\boldsymbol{Q} = \boldsymbol{Q}$, which can be rewritten as $\boldsymbol{Q}\left(\boldsymbol{Q}^\top\boldsymbol{Q} - \mathbb{I}_n\right) = \mathbf{0}$. From this last equality, the fact that $\boldsymbol{Q}^\top\boldsymbol{Q} - \mathbb{I}_n = \left(\boldsymbol{Q}^\top\boldsymbol{Q} - \mathbb{I}_n\right)^\top$, and recalling that the vector $\mathbf{1}_n$ spans the null space of matrix $\boldsymbol{Q}$, it follows that

$$\boldsymbol{Q}^\top\boldsymbol{Q} - \mathbb{I}_n = \beta \mathbf{1}_n \mathbf{1}_n^\top, \qquad \text{for some } \beta \in \mathbb{R}.$$

Finally, multiplying the equation above on the right by $\mathbf{1}_n$ yields

$$\begin{aligned} \left(\boldsymbol{Q}^\top\boldsymbol{Q} - \mathbb{I}_n\right)\mathbf{1}_n &= \beta \mathbf{1}_n \mathbf{1}_n^\top \mathbf{1}_n, \\ -\mathbf{1}_n &= n\beta \mathbf{1}_n, \end{aligned}$$

which implies that $\beta = -\frac{1}{n}$ and, hence, we obtain

$$\boldsymbol{Q}^\top\boldsymbol{Q} = \mathbb{I}_n - \frac{1}{n}\mathbf{1}_n\mathbf{1}_n^\top. \qquad \square$$

B.2 Proofs and Derivations in Part II

B.2.1 Time-Derivative of the Coordination States

The time-derivative of the ith coordination state is given by

$$\dot{\xi}_i(t) = \frac{\mathrm{d}}{\mathrm{d}t}\left(\eta_i\left(\ell_i'(t)\right)\right) = \left.\frac{\mathrm{d}\eta_i}{\mathrm{d}\ell_i'}\right|_{\ell_i'(t)} \dot{\ell}_i'(t).$$

From the definitions of $\ell_{d,i}'(\cdot)$ and $\eta_i(\cdot)$ in Section 4.1, we have that the following equality holds for all $\ell_i' \in [0, 1]$:

$$\ell_{d,i}'\left(\eta_i(\ell_i')\right) = \ell_i'$$

and, therefore, taking the derivative with respect to ℓ_i' on both sides yields

$$\left.\frac{\mathrm{d}\ell_{d,i}'}{\mathrm{d}\eta_i}\right|_{\eta_i(\ell_i')} \left.\frac{\mathrm{d}\eta_i}{\mathrm{d}\ell_i'}\right|_{\ell_i'} = 1. \tag{B.1}$$

From the definition of $\ell_{d,i}'(\cdot)$, it follows that

$$\left.\frac{\mathrm{d}\ell_{d,i}'}{\mathrm{d}\eta_i}\right|_{\eta_i(\ell_i')} = \frac{1}{\ell_{fi}} v_{d,i}(\eta_i(\ell_i')), \qquad \text{for all } \ell_i' \in [0, 1],$$

which, along with equality (B.1), implies that

$$\left.\frac{\mathrm{d}\eta_i}{\mathrm{d}\ell_i'}\right|_{\ell_i'} = \frac{1}{\frac{1}{\ell_{fi}} v_{d,i}(\eta_i(\ell_i'))}, \qquad \text{for all } \ell_i' \in [0, 1].$$

From the above equations, the evolution of the ith coordination state can be expressed as

$$\dot{\xi}_i(t) = \frac{1}{\frac{1}{\ell_{fi}} v_{d,i}(\eta_i(\ell_i'(t)))} \dot{\ell}_i'(t),$$

which can be simplified to

$$\dot{\xi}_i(t) = \frac{\dot{\ell}_i(t)}{v_{d,i}(\xi_i(t))}.$$

B.2.2 Closed-Loop Coordination Error Dynamics

From the definition of $\boldsymbol{\zeta}_1(t)$ and $\boldsymbol{\zeta}_2(t)$ and the closed-loop coordination dynamics (4.7), it follows that

$$\begin{aligned}\dot{\boldsymbol{\zeta}}_1(t) &= -k_P \boldsymbol{Q}\boldsymbol{L}(t)\boldsymbol{\xi}(t) + \boldsymbol{Q}\mathbf{1}_n + \boldsymbol{Q}\begin{bmatrix} \mathbf{0} \\ \boldsymbol{\zeta}_2(t) \end{bmatrix} \\ &= -k_P \boldsymbol{Q}\boldsymbol{L}(t)\boldsymbol{\xi}(t) + \boldsymbol{Q}\boldsymbol{C}\boldsymbol{\zeta}_2(t).\end{aligned}$$

The properties of the projection matrix $\boldsymbol{\Pi}_\xi$ imply that

$$\begin{aligned}\dot{\boldsymbol{\zeta}}_1(t) &= -k_P \boldsymbol{Q}\boldsymbol{L}(t)\boldsymbol{\Pi}_\xi\boldsymbol{\xi}(t) + \boldsymbol{Q}\boldsymbol{C}\boldsymbol{\zeta}_2(t) \\ &= -k_P \boldsymbol{Q}\boldsymbol{L}(t)\boldsymbol{Q}^\top\boldsymbol{Q}\boldsymbol{\xi}(t) + \boldsymbol{Q}\boldsymbol{C}\boldsymbol{\zeta}_2(t) \\ &= -k_P \bar{\boldsymbol{L}}(t)\boldsymbol{\zeta}_1(t) + \boldsymbol{Q}\boldsymbol{C}\boldsymbol{\zeta}_2(t).\end{aligned} \tag{B.2}$$

Similarly,

$$\dot{\boldsymbol{\zeta}}_2(t) = -k_I \boldsymbol{C}^\top \boldsymbol{L}(t)\boldsymbol{\xi}(t) = -k_I \boldsymbol{C}^\top \boldsymbol{Q}^\top \bar{\boldsymbol{L}}(t)\boldsymbol{\zeta}_1(t). \tag{B.3}$$

Eqs. (B.2) and (B.3) lead to the dynamics (4.9).

B.2.3 Proof of Lemma 3.1

We start by noting that the following upper bounds hold over the compact set Ω_{pf} introduced in (3.13):

$$\|\boldsymbol{p}_F\| \le c c_1 < \frac{c_1}{\sqrt{2}}, \tag{B.4}$$

$$\Psi(\tilde{\boldsymbol{R}}) \le c^2 < \frac{1}{2}. \tag{B.5}$$

Consider now the Lyapunov function candidate

$$V_{pf}(\boldsymbol{p}_F, \tilde{\boldsymbol{R}}) = \Psi(\tilde{\boldsymbol{R}}) + \frac{1}{c_1^2}\|\boldsymbol{p}_F\|^2 .$$

This function is locally positive-definite about $(\boldsymbol{p}_F, \tilde{R}_{11}) = (\mathbf{0}, 1)$ within the set Ω_{pf}. Moreover, we note that $\|\boldsymbol{e}_{\tilde{\boldsymbol{R}}}\|$ can be related to the function $\Psi(\tilde{\boldsymbol{R}})$ as follows:

$$\begin{aligned}\|\boldsymbol{e}_{\tilde{\boldsymbol{R}}}\|^2 &= \frac{1}{4}\left(\tilde{R}_{12}^2 + \tilde{R}_{13}^2\right) = \frac{1}{4}\left(1 - \tilde{R}_{11}^2\right) = \frac{1}{4}\left(1 - \tilde{R}_{11}\right)\left(1 + \tilde{R}_{11}\right) \\ &= \Psi(\tilde{\boldsymbol{R}})\left(1 - \Psi(\tilde{\boldsymbol{R}})\right) .\end{aligned}$$

Then, the bound in (B.5) implies that, inside the set Ω_{pf}, the function $\Psi(\tilde{\boldsymbol{R}})$ satisfies

$$\|\boldsymbol{e}_{\tilde{\boldsymbol{R}}}\|^2 \le \Psi(\tilde{\boldsymbol{R}}) \le \frac{1}{1-c^2}\|\boldsymbol{e}_{\tilde{\boldsymbol{R}}}\|^2 .$$

It thus follows that, within the set Ω_{pf}, the Lyapunov function V_{pf} can be bounded as

$$\|\boldsymbol{e}_{\tilde{\boldsymbol{R}}}\|^2 + \frac{1}{c_1^2}\|\boldsymbol{p}_F\|^2 \le V_{pf} \le \frac{1}{1-c^2}\|\boldsymbol{e}_{\tilde{\boldsymbol{R}}}\|^2 + \frac{1}{c_1^2}\|\boldsymbol{p}_F\|^2 . \tag{B.6}$$

From the dynamics (3.6), the time-derivative of V_{pf} is given by

$$\begin{aligned}\dot{V}_{pf} &= \dot{\Psi}(\tilde{\boldsymbol{R}}) + \frac{2}{c_1^2}\boldsymbol{p}_F \cdot \dot{\boldsymbol{p}}_F \\ &= \boldsymbol{e}_{\tilde{\boldsymbol{R}}} \cdot \left(\begin{bmatrix} q \\ r \end{bmatrix} - \boldsymbol{\Pi}_{\boldsymbol{R}}\tilde{\boldsymbol{R}}^\top\left(\boldsymbol{R}_F^D\{\boldsymbol{\omega}_{F/I}\}_F + \{\boldsymbol{\omega}_{D/F}\}_D\right)\right) \\ &\quad + \frac{2}{c_1^2}\boldsymbol{p}_F \cdot \left(-\dot{\ell}\,\hat{\mathbf{t}} - \boldsymbol{\omega}_{F/I} \times \boldsymbol{p}_F + v\,\hat{\mathbf{w}}_1\right) .\end{aligned}$$

In the equation above, $\dot{\boldsymbol{p}}_F$ denotes the componentwise derivative of vector $\boldsymbol{p}_F$ and therefore $\dot{\boldsymbol{p}}_F = \dot{\boldsymbol{p}}_F]_F$. The rate commands (3.8), together with the law (3.7) for the rate of progression of the virtual target along the path, lead to

$$\begin{aligned}\dot{V}_{pf} = &-2k_{\tilde{R}}\,\boldsymbol{e}_{\tilde{\boldsymbol{R}}} \cdot \boldsymbol{e}_{\tilde{\boldsymbol{R}}} + \frac{2}{c_1^2}\left(-k_\ell\left(\boldsymbol{p}_F \cdot \hat{\mathbf{t}}\right)^2 - \boldsymbol{p}_F \cdot \left(\boldsymbol{\omega}_{F/I} \times \boldsymbol{p}_F\right)\right. \\ &\left. + v\,\boldsymbol{p}_F \cdot \left(\hat{\mathbf{w}}_1 - \left(\hat{\mathbf{w}}_1 \cdot \hat{\mathbf{t}}\right)\hat{\mathbf{t}}\right)\right).\end{aligned} \tag{B.7}$$

Since $(\boldsymbol{p}_F \cdot \hat{\mathbf{t}}) = x_F$ and $(\boldsymbol{p}_F \cdot (\boldsymbol{\omega}_{F/I} \times \boldsymbol{p}_F)) = 0$, (B.7) reduces to

$$\dot{V}_{pf} = -2k_{\tilde{R}}\,\boldsymbol{e}_{\tilde{\boldsymbol{R}}} \cdot \boldsymbol{e}_{\tilde{\boldsymbol{R}}} - \frac{2k_\ell}{c_1^2}x_F^2 + \frac{2v}{c_1^2}\left(\boldsymbol{p}_F \cdot \left(\hat{\mathbf{w}}_1 - \left(\hat{\mathbf{w}}_1 \cdot \hat{\mathbf{t}}\right)\hat{\mathbf{t}}\right)\right) . \tag{B.8}$$

Letting $\boldsymbol{p}_{\mathbf{x}}(t)$ denote the path-following cross-track error, which can be expressed as

$$\boldsymbol{p}_{\mathbf{x}} = \left(\boldsymbol{p}_F \cdot \hat{\mathbf{n}}_1\right)\hat{\mathbf{n}}_1 + \left(\boldsymbol{p}_F \cdot \hat{\mathbf{n}}_2\right)\hat{\mathbf{n}}_2 = y_F\,\hat{\mathbf{n}}_1 + z_F\,\hat{\mathbf{n}}_2\,, \tag{B.9}$$

we have the following equality:

$$\boldsymbol{p}_F \cdot \left(\hat{\mathbf{w}}_1 - \left(\hat{\mathbf{w}}_1 \cdot \hat{\mathbf{t}}\right)\hat{\mathbf{t}}\right) = \boldsymbol{p}_F \cdot \left(\left(\hat{\mathbf{w}}_1 \cdot \hat{\mathbf{n}}_1\right)\hat{\mathbf{n}}_1 + \left(\hat{\mathbf{w}}_1 \cdot \hat{\mathbf{n}}_2\right)\hat{\mathbf{n}}_2\right) = \boldsymbol{p}_{\mathbf{x}} \cdot \hat{\mathbf{w}}_1\,. \tag{B.10}$$

Substituting (B.10) into (B.8), we obtain

$$\dot{V}_{pf} = -2k_{\tilde{R}}\,\boldsymbol{e}_{\tilde{\boldsymbol{R}}} \cdot \boldsymbol{e}_{\tilde{\boldsymbol{R}}} - \frac{2k_\ell}{c_1^2}x_F^2 + \frac{2v}{c_1^2}\left(\boldsymbol{p}_{\mathbf{x}} \cdot \hat{\mathbf{w}}_1\right). \tag{B.11}$$

Consider now the quantity $(\hat{\mathbf{w}}_1 \cdot \hat{\mathbf{b}}_{1\mathbf{D}})$, which represents the cosine of the angle ψ_e between the desired direction of the velocity vector $\hat{\mathbf{b}}_{1\mathbf{D}}$ and the actual direction of the vehicle's velocity vector $\hat{\mathbf{w}}_1$. From the definition of $\Psi(\tilde{\boldsymbol{R}})$ in (3.3), we have that

$$\hat{\mathbf{w}}_1 \cdot \hat{\mathbf{b}}_{1\mathbf{D}} = \cos\psi_e = \tilde{R}_{11} = 1 - 2\Psi(\tilde{\boldsymbol{R}})\,.$$

The bound in (B.5) implies that, within the set Ω_{pf}, the quantity $(\hat{\mathbf{w}}_1 \cdot \hat{\mathbf{b}}_{1\mathbf{D}})$ is bounded away from zero:

$$\hat{\mathbf{w}}_1 \cdot \hat{\mathbf{b}}_{1\mathbf{D}} = 1 - 2\Psi(\tilde{\boldsymbol{R}}) \geq 1 - 2c^2 > 0\,.$$

The quantity $\frac{1}{(\hat{\mathbf{w}}_1 \cdot \hat{\mathbf{b}}_{1\mathbf{D}})}$ is therefore well defined within the set Ω_{pf}. Next, we add and subtract the term $\frac{2v}{c_1^2}\frac{(\boldsymbol{p}_{\mathbf{x}} \cdot \hat{\mathbf{b}}_{1\mathbf{D}})}{(\hat{\mathbf{w}}_1 \cdot \hat{\mathbf{b}}_{1\mathbf{D}})}$ to (B.11) to obtain

$$\dot{V}_{pf} = -2k_{\tilde{R}}\,\boldsymbol{e}_{\tilde{\boldsymbol{R}}} \cdot \boldsymbol{e}_{\tilde{\boldsymbol{R}}} - \frac{2k_\ell}{c_1^2}x_F^2 + \frac{2v}{c_1^2}\frac{(\boldsymbol{p}_{\mathbf{x}} \cdot \hat{\mathbf{b}}_{1\mathbf{D}})}{(\hat{\mathbf{w}}_1 \cdot \hat{\mathbf{b}}_{1\mathbf{D}})} + \frac{2v}{c_1^2}\frac{\boldsymbol{p}_{\mathbf{x}} \cdot (\hat{\mathbf{w}}_1 \times (\hat{\mathbf{w}}_1 \times \hat{\mathbf{b}}_{1\mathbf{D}}))}{(\hat{\mathbf{w}}_1 \cdot \hat{\mathbf{b}}_{1\mathbf{D}})}\,.$$

The definitions of $\hat{\mathbf{b}}_{1\mathbf{D}}(t)$ and $\boldsymbol{p}_{\mathbf{x}}(t)$ in (3.2) and (B.9) lead to

$$\begin{aligned}\dot{V}_{pf} = &-2k_{\tilde{R}}\,\boldsymbol{e}_{\tilde{\boldsymbol{R}}} \cdot \boldsymbol{e}_{\tilde{\boldsymbol{R}}} - \frac{2k_\ell}{c_1^2}x_F^2 - \frac{2v}{c_1^2(\hat{\mathbf{w}}_1 \cdot \hat{\mathbf{b}}_{1\mathbf{D}})\left(d^2 + \boldsymbol{p}_{\mathbf{x}} \cdot \boldsymbol{p}_{\mathbf{x}}\right)^{\frac{1}{2}}}\,\boldsymbol{p}_{\mathbf{x}} \cdot \boldsymbol{p}_{\mathbf{x}} \\ &+ \frac{2v}{c_1^2}\frac{\boldsymbol{p}_{\mathbf{x}} \cdot (\hat{\mathbf{w}}_1 \times (\hat{\mathbf{w}}_1 \times \hat{\mathbf{b}}_{1\mathbf{D}}))}{(\hat{\mathbf{w}}_1 \cdot \hat{\mathbf{b}}_{1\mathbf{D}})}\,.\end{aligned}$$

We now note that, within the set Ω_{pf}, the following bounds hold:

$$0 < 1 - 2c^2 \leq (\hat{\mathbf{w}}_1 \cdot \hat{\mathbf{b}}_{1\mathbf{D}}) \leq 1\,, \qquad \|\boldsymbol{p}_{\mathbf{x}}\| \leq \|\boldsymbol{p}_F\| \leq cc_1\,.$$

These bounds, together with the assumption on the vehicle speed in (3.9), yield the following bound for $\dot{V}_{pf}$:

$$\dot{V}_{pf} \leq -2k_{\tilde{R}}\|\boldsymbol{e}_{\tilde{\boldsymbol{R}}}\|^2 - \frac{2k_\ell}{c_1^2}x_F^2 - \frac{2v_{\min}}{c_1^2\left(d^2+c^2c_1^2\right)^{\frac{1}{2}}}\|\boldsymbol{p}_{\mathbf{x}}\|^2 + \frac{2v_{\max}}{c_1^2(1-2c^2)}\|\boldsymbol{p}_{\mathbf{x}}\|\,\|\hat{\mathbf{w}}_{\mathbf{1}}\times(\hat{\mathbf{w}}_{\mathbf{1}}\times\hat{\mathbf{b}}_{\mathbf{1D}})\|\,.$$

The term $\|\hat{\mathbf{w}}_{\mathbf{1}}\times(\hat{\mathbf{w}}_{\mathbf{1}}\times\hat{\mathbf{b}}_{\mathbf{1D}})\|$ represents the absolute value of the sine of the angle ψ_e. Therefore, we can write

$$\|\hat{\mathbf{w}}_{\mathbf{1}}\times(\hat{\mathbf{w}}_{\mathbf{1}}\times\hat{\mathbf{b}}_{\mathbf{1D}})\| = |\sin(\psi_e)| = \sqrt{1-\cos^2(\psi_e)} = \sqrt{1-\tilde{R}_{11}^2} = \sqrt{\tilde{R}_{12}^2+\tilde{R}_{13}^2} = 2\|\boldsymbol{e}_{\tilde{\boldsymbol{R}}}\|\,,$$

which yields

$$\dot{V}_{pf} \leq -2k_{\tilde{R}}\|\boldsymbol{e}_{\tilde{\boldsymbol{R}}}\|^2 - \frac{2k_\ell}{c_1^2}x_F^2 - \frac{2v_{\min}}{c_1^2\left(d^2+c^2c_1^2\right)^{\frac{1}{2}}}\|\boldsymbol{p}_{\mathbf{x}}\|^2 + \frac{4v_{\max}}{c_1^2(1-2c^2)}\|\boldsymbol{p}_{\mathbf{x}}\|\,\|\boldsymbol{e}_{\tilde{\boldsymbol{R}}}\|\,.$$

Letting $\tilde{k}_\ell := \min\left\{k_\ell\,,\ \frac{v_{\min}}{(d^2+c^2c_1^2)^{\frac{1}{2}}}\right\}$ and noting that $\|\boldsymbol{p}_{\mathbf{x}}\| \leq \|\boldsymbol{p}_{\boldsymbol{F}}\|$, we have

$$\dot{V}_{pf} \leq -2k_{\tilde{R}}\|\boldsymbol{e}_{\tilde{\boldsymbol{R}}}\|^2 - \frac{2\tilde{k}_\ell}{c_1^2}\|\boldsymbol{p}_{\boldsymbol{F}}\|^2 + \frac{4v_{\max}}{c_1^2(1-2c^2)}\|\boldsymbol{p}_{\boldsymbol{F}}\|\,\|\boldsymbol{e}_{\tilde{\boldsymbol{R}}}\|\,.$$

From the choice for the characteristic distance d and the path-following control parameters k_ℓ and $k_{\tilde{R}}$ in (3.10), and the definition of $\bar{\lambda}_{pf}$ in (3.12), it follows that:

$$\begin{bmatrix} k_{\tilde{R}} & -\frac{v_{\max}}{c_1^2(1-2c^2)} \\ -\frac{v_{\max}}{c_1^2(1-2c^2)} & \frac{\tilde{k}_\ell}{c_1^2} \end{bmatrix} \geq \bar{\lambda}_{pf}\begin{bmatrix} \frac{1}{1-c^2} & 0 \\ 0 & \frac{1}{c_1^2} \end{bmatrix},$$

which implies that, within the set Ω_{pf}, the following bound holds:

$$\dot{V}_{pf} \leq -2\bar{\lambda}_{pf}\left(\frac{1}{1-c^2}\|\boldsymbol{e}_{\tilde{\boldsymbol{R}}}\|^2 + \frac{1}{c_1^2}\|\boldsymbol{p}_{\boldsymbol{F}}\|^2\right) \leq -2\bar{\lambda}_{pf}V_{pf}\,.$$

It follows from [2, Theorem 4.10] that both $\|\boldsymbol{e}_{\tilde{\boldsymbol{R}}}\|$ and $\|\boldsymbol{p}_{\boldsymbol{F}}\|$ converge exponentially fast to zero for all initial conditions inside the compact set Ω_{pf}. □

B.2.4 *Proof of Lemma 3.2*

First, we show that the rate commands $q_c(t)$ and $r_c(t)$ are bounded for all $(\boldsymbol{p}_{\boldsymbol{F}}, \tilde{\boldsymbol{R}}) \in \Omega_{pf}$. To this end, we note that over the compact set Ω_{pf}, introduced in (3.13), the following inequalities hold:

$$\|\boldsymbol{p}_F\| \leq c c_1 \,, \tag{B.12}$$

$$\Psi(\tilde{\boldsymbol{R}}) \leq c^2 \,. \tag{B.13}$$

The first inequality above, together with the bound on the vehicle speed in (3.9), implies that $\dot{\ell}(t)$ satisfies

$$|\dot{\ell}| \leq v_{\max} + k_\ell \, c c_1 \,.$$

From the assumption on the feasibility of the path, we can conclude that both $k_1(\ell)$ and $k_2(\ell)$ are bounded, and therefore the bound on $\dot{\ell}(t)$ implies that $\boldsymbol{\omega}_{\boldsymbol{F}/\boldsymbol{I}}(t)$ is also bounded. It then follows from (3.6) that $\dot{\boldsymbol{p}}_F(t)$ is bounded, which, along with inequality (B.12), implies that the entries of $\dot{\boldsymbol{R}}_{\boldsymbol{D}}^{\boldsymbol{F}}(t)$ are bounded. From the kinematic equation

$$\left(\{\boldsymbol{\omega}_{\boldsymbol{D}/\boldsymbol{F}}\}_D\right)^\wedge = \boldsymbol{R}_{\boldsymbol{F}}^{\boldsymbol{D}} \dot{\boldsymbol{R}}_{\boldsymbol{D}}^{\boldsymbol{F}} \,,$$

it follows that $\boldsymbol{\omega}_{\boldsymbol{D}/\boldsymbol{F}}(t)$ is also bounded. Moreover, since $\|\boldsymbol{e}_{\tilde{\boldsymbol{R}}}\| \leq \Psi(\tilde{\boldsymbol{R}})$, inequality (B.13) implies that the attitude error $\boldsymbol{e}_{\tilde{\boldsymbol{R}}}(t)$ satisfies

$$\|\boldsymbol{e}_{\tilde{\boldsymbol{R}}}\| \leq c^2 \,.$$

From the bounds on $\boldsymbol{\omega}_{\boldsymbol{F}/\boldsymbol{I}}(t)$, $\boldsymbol{\omega}_{\boldsymbol{D}/\boldsymbol{F}}(t)$, and $\boldsymbol{e}_{\tilde{\boldsymbol{R}}}(t)$ it follows that, for all $(\boldsymbol{p}_F, \tilde{\boldsymbol{R}}) \in \Omega_{pf}$, the rate commands $q_c(t)$ and $r_c(t)$ are bounded. Then, based on the assumption made in Section 2.2.4 on the tracking capabilities of the vehicle with its autopilot, we have that, for all $(\boldsymbol{p}_F, \tilde{\boldsymbol{R}}) \in \Omega_{pf}$, the following performance bounds hold:

$$|q_c - q| \leq \gamma_q \,, \qquad |r_c - r| \leq \gamma_r \,. \tag{B.14}$$

Next, we consider again the Lyapunov function candidate

$$V_{pf}(\boldsymbol{p}_F, \tilde{\boldsymbol{R}}) = \Psi(\tilde{\boldsymbol{R}}) + \frac{1}{c_1^2}\|\boldsymbol{p}_F\|^2 \,.$$

From the dynamics (3.6), the time-derivative of V_{pf} is given by

$$\begin{aligned}\dot{V}_{pf} = \boldsymbol{e}_{\tilde{\boldsymbol{R}}} \cdot & \left(\begin{bmatrix} q \\ r \end{bmatrix} - \boldsymbol{\Pi}_{\boldsymbol{R}} \tilde{\boldsymbol{R}}^\top \left(\boldsymbol{R}_{\boldsymbol{F}}^{\boldsymbol{D}} \{\boldsymbol{\omega}_{\boldsymbol{F}/\boldsymbol{I}}\}_F + \{\boldsymbol{\omega}_{\boldsymbol{D}/\boldsymbol{F}}\}_D \right) \right) \\ & + \frac{2}{c_1^2} \boldsymbol{p}_F \cdot \left(-\dot{\ell}\,\hat{\mathbf{t}} - \boldsymbol{\omega}_{F/I} \times \boldsymbol{p}_F + v\,\hat{\mathbf{w}}_1 \right) .\end{aligned}$$

We add and subtract the term $\boldsymbol{e}_{\tilde{\boldsymbol{R}}} \cdot \left[\begin{smallmatrix} q_c \\ r_c \end{smallmatrix}\right]$ to the above equation to obtain

$$\begin{aligned}\dot{V}_{pf} = \boldsymbol{e}_{\tilde{\boldsymbol{R}}} \cdot & \left(\begin{bmatrix} q_c \\ r_c \end{bmatrix} - \boldsymbol{\Pi}_{\boldsymbol{R}} \tilde{\boldsymbol{R}}^\top \left(\boldsymbol{R}_{\boldsymbol{F}}^{\boldsymbol{D}} \{\boldsymbol{\omega}_{\boldsymbol{F}/\boldsymbol{I}}\}_F + \{\boldsymbol{\omega}_{\boldsymbol{D}/\boldsymbol{F}}\}_D \right) \right) \\ & + \frac{2}{c_1^2} \boldsymbol{p}_F \cdot \left(-\dot{\ell}\,\hat{\mathbf{t}} - \boldsymbol{\omega}_{F/I} \times \boldsymbol{p}_F + v\,\hat{\mathbf{w}}_1 \right) - \boldsymbol{e}_{\tilde{\boldsymbol{R}}} \cdot \begin{bmatrix} q_c - q \\ r_c - r \end{bmatrix} .\end{aligned}$$

Similarly to the proof of Lemma 3.1, we have that, inside the set Ω_{pf}, the following bound holds:

$$\dot{V}_{pf} \leq -2\bar{\lambda}_{pf}\left(\frac{1}{1-c^2}\,\|\boldsymbol{e}_{\tilde{\boldsymbol{R}}}\|^2 + \frac{1}{c_1^2}\,\|\boldsymbol{p}_{\boldsymbol{F}}\|^2\right) + \|\boldsymbol{e}_{\tilde{\boldsymbol{R}}}\|\left\|\begin{bmatrix} q_c - q \\ r_c - r \end{bmatrix}\right\|,$$

where $\bar{\lambda}_{pf}$ was defined in (3.12). From the performance bounds (B.14) and the definition of γ_ω in (3.14), it follows that

$$\left\|\begin{bmatrix} q_c - q \\ r_c - r \end{bmatrix}\right\| \leq \gamma_\omega\,,$$

yielding

$$\dot{V}_{pf} \leq -2\bar{\lambda}_{pf}\left(\frac{1}{1-c^2}\,\|\boldsymbol{e}_{\tilde{\boldsymbol{R}}}\|^2 + \frac{1}{c_1^2}\,\|\boldsymbol{p}_{\boldsymbol{F}}\|^2\right) + \|\boldsymbol{e}_{\tilde{\boldsymbol{R}}}\|\gamma_\omega\,.$$

We now rewrite the above inequality as

$$\begin{aligned}\dot{V}_{pf} \leq\; & -2\bar{\lambda}_{pf}(1-\delta_\lambda)\left(\frac{1}{1-c^2}\,\|\boldsymbol{e}_{\tilde{\boldsymbol{R}}}\|^2 + \frac{1}{c_1^2}\,\|\boldsymbol{p}_{\boldsymbol{F}}\|^2\right) \\ & - 2\bar{\lambda}_{pf}\delta_\lambda\left(\frac{1}{1-c^2}\,\|\boldsymbol{e}_{\tilde{\boldsymbol{R}}}\|^2 + \frac{1}{c_1^2}\,\|\boldsymbol{p}_{\boldsymbol{F}}\|^2\right) + \|\boldsymbol{e}_{\tilde{\boldsymbol{R}}}\|\gamma_\omega\,,\end{aligned}$$

where $0 < \delta_\lambda < 1$. Then, for all $\boldsymbol{p}_{\boldsymbol{F}}(t)$ and $\boldsymbol{e}_{\tilde{\boldsymbol{R}}}(t)$ satisfying

$$-2\bar{\lambda}_{pf}\delta_\lambda\left(\frac{1}{1-c^2}\,\|\boldsymbol{e}_{\tilde{\boldsymbol{R}}}\|^2 + \frac{1}{c_1^2}\,\|\boldsymbol{p}_{\boldsymbol{F}}\|^2\right) + \|\boldsymbol{e}_{\tilde{\boldsymbol{R}}}\|\gamma_\omega \leq 0\,, \tag{B.15}$$

we have that

$$\dot{V}_{pf} \leq -2\bar{\lambda}_{pf}(1-\delta_\lambda)\left(\frac{1}{1-c^2}\,\|\boldsymbol{e}_{\tilde{\boldsymbol{R}}}\|^2 + \frac{1}{c_1^2}\,\|\boldsymbol{p}_{\boldsymbol{F}}\|^2\right) \leq -2\bar{\lambda}_{pf}(1-\delta_\lambda)V_{pf}\,.$$

Inequality (B.15) is satisfied outside the bounded set D defined by

$$\begin{aligned}D := \Bigg\{(\boldsymbol{p}_{\boldsymbol{F}}, \tilde{\boldsymbol{R}}) \in \mathbb{R}^3 \times \mathsf{SO(3)} \;\mid\; & \frac{1}{1-c^2}\left(\|\boldsymbol{e}_{\tilde{\boldsymbol{R}}}\| - \frac{(1-c^2)\,\gamma_\omega}{4\bar{\lambda}_{pf}\delta_\lambda}\right)^2 + \frac{1}{c_1^2}\|\boldsymbol{p}_{\boldsymbol{F}}\|^2 \\ & < \frac{(1-c^2)\,\gamma_\omega^2}{16\bar{\lambda}_{pf}^2\delta_\lambda^2}\Bigg\}.\end{aligned}$$

The set D is in the interior of the compact set F given by

$$F := \left\{(\boldsymbol{p}_{\boldsymbol{F}}, \tilde{\boldsymbol{R}}) \in \mathbb{R}^3 \times \mathsf{SO(3)} \;\mid\; \frac{1}{1-c^2}\,\|\boldsymbol{e}_{\tilde{\boldsymbol{R}}}\|^2 + \frac{1}{c_1^2}\|\boldsymbol{p}_{\boldsymbol{F}}\|^2 \leq \frac{(1-c^2)\,\gamma_\omega^2}{4\bar{\lambda}_{pf}^2\delta_\lambda^2}\right\},$$

which in its turn is contained in the compact set Ω_b defined by

$$\Omega_b := \left\{ (\boldsymbol{p_F}, \tilde{\boldsymbol{R}}) \in \mathbb{R}^3 \times \mathsf{SO(3)} \mid \Psi(\tilde{\boldsymbol{R}}) + \frac{1}{c_1^2}\|\boldsymbol{p_F}\|^2 \leq \frac{(1-c^2)\gamma_\omega^2}{4\bar{\lambda}_{pf}^2\delta_\lambda^2} \right\}.$$

Then, the design constraint for the performance bounds γ_q and γ_r in (3.14) implies that the set Ω_b is in the interior of the set Ω_{pf} introduced in (3.13), that is, $\Omega_b \subset \Omega_{pf}$.

Equipped with the above results and using a proof similar to that of Theorem 4.18 in [2], it can be shown that, for every initial state $(\boldsymbol{p_F}(0), \tilde{\boldsymbol{R}}(0)) \in \Omega_{pf}$, there is a time $T_b \geq 0$ such that the following bounds are satisfied:

$$V_{pf}(t) \leq V_{pf}(0)\mathrm{e}^{-2\bar{\lambda}_{pf}(1-\delta_\lambda)t}, \qquad \text{for all } 0 \leq t < T_b,$$

$$V_{pf}(t) \leq \frac{(1-c^2)\gamma_\omega^2}{4\bar{\lambda}_{pf}^2\delta_\lambda^2}, \qquad \text{for all } t \geq T_b.$$

The bounds in (3.15) follow immediately from the two bounds above and inequalities (B.6). □

B.2.5 Proof of Lemma 4.1

To prove that the origin of the closed-loop kinematic coordination error dynamics (4.9) is globally uniformly exponentially stable (GUES) under the connectivity condition (2.8), we first consider the system

$$\dot{\boldsymbol{\phi}}(t) = -k_P\bar{\boldsymbol{L}}(t)\boldsymbol{\phi}(t), \qquad \boldsymbol{\phi}(t) \in \mathbb{R}^{n-1}, \tag{B.16}$$

where k_P is the proportional coordination control gain introduced in (4.6). Letting $\boldsymbol{D}(t)$ be the time-varying incidence matrix, $\boldsymbol{L}(t) = \boldsymbol{D}(t)\,\boldsymbol{D}^\top(t)$ and we can rewrite the system above as

$$\dot{\boldsymbol{\phi}}(t) = -k_P(\boldsymbol{QD}(t))(\boldsymbol{QD}(t))^\top\boldsymbol{\phi}(t).$$

Then, since $\boldsymbol{QD}(t)$ is piecewise constant in time and, in addition, $\|\boldsymbol{QD}(t)\|^2 \leq n$, one can prove that system (B.16) is GUES and the following bound holds:

$$\|\boldsymbol{\phi}(t)\| \leq \kappa_\phi\|\boldsymbol{\phi}(0)\|\mathrm{e}^{-\lambda_{cd}^P t}$$

with $\kappa_\phi = 1$ and $\lambda_{cd}^P \geq \bar{\lambda}_{cd}^P := \frac{k_P n\mu}{(1+k_P nT)^2}$. This result can be proven along the same lines as Lemma 5 in [4] or Lemma 3 in [5]. Since $\bar{\boldsymbol{L}}(t)$ is continuous for almost all $t \geq 0$ and uniformly bounded, and system (B.16) is GUES, then Lemma 1 in [5] and a similar argument as in Theorem 4.12 in [2] imply that, for any constants $\bar{c}_3$ and $\bar{c}_4$ satisfying $0 < \bar{c}_3 \leq \bar{c}_4$, there exists a continuous, piecewise-differentiable matrix $\boldsymbol{P}_{\boldsymbol{cd_0}}(t) = \boldsymbol{P}_{\boldsymbol{cd_0}}^\top(t)$, such that

$$\bar{c}_1 \mathbb{I}_{\boldsymbol{n-1}} := \frac{\bar{c}_3}{2k_P n} \mathbb{I}_{\boldsymbol{n-1}} \leq \boldsymbol{P}_{\boldsymbol{cd_0}}(t) \leq \frac{\bar{c}_4}{2\bar{\lambda}_{cd}^P} \mathbb{I}_{\boldsymbol{n-1}} =: \bar{c}_2 \mathbb{I}_{\boldsymbol{n-1}}\,, \tag{B.17a}$$

$$\dot{\boldsymbol{P}}_{\boldsymbol{cd_0}}(t) - k_P \bar{\boldsymbol{L}}(t) \boldsymbol{P}_{\boldsymbol{cd_0}}(t) - k_P \boldsymbol{P}_{\boldsymbol{cd_0}}(t) \bar{\boldsymbol{L}}(t) \leq -\bar{c}_3 \mathbb{I}_{\boldsymbol{n-1}}\,. \tag{B.17b}$$

Next, we apply the change of variables

$$z(t) := \boldsymbol{S}_\zeta \boldsymbol{\zeta}(t)\,, \qquad \boldsymbol{S}_\zeta := \begin{bmatrix} \mathbb{I}_{\boldsymbol{n-1}} & \boldsymbol{0} \\ -\frac{k_I}{k_P} \boldsymbol{C}^\top \boldsymbol{Q}^\top & \mathbb{I}_{\boldsymbol{n-n_\ell}} \end{bmatrix}, \tag{B.18}$$

to the kinematic coordination error dynamics (4.9), which leads to

$$\dot{z}(t) = \boldsymbol{S}_\zeta \boldsymbol{A}_\zeta(t) \boldsymbol{S}_\zeta^{-1} z(t) = \begin{bmatrix} -k_P \bar{\boldsymbol{L}}(t) + \frac{k_I}{k_P} \boldsymbol{Q}\boldsymbol{C}\boldsymbol{C}^\top \boldsymbol{Q}^\top & \boldsymbol{Q}\boldsymbol{C} \\ -\frac{k_I^2}{k_P^2} \boldsymbol{C}^\top \boldsymbol{Q}^\top \boldsymbol{Q}\boldsymbol{C}\boldsymbol{C}^\top \boldsymbol{Q}^\top & -\frac{k_I}{k_P} \boldsymbol{C}^\top \boldsymbol{Q}^\top \boldsymbol{Q}\boldsymbol{C} \end{bmatrix} z(t)\,. \tag{B.19}$$

Consider now the Lyapunov function candidate

$$V_{cd}(t, z) := z^\top \boldsymbol{P}_{\boldsymbol{cd}}(t)\, z\,, \tag{B.20}$$

where $\boldsymbol{P}_{\boldsymbol{cd}}(t)$ is defined as

$$\boldsymbol{P}_{\boldsymbol{cd}}(t) := \begin{bmatrix} \boldsymbol{P}_{\boldsymbol{cd_0}}(t) & \boldsymbol{0} \\ \boldsymbol{0} & \frac{k_P^3}{k_I^3} \left(\boldsymbol{C}^\top \boldsymbol{Q}^\top \boldsymbol{Q}\boldsymbol{C}\right)^{-1} \end{bmatrix}.$$

The time derivative of V_{cd} along the trajectories of the system (B.19) is given by

$$\dot{V}_{cd}(t) =$$

$$z^\top(t) \begin{bmatrix} \dot{\boldsymbol{P}}_{\boldsymbol{cd_0}}(t) - k_P \bar{\boldsymbol{L}}(t) \boldsymbol{P}_{\boldsymbol{cd_0}}(t) - k_P \boldsymbol{P}_{\boldsymbol{cd_0}}(t) \bar{\boldsymbol{L}}(t) + \frac{k_I}{k_P}\left(\boldsymbol{Q}\boldsymbol{C}\boldsymbol{C}^\top \boldsymbol{Q}^\top \boldsymbol{P}_{\boldsymbol{cd_0}}(t) + \boldsymbol{P}_{\boldsymbol{cd_0}}(t) \boldsymbol{Q}\boldsymbol{C}\boldsymbol{C}^\top \boldsymbol{Q}^\top\right) & \left(\boldsymbol{P}_{\boldsymbol{cd_0}}(t) - \frac{k_P}{k_I} \mathbb{I}_{\boldsymbol{n-1}}\right) \boldsymbol{Q}\boldsymbol{C} \\ \boldsymbol{C}^\top \boldsymbol{Q}^\top \left(\boldsymbol{P}_{\boldsymbol{cd_0}}(t) - \frac{k_P}{k_I} \mathbb{I}_{\boldsymbol{n-1}}\right) & -2\frac{k_P^2}{k_I^2} \mathbb{I}_{\boldsymbol{n-n_\ell}} \end{bmatrix} z(t)\,.$$

Inequality (B.17b) implies that

$$\dot{V}_{cd}(t) \leq z^\top(t) \begin{bmatrix} -\bar{c}_3 \mathbb{I}_{\boldsymbol{n-1}} + \frac{k_I}{k_P}\left(\boldsymbol{Q}\boldsymbol{C}\boldsymbol{C}^\top \boldsymbol{Q}^\top \boldsymbol{P}_{\boldsymbol{cd_0}}(t) + \boldsymbol{P}_{\boldsymbol{cd_0}}(t) \boldsymbol{Q}\boldsymbol{C}\boldsymbol{C}^\top \boldsymbol{Q}^\top\right) & \left(\boldsymbol{P}_{\boldsymbol{cd_0}}(t) - \frac{k_P}{k_I} \mathbb{I}_{\boldsymbol{n-1}}\right) \boldsymbol{Q}\boldsymbol{C} \\ \boldsymbol{C}^\top \boldsymbol{Q}^\top \left(\boldsymbol{P}_{\boldsymbol{cd_0}}(t) - \frac{k_P}{k_I} \mathbb{I}_{\boldsymbol{n-1}}\right) & -2\frac{k_P^2}{k_I^2} \mathbb{I}_{\boldsymbol{n-n_\ell}} \end{bmatrix} z(t)\,.$$

Now, for any $\rho_k \geq 2$, define

$$\bar{\lambda}_{cd} := \frac{\bar{\lambda}_{cd}^P}{1 + \rho_k \frac{n}{n_\ell}}\,.$$

Then, letting

$$k_P > 0\,, \qquad k_I = k_P \bar{\lambda}_{cd} \frac{n}{n_\ell} \rho_k\,, \qquad \bar{c}_3 = \bar{c}_4 = \frac{\bar{\lambda}_{cd}^P}{\bar{\lambda}_{cd}} \frac{2n_\ell}{\rho_k n}\,, \tag{B.21}$$

and noting that $\|\boldsymbol{QC}\| \leq 1$ and $\lambda_{\min}(\boldsymbol{C}^\top \boldsymbol{Q}^\top \boldsymbol{QC}) = \frac{n_\ell}{n}$, one can use inequalities (B.17) and Schur complements to prove that the following inequality holds for all $t \geq 0$:[1]

$$\begin{bmatrix} -\bar{c}_3 \mathbb{I}_{n-1} + \frac{k_I}{k_P}\left(\boldsymbol{QCC}^\top \boldsymbol{Q}^\top \boldsymbol{P}_{cd_0}(t) + \boldsymbol{P}_{cd_0}(t)\boldsymbol{QCC}^\top \boldsymbol{Q}^\top\right) & \left(\boldsymbol{P}_{cd_0}(t) - \frac{k_P}{k_I}\mathbb{I}_{n-1}\right)\boldsymbol{QC} \\ \boldsymbol{C}^\top \boldsymbol{Q}^\top \left(\boldsymbol{P}_{cd_0}(t) - \frac{k_P}{k_I}\mathbb{I}_{n-1}\right) & -2\frac{k_P^2}{k_I^2}\mathbb{I}_{n-n_\ell} \end{bmatrix} \leq -2\bar{\lambda}_{cd} \begin{bmatrix} \bar{c}_2 \mathbb{I}_{n-1} & \boldsymbol{0} \\ \boldsymbol{0} & \frac{k_P^3}{k_I^3}\left(\boldsymbol{C}^\top \boldsymbol{Q}^\top \boldsymbol{QC}\right)^{-1} \end{bmatrix}. \tag{B.22}$$

Then, for the choice of parameters in (B.21), inequality (B.22) implies that

$$\dot{V}_{cd}(t) \leq -2\bar{\lambda}_{cd}\, z^\top(t) \begin{bmatrix} \boldsymbol{P}_{cd_0}(t) & \boldsymbol{0} \\ \boldsymbol{0} & \frac{k_P^3}{k_I^3}\left(\boldsymbol{C}^\top \boldsymbol{Q}^\top \boldsymbol{QC}\right)^{-1} \end{bmatrix} z(t) = -2\bar{\lambda}_{cd} V_{cd}(t).$$

Application of the comparison lemma (see [2, Lemma 3.4]) yields

$$V_{cd}(t) \leq V_{cd}(0)\mathrm{e}^{-2\bar{\lambda}_{cd}t},$$

and, since

$$\min\left\{\bar{c}_1, \frac{k_P^3}{k_I^3}\right\} \|z(t)\|^2 \leq V_{cd}(t) \leq \max\left\{\bar{c}_2, \frac{k_P^3}{k_I^3}\frac{n}{n_\ell}\right\} \|z(t)\|^2,$$

it follows that

$$\|z(t)\| \leq \left(\frac{\max\left\{\bar{c}_2, \frac{k_P^3}{k_I^3}\frac{n}{n_\ell}\right\}}{\min\left\{\bar{c}_1, \frac{k_P^3}{k_I^3}\right\}} \right)^{\frac{1}{2}} \|z(0)\|\mathrm{e}^{-\bar{\lambda}_{cd}t}.$$

The similarity transformation in (B.18) implies that

$$\|\boldsymbol{\zeta}(t)\| \leq \kappa_{\zeta 0}\|\boldsymbol{\zeta}(0)\|\mathrm{e}^{-\bar{\lambda}_{cd}t}, \qquad \kappa_{\zeta 0} := \|\boldsymbol{S}_\zeta^{-1}\| \left(\frac{\max\left\{\bar{c}_2, \frac{k_P^3}{k_I^3}\frac{n}{n_\ell}\right\}}{\min\left\{\bar{c}_1, \frac{k_P^3}{k_I^3}\right\}} \right)^{\frac{1}{2}} \|\boldsymbol{S}_\zeta\|, \tag{B.23}$$

and, consequently, system (4.9) is GUES with (guaranteed) rate of convergence $\bar{\lambda}_{cd}$.

[1] The proof of this result can be found at the end of this section.

To prove the bounds in (4.12) and (4.13), we first note that from the kinematic equations (4.7) and the definition of $\boldsymbol{\zeta}_1(t)$ and $\boldsymbol{\zeta}_2(t)$ in (4.8) it follows that

$$\boldsymbol{Q}^\top \boldsymbol{\zeta}_1(t) = \boldsymbol{Q}^\top \boldsymbol{Q}\boldsymbol{\xi}(t) = \boldsymbol{\Pi}_{\xi}\boldsymbol{\xi}(t) = \boldsymbol{\xi}(t) - \left(\frac{1}{n}\mathbf{1}_n^\top \boldsymbol{\xi}(t)\right)\mathbf{1}_n\,, \tag{B.24}$$

$$\dot{\boldsymbol{\xi}}(t) - \mathbf{1}_n = -k_P \boldsymbol{L}(t)\boldsymbol{Q}^\top \boldsymbol{\zeta}_1(t) + \begin{bmatrix} \mathbf{0} \\ \zeta_2(t) \end{bmatrix}. \tag{B.25}$$

Partitioning the matrix $\boldsymbol{Q}$ as $\boldsymbol{Q} = [\boldsymbol{q}_1, \ldots, \boldsymbol{q}_n]$, $\boldsymbol{q}_i \in \mathbb{R}^{n-1}$, equality (B.24) above yields

$$\xi_i(t) - \xi_j(t) = \boldsymbol{q}_i^\top \boldsymbol{\zeta}_1(t) - \boldsymbol{q}_j^\top \boldsymbol{\zeta}_1(t)\,.$$

Then, recalling that $\boldsymbol{Q}^\top \boldsymbol{Q} = \boldsymbol{\Pi}_{\xi}$, which implies that $\|\boldsymbol{q}_i\|^2 = 1 - \frac{1}{n}$, one obtains the following bound:

$$|\xi_i(t) - \xi_j(t)| \leq 2\left(1 - \frac{1}{n}\right)^{\frac{1}{2}} \|\boldsymbol{\zeta}_1(t)\| \leq 2\left(1 - \frac{1}{n}\right)^{\frac{1}{2}} \|\boldsymbol{\zeta}(t)\|\,, \qquad i, j = 1, \ldots, n\,. \tag{B.26}$$

Equality (B.25) leads to

$$|\dot{\xi}_i(t) - 1| \leq k_P \|\boldsymbol{L}(t)\| \|\boldsymbol{Q}^\top\| \|\boldsymbol{\zeta}_1(t)\| + \|\boldsymbol{\zeta}_2(t)\|\,,$$

and recalling that $\lambda_{\max}(\boldsymbol{L}(t)) \leq n$ and $\boldsymbol{Q}\boldsymbol{Q}^\top = \mathbb{I}_{n-1}$, it follows that

$$|\dot{\xi}_i(t) - 1| \leq (k_P n + 1)\, \|\boldsymbol{\zeta}(t)\|\,. \tag{B.27}$$

Inequalities (B.23), (B.26), and (B.27) lead to the bounds in (4.12) and (4.13) with

$$\kappa_{\xi 0} = 2\left(1 - \frac{1}{n}\right)^{\frac{1}{2}} \kappa_{\zeta 0}\,, \qquad \kappa_{\dot{\xi} 0} = (k_P n + 1)\,\kappa_{\zeta 0}\,. \qquad \square$$

Proof of Inequality (B.22)

To prove inequality (B.22), we start by showing that the following inequality

$$\begin{bmatrix} \bar{c}_3 \mathbb{I}_{n-1} - \frac{k_I}{k_P}\left(\boldsymbol{Q}\boldsymbol{C}\boldsymbol{C}^\top \boldsymbol{Q}^\top \boldsymbol{P}_{cd_0}(t) + \boldsymbol{P}_{cd_0}(t)\boldsymbol{Q}\boldsymbol{C}\boldsymbol{C}^\top \boldsymbol{Q}^\top\right) & \left(\frac{k_P}{k_I}\mathbb{I}_{n-1} - \boldsymbol{P}_{cd_0}(t)\right)\boldsymbol{Q}\boldsymbol{C} \\ \boldsymbol{C}^\top \boldsymbol{Q}^\top \left(\frac{k_P}{k_I}\mathbb{I}_{n-1} - \boldsymbol{P}_{cd_0}(t)\right) & 2\frac{k_P^2}{k_I^2}\mathbb{I}_{n-n_\ell} \end{bmatrix} \geq 2\bar{\lambda}_{cd} \begin{bmatrix} \bar{c}_2 \mathbb{I}_{n-1} & \mathbf{0} \\ \mathbf{0} & \frac{k_P^3}{k_I^3}\frac{n}{n_\ell}\mathbb{I}_{n-n_\ell} \end{bmatrix}, \tag{B.28}$$

or equivalently

$$\begin{bmatrix} (\bar{c}_3-2\bar{\lambda}_{cd}\bar{c}_2)\mathbb{I}_{n-1}-\frac{k_I}{k_P}\left(QCC^\top Q^\top P_{cd_0}(t)+P_{cd_0}(t)QCC^\top Q^\top\right) & \left(\frac{k_P}{k_I}\mathbb{I}_{n-1}-P_{cd_0}(t)\right)QC \\ C^\top Q^\top\left(\frac{k_P}{k_I}\mathbb{I}_{n-1}-P_{cd_0}(t)\right) & 2\frac{k_P^2}{k_I^2}\left(1-\bar{\lambda}_{cd}\frac{k_P}{k_I}\frac{n}{n_\ell}\right)\mathbb{I}_{n-n_\ell} \end{bmatrix} \geq \mathbf{0}, \tag{B.29}$$

holds for all $t \geq 0$. To this end, we note that Schur complements can be used to prove that inequality (B.29) holds for all $t \geq 0$ if and only if the following set of inequalities also holds for all $t \geq 0$:

$$\left(\bar{c}_3 - 2\bar{\lambda}_{cd}\bar{c}_2\right)\mathbb{I}_{n-1} - \frac{k_I}{k_P}\left(QCC^\top Q^\top P_{cd_0}(t) + P_{cd_0}(t)QCC^\top Q^\top\right) \geq \mathbf{0}, \tag{B.30a}$$

$$1 - \bar{\lambda}_{cd}\frac{k_P}{k_I}\frac{n}{n_\ell} > 0, \tag{B.30b}$$

$$\begin{aligned} &\left(\bar{c}_3 - 2\bar{\lambda}_{cd}\bar{c}_2\right)\mathbb{I}_{n-1} - \frac{k_I}{k_P}\left(QCC^\top Q^\top P_{cd_0}(t) + P_{cd_0}(t)QCC^\top Q^\top\right) \\ &\quad - \left(\frac{k_P}{k_I}\mathbb{I}_{n-1} - P_{cd_0}(t)\right)QC\frac{1}{2}\frac{k_I^2}{k_P^2}\left(1 - \bar{\lambda}_{cd}\frac{k_P}{k_I}\frac{n}{n_\ell}\right)^{-1} \\ &\quad \times C^\top Q^\top\left(\frac{k_P}{k_I}\mathbb{I}_{n-1} - P_{cd_0}(t)\right) \geq \mathbf{0}. \end{aligned} \tag{B.30c}$$

The last inequality above can be rewritten as

$$\begin{aligned} &\left(\bar{c}_3 - 2\bar{\lambda}_{cd}\bar{c}_2\right)\mathbb{I}_{n-1} - \frac{k_I}{k_P}\left(1 - \frac{\alpha}{2}\right)\left(QCC^\top Q^\top P_{cd_0}(t) + P_{cd_0}(t)QCC^\top Q^\top\right) \\ &\quad - \frac{\alpha}{2}QCC^\top Q^\top - \frac{k_I^2}{k_P^2}\frac{\alpha}{2}P_{cd_0}(t)QCC^\top Q^\top P_{cd_0}(t) \geq \mathbf{0}, \end{aligned}$$

where we have defined

$$\alpha := \left(1 - \bar{\lambda}_{cd}\frac{k_P}{k_I}\frac{n}{n_\ell}\right)^{-1}.$$

Recalling now that $P_{cd_0}(t) \leq \bar{c}_2\mathbb{I}_{n-1}$, $\bar{c}_2 = \frac{\bar{c}_4}{2\bar{\lambda}_{cd}^P}$, and $\|QC\| \leq 1$, it is simple to show that inequalities (B.30) hold for any k_P, k_I, $\bar{\lambda}_{cd}$, $\bar{c}_3$, and $\bar{c}_4$ satisfying

$$\left(\bar{c}_3 - \frac{\bar{\lambda}_{cd}}{\bar{\lambda}_{cd}^P}\bar{c}_4\right) - \frac{k_I}{k_P}\frac{\bar{c}_4}{\bar{\lambda}_{cd}^P} \geq 0, \tag{B.31a}$$

$$\alpha > 0, \tag{B.31b}$$

$$2 - \alpha \geq 0\,, \tag{B.31c}$$

$$\left(\bar{c}_3 - \frac{\bar{\lambda}_{cd}}{\bar{\lambda}^P_{cd}}\bar{c}_4\right) - \frac{k_I}{k_P}\left(1 - \frac{\alpha}{2}\right)\frac{\bar{c}_4}{\bar{\lambda}^P_{cd}} - \frac{\alpha}{2} - \frac{k_I^2}{k_P^2}\frac{\alpha}{8}\left(\frac{\bar{c}_4}{\bar{\lambda}^P_{cd}}\right)^2 \geq 0\,, \tag{B.31d}$$

$$\bar{c}_4 - \bar{c}_3 \geq 0\,, \tag{B.31e}$$

where the last inequality is needed to ensure existence of $\boldsymbol{P_{cd_0}}(t)$ satisfying (B.17).

Next, we let $X := \frac{\bar{\lambda}_{cd}}{\bar{\lambda}^P_{cd}}\bar{c}_4$ and $Y := \frac{k_I}{k_P}\frac{1}{\bar{\lambda}_{cd}}$, and note that

$$XY = \frac{k_I}{k_P}\frac{\bar{c}_4}{\bar{\lambda}^P_{cd}}\,, \qquad \alpha = \frac{Yn_\ell}{Yn_\ell - n}\,.$$

Then, inequalities (B.31) can be rewritten as

$$(\bar{c}_3 - X) - XY \geq 0\,, \tag{B.32a}$$

$$Y - \frac{n}{n_\ell} > 0\,, \tag{B.32b}$$

$$Y - 2\frac{n}{n_\ell} \geq 0\,, \tag{B.32c}$$

$$(\bar{c}_3 - X) - \left(1 - \frac{\alpha}{2}\right)XY - \frac{\alpha}{2} - \frac{\alpha}{8}(XY)^2 \geq 0\,, \tag{B.32d}$$

$$\frac{\bar{\lambda}^P_{cd}}{\bar{\lambda}_{cd}}X - \bar{c}_3 \geq 0\,. \tag{B.32e}$$

Clearly, inequality (B.32c) is more strict than inequality (B.32b). It can also be proven that inequality (B.32d) is more strict than inequality (B.32a). Therefore, inequalities (B.32) can be simplified to

$$Y - 2\frac{n}{n_\ell} \geq 0\,, \tag{B.33a}$$

$$(\bar{c}_3 - X) - \left(1 - \frac{\alpha}{2}\right)XY - \frac{\alpha}{2} - \frac{\alpha}{8}(XY)^2 \geq 0\,, \tag{B.33b}$$

$$\frac{\bar{\lambda}^P_{cd}}{\bar{\lambda}_{cd}}X - \bar{c}_3 \geq 0\,. \tag{B.33c}$$

In particular, we note that the above inequalities have a solution if and only if

$$2\frac{n}{n_\ell} \leq Y \leq \frac{\bar{\lambda}^P_{cd}}{\bar{\lambda}_{cd}} - 1\,.$$

Let now ρ_k be a positive constant satisfying $\rho_k \geq 2$, and set

$$Y = \rho_k\frac{n}{n_\ell}\,, \qquad X = \frac{2}{Y}\,, \qquad \bar{\lambda}_{cd} = \frac{\bar{\lambda}^P_{cd}}{1 + Y}\,, \qquad \bar{c}_3 = \frac{\bar{\lambda}^P_{cd}}{\bar{\lambda}_{cd}}X\,.$$

It is straightforward to verify that inequalities (B.33) hold for this particular choice of parameters X, Y, $\bar{\lambda}_{cd}$, and $\bar{c}_3$. This implies that inequality (B.28) holds for all $t \geq 0$ if k_P, k_I, $\bar{\lambda}_{cd}$, $\bar{c}_3$, and $\bar{c}_4$ are set to satisfy

$$\bar{\lambda}_{cd} = \frac{\bar{\lambda}^p_{cd}}{1 + \rho_k \frac{n}{n_\ell}}, \qquad \frac{k_I}{k_P} = \bar{\lambda}_{cd} \frac{n}{n_\ell} \rho_k, \qquad \bar{c}_3 = \bar{c}_4 = \frac{\bar{\lambda}^p_{cd}}{\bar{\lambda}_{cd}} \frac{2n_\ell}{\rho_k n}.$$

Finally, we note that $\left(\boldsymbol{C}^\top \boldsymbol{Q}^\top \boldsymbol{Q}\boldsymbol{C}\right)^{-1} \leq \frac{n}{n_\ell} \mathbb{I}_{n-n_\ell}$, which leads to

$$\begin{bmatrix} \bar{c}_2 \mathbb{I}_{n-1} & \mathbf{0} \\ \mathbf{0} & \frac{k_P^3}{k_I^3} \left(\boldsymbol{C}^\top \boldsymbol{Q}^\top \boldsymbol{Q}\boldsymbol{C}\right)^{-1} \end{bmatrix} \leq \begin{bmatrix} \bar{c}_2 \mathbb{I}_{n-1} & \mathbf{0} \\ \mathbf{0} & \frac{k_P^3}{k_I^3} \frac{n}{n_\ell} \mathbb{I}_{n-n_\ell} \end{bmatrix}.$$

The inequality above, along with (B.28), imply that inequality (B.22) holds for all $t \geq 0$. □

B.2.6 Proof of Lemma 4.2

Input-to-state stability (ISS) can be proven along the same lines as Lemma 4.6 in [2]. In fact, we can conclude that system (4.14) is ISS because it is a linear system, the Laplacian $\boldsymbol{L}(t)$ is bounded, the unforced system has a globally exponentially stable equilibrium point at the origin (see Lemma 4.1), and the speed tracking error vector $\boldsymbol{e}_v(t)$ is assumed to be piecewise continuous in t and bounded for all $t \geq 0$. The constants $\kappa_{\zeta 0}$ and $\kappa_{\zeta 1}$ in (4.15) can be derived using a proof similar to that of Lemma 4.6 in [2], and are given by

$$\kappa_{\zeta 0} = \|\boldsymbol{S}_\zeta^{-1}\| \left(\frac{\max\left\{\bar{c}_2, \frac{k_P^3}{k_I^3} \frac{n}{n_\ell}\right\}}{\min\left\{\bar{c}_1, \frac{k_P^3}{k_I^3}\right\}} \right)^{\frac{1}{2}} \|\boldsymbol{S}_\zeta\|,$$

$$\kappa_{\zeta 1} = \frac{1}{v_{d\,\min}} \|\boldsymbol{S}_\zeta^{-1}\| \left(\frac{\max\left\{\bar{c}_2, \frac{k_P^3}{k_I^3} \frac{n}{n_\ell}\right\}}{\min\left\{\bar{c}_1, \frac{k_P^3}{k_I^3}\right\}} \right)^{\frac{3}{2}} \frac{\|\boldsymbol{S}_\zeta\|}{\bar{\lambda}_{cd} \theta_\lambda},$$

where $v_{d\,\min}$ was introduced in (2.3).

Finally, the bounds in (4.16) and (4.17) follow from the bound in (4.15) and the following inequalities:

$$|\xi_i(t) - \xi_j(t)| \leq 2\left(1 - \frac{1}{n}\right)^{\frac{1}{2}} \|\boldsymbol{\zeta}(t)\|, \qquad i, j = 1, \ldots, n, \tag{B.34}$$

$$|\dot{\xi}_i(t) - 1| \leq (k_P n + 1) \|\boldsymbol{\zeta}(t)\| + \frac{1}{v_{d\,\min}} \|\boldsymbol{e}_v(t)\|, \qquad i = 1, \ldots, n. \tag{B.35}$$

□

B.2.7 Proof of Theorem 4.1

We prove the claims of the theorem by contradiction. Consider one of the vehicles involved in the mission and assume that, at some time, it is not able to remain inside the pre-specified tube centered at its desired path. We assume that all other vehicles satisfy the claims of the theorem. In addition, we assume that no vehicle has yet reached its final destination. In what follows, we establish the validity of the theorem by showing that the hypotheses above imply a contradiction.

More precisely, consider the ith vehicle and assume that, at a given time $t > 0$, its path-following error is such that $(\boldsymbol{p}_{F,i}(t), \tilde{\boldsymbol{R}}_i(t)) \notin \Omega_{pf}$. For all other vehicles j, $j = 1, \ldots, n$, $j \neq i$, we assume that $(\boldsymbol{p}_{F,j}(\tau), \tilde{\boldsymbol{R}}_j(\tau)) \in \Omega_{pf}$ for all $\tau \in [0, t]$. Next, for the ith vehicle, consider the path-following Lyapunov function candidate:

$$V_{pf,i}(\boldsymbol{p}_{F,i}, \tilde{\boldsymbol{R}}_i) = \Psi(\tilde{\boldsymbol{R}}_i) + \frac{1}{c_1^2}\|\boldsymbol{p}_{F,i}\|^2 .$$

Since $(\boldsymbol{p}_{F,i}(0), \tilde{\boldsymbol{R}}_i(0)) \in \Omega_{pf}$ by assumption, and $V_{pf,i}$ evaluated along the system trajectories is continuous and differentiable, we have that, if $(\boldsymbol{p}_{F,i}(t), \tilde{\boldsymbol{R}}_i(t)) \notin \Omega_{pf}$ for some $t > 0$, then there exists a time t', $0 \leq t' < t$, such that

$$V_{pf,i}(t') = c^2 , \tag{B.36}$$

$$\dot{V}_{pf,i}(t') > 0 , \tag{B.37}$$

while

$$V_{pf,i}(\tau) \leq c^2 , \qquad \text{for all } \tau \in [0, t') . \tag{B.38}$$

Equality (B.36) and the bound in (B.38) imply that the following inequalities hold for all $\tau \in [0, t']$:

$$\|\boldsymbol{p}_{F,i}(\tau)\| \leq cc_1 , \qquad \Psi(\tilde{\boldsymbol{R}}_i(\tau)) \leq c^2 . \tag{B.39}$$

These two bounds, along with the choice for the characteristic distance d in (4.18), yield

$$\hat{\mathbf{w}}_{1,i}(\tau) \cdot \hat{\mathbf{t}}_i(\tau) \geq \frac{(1 - 2c^2)d - 2c(1 - c^2)^{\frac{1}{2}} cc_1}{(d^2 + (cc_1)^2)^{\frac{1}{2}}} =: c_2 > 0 , \qquad \text{for all } \tau \in [0, t'] . \tag{B.40}$$

The quantity $\frac{1}{\hat{\mathbf{w}}_{1,i}(\tau)\cdot\hat{\mathbf{t}}_i(\tau)}$ is thus well defined for all $\tau \in [0, t']$, which implies that the speed command $v_{c,i}(\tau)$ in (4.4) is also well defined for all $\tau \in [0, t']$. It follows from Lemma 4.1 that

$$\|\boldsymbol{\zeta}(\tau)\| \leq \kappa_{\zeta 0}\|\boldsymbol{\zeta}(0)\|e^{-\bar{\lambda}_{cd}\tau} , \qquad \text{for all } \tau \in [0, t'] ,$$

where $\kappa_{\zeta 0}$ was defined in (B.23). This bound, the speed command (4.4), the coordination law (4.6), inequality (B.26), and the bounds in (B.39) and (B.40) lead to

$$
\begin{aligned}
v_{d\,\min}\left(1-\kappa_1\kappa_{\zeta 0}\|\boldsymbol{\zeta}(0)\|\right)-k_\ell c c_1 \le v_{c,i}(\tau) \\
\le \frac{1}{c_2}\left(v_{d\,\max}\left(1+\kappa_1\kappa_{\zeta 0}\|\boldsymbol{\zeta}(0)\|\right)+k_\ell c c_1\right), \\
\text{for all } \tau \in [0,t'],
\end{aligned}
$$

where $\kappa_1 := 2k_P\left(\frac{(n-1)^3}{n}\right)^{\frac{1}{2}}+1$. The assumption on the initial condition in (4.21) implies that

$$
v_{\min} \le v_{c,i}(\tau) \le v_{\max}, \qquad \text{for all } \tau \in [0,t']. \tag{B.41}
$$

We can now use a proof similar to the one of Lemma 3.1 to show that, for all $\tau \in [0,t']$, $\dot{V}_{pf,i} < 0$ on the boundary of Ω_{pf}, which contradicts the claim in (B.36)-(B.37). Therefore, we have that, for all $t \ge 0$ and all $i \in \{1,\ldots,n\}$, the path-following errors $\boldsymbol{p}_{F,i}(t)$ and $\tilde{\boldsymbol{R}}_i(t)$ satisfy $(\boldsymbol{p}_{F,i}(t), \tilde{\boldsymbol{R}}_i(t)) \in \Omega_{pf}$. At the kinematic level, the bounds in (4.22) follow directly from (B.41). Then, Lemmas 3.1 and 4.1 can be used to prove exponential stability of the origin of dynamics (3.6) and (4.9), with guaranteed rates of convergence $\bar{\lambda}_{pf}$ and $\bar{\lambda}_{cd}$, respectively. □

B.2.8 Proof of Theorem 4.2

First, in order to simplify the notation in this proof, we define the positive constants $v_{c\,\min}$ and $v_{c\,\max}$ as

$$
v_{c\,\min} := v_{\min} + \gamma_v, \qquad v_{c\,\max} := v_{\max} - \gamma_v,
$$

which, as will become clear later in the proof, characterize respectively lower and upper bounds on the vehicle speed commands.

We now prove the claims of the theorem by contradiction. To this end, similarly to the proof of Theorem 4.1, we assume that one of the vehicles involved in the mission is not able to remain inside the pre-specified tube centered at its desired path. We assume that all other vehicles satisfy the claims of the theorem. In addition, we assume that no vehicle has yet reached its final destination. In what follows, we establish the validity of the theorem by showing that the hypotheses above imply a contradiction.

More precisely, consider the ith vehicle and assume that, at a given time $t > 0$, its path-following error is such that $(\boldsymbol{p}_{F,i}(t), \tilde{\boldsymbol{R}}_i(t)) \notin \Omega_{pf}$. For all other vehicles j, $j = 1,\ldots,n$, $j \ne i$, we assume that $(\boldsymbol{p}_{F,j}(\tau), \tilde{\boldsymbol{R}}_j(\tau)) \in \Omega_{pf}$ for all $\tau \in [0,t]$. Next, for the ith vehicle, consider the path-following Lyapunov function candidate

$$
V_{pf,i}(\boldsymbol{p}_{F,i}, \tilde{\boldsymbol{R}}_i) = \Psi(\tilde{\boldsymbol{R}}_i) + \frac{1}{c_1^2}\|\boldsymbol{p}_{F,i}\|^2 .
$$

Since $(\boldsymbol{p}_{F,i}(0), \tilde{\boldsymbol{R}}_i(0)) \in \Omega_{pf}$ by assumption, and $V_{pf,i}$ evaluated along the system trajectories is continuous and differentiable, we have that, if $(\boldsymbol{p}_{F,i}(t), \tilde{\boldsymbol{R}}_i(t)) \notin \Omega_{pf}$ for some $t > 0$, then there exists a time t', $0 \le t' < t$, such that

$$V_{pf,i}(t') = c^2 , \tag{B.42}$$

$$\dot{V}_{pf,i}(t') > 0 , \tag{B.43}$$

while

$$V_{pf,i}(\tau) \le c^2 , \qquad \text{for all } \tau \in [0, t') . \tag{B.44}$$

Equality (B.42) and the bound in (B.44) imply that the following inequalities hold for all $\tau \in [0, t']$:

$$\|\boldsymbol{p}_{F,i}(\tau)\| \le c c_1 , \qquad \Psi(\tilde{\boldsymbol{R}}_i(\tau)) \le c^2 . \tag{B.45}$$

These two bounds, along with the choice for the characteristic distance d in (4.18), yield

$$\hat{\mathbf{w}}_{1,i}(\tau) \cdot \hat{\mathbf{t}}_i(\tau) \ge c_2 > 0 , \qquad \text{for all } \tau \in [0, t'] . \tag{B.46}$$

The quantity $\frac{1}{\hat{\mathbf{w}}_{1,i}(\tau)\cdot\hat{\mathbf{t}}_i(\tau)}$ is thus well defined for all $\tau \in [0, t']$, which implies that the speed command $v_{c,i}(\tau)$ in (4.4) is also well defined for all $\tau \in [0, t']$.

At this point, we prove (by contradiction) that, with the assumptions made and the results derived so far, the speed commands of all n vehicles satisfy

$$v_{c\,\min} \le v_{c,j}(\tau) \le v_{c\,\max} , \qquad \text{for all } \tau \in [0, t'] , \quad \text{and all } j \in \{1, \dots, n\} .$$

To show this, let t'', $0 \le t'' \le t'$, be the first time at which one of the vehicles, say vehicle k, violates one of the bounds above. This implies that at time t'' one of the following inequalities is satisfied:

$$v_{c\,\min} > v_{c,k}(t'') , \qquad \text{or} \qquad v_{c,k}(t'') < v_{c\,\max} , \tag{B.47}$$

while

$$v_{c\,\min} \le v_{c,k}(\tau) \le v_{c\,\max} , \quad \text{for all } \tau \in [0, t'') , \tag{B.48}$$

$$v_{c\,\min} \le v_{c,j}(\tau) \le v_{c\,\max} , \quad \text{for all } \tau \in [0, t''] , \quad \text{and all } j = \{1, \dots, n\}, j \ne k . \tag{B.49}$$

The bounds in (B.48) and (B.49), along with the assumption on the vehicle dynamics in (2.10)-(2.11), yield

$$|v_{c,k}(\tau) - v_k(\tau)| \le \gamma_v \qquad \text{for all } \tau \in [0, t'') , \tag{B.50}$$

$$|v_{c,j}(\tau) - v_j(\tau)| \le \gamma_v \qquad \text{for all } \tau \in [0, t''] , \quad \text{and all } j = \{1, \dots, n\}, j \ne k , \tag{B.51}$$

which, in turn, lead to

$$v_{\min} \leq v_k(\tau) \leq v_{\max}\,, \qquad \text{for all } \tau \in [0, t'')\,, \tag{B.52}$$

$$v_{\min} \leq v_j(\tau) \leq v_{\max}\,, \qquad \text{for all } \tau \in [0, t'']\,, \quad \text{and all } j = \{1, \ldots, n\},\ j \neq k\,. \tag{B.53}$$

Continuity of $v_k(\cdot)$ and the bound in (B.52) above imply that $v_k(t'')$ is bounded. Moreover, since we have assumed that no vehicle has yet reached its final destination at time t, we have that the coordination states $\xi_j(\tau)$, $j = 1, \ldots, n$, are bounded for all $\tau \in [0, t'']$. Boundedness of $\xi_j(\tau)$ for all $\tau \in [0, t'']$ and all $j = 1, \ldots, n$ implies that, in particular, $u_{\text{coord},k}(t'')$ is bounded, which, together with inequalities (B.45) and (B.46), implies that $v_{c,k}(t'')$ is also bounded. From the boundedness of both $v_k(t'')$ and $v_{c,k}(t'')$ we can conclude that $e_{v,k}(t'')$ is bounded. A proof similar to the one of Lemma 4.2 can now be used to show that the choice of the coordination control gains k_P and k_I in (4.19) ensures that the following bound holds:

$$\|\boldsymbol{\zeta}(\tau)\| \leq \kappa_{\zeta 0}\|\boldsymbol{\zeta}(0)\| \mathrm{e}^{-\lambda_{cd}\tau} + \kappa_{\zeta 1} \sup_{s \in [0,\tau)} \|\boldsymbol{e}_v(s)\|\,, \qquad \text{for all } \tau \in [0, t'']\,,$$

which, along with the bounds in (B.50) and (B.51), leads to

$$\|\boldsymbol{\zeta}(t'')\| \leq \kappa_{\zeta 0}\|\boldsymbol{\zeta}(0)\| + \kappa_{\zeta 1}\sqrt{n}\gamma_v\,.$$

This bound, the speed command (4.4), the coordination law (4.6), inequality (B.34), the bounds in (B.45) and (B.46), and the assumption that $(\boldsymbol{p}_{F,j}(\tau), \tilde{\boldsymbol{R}}_j(\tau)) \in \Omega_{pf}$ for all $\tau \in [0, t]$ and all $j \in \{1, \ldots, n\}$, $j \neq i$, lead to

$$v_{d\,\min}\left(1 - \kappa_1\left(\kappa_{\zeta 0}\|\boldsymbol{\zeta}(0)\| + \kappa_{\zeta 1}\sqrt{n}\gamma_v\right)\right) - k_\ell c c_1 \leq v_{c,k}(t'') \leq \frac{1}{c_2}\left(v_{d\,\max}\left(1 + \kappa_1\left(\kappa_{\zeta 0}\|\boldsymbol{\zeta}(0)\| + \kappa_{\zeta 1}\sqrt{n}\gamma_v\right)\right) + k_\ell c c_1\right)\,.$$

The assumption on the initial condition in (4.27) implies that

$$v_{c\,\min} \leq v_{c,k}(t'') \leq v_{c\,\max}\,,$$

which contradicts the claim in (B.47). Therefore we have that

$$v_{c\,\min} \leq v_{c,j}(\tau) \leq v_{c\,\max}\,, \qquad \text{for all } \tau \in [0, t']\,, \quad \text{and all } j = \{1, \ldots, n\}\,.$$

Then, the assumption on the vehicle dynamics in (2.10)-(2.11) yields

$$|v_{c,j}(\tau) - v_j(\tau)| \leq \gamma_v \qquad \text{for all } \tau \in [0, t']\,, \quad \text{and all } j = \{1, \ldots, n\}\,,$$

which, in turn, leads to

$$v_{\min} \leq v_j(\tau) \leq v_{\max}\,, \qquad \text{for all } \tau \in [0, t']\,, \quad \text{and all } j = \{1, \ldots, n\}\,. \tag{B.54}$$

We can now use a proof similar to the one of Lemma 3.2 to show that, for all $\tau \in [0, t']$, $\dot{V}_{pf,i} < 0$ on the boundary of Ω_{pf}, which contradicts the claim in (B.42)-(B.43). Therefore, we have that, for all $t \geq 0$ and all $j \in \{1, \ldots, n\}$, the path-following errors $\boldsymbol{p}_{F,j}(t)$ and $\tilde{\boldsymbol{R}}_j(t)$ satisfy $(\boldsymbol{p}_{F,j}(t), \tilde{\boldsymbol{R}}_j(t)) \in \Omega_{pf}$. The bounds in (4.28) follow directly from (B.54), while the bounds in (4.29) and (4.30) can be derived from proofs similar to those of Lemmas 3.2 and 4.2. □

B.2.9 Proof of Lemma 6.1

Let $\hat{\boldsymbol{\xi}} \in \mathbb{R}^n$, and note that $\mathbf{q}(\hat{\boldsymbol{\xi}}) = \boldsymbol{k}\Delta$, for some $\boldsymbol{k} \in \mathbb{Z}^n$. Also, for any $\boldsymbol{w} \in \mathrm{K}(\mathbf{q}(\hat{\boldsymbol{\xi}}))$, we have that

$$w_i \quad \begin{cases} = \quad k_i\Delta\,, & \hat{\xi}_i \neq k_i\Delta - \frac{\Delta}{2} \\ \in \quad [(k_i-1)\Delta, k_i\Delta]\,, & \hat{\xi}_i = k_i\Delta - \frac{\Delta}{2} \end{cases},$$

where $w_i \in \mathbb{R}$, $\hat{\xi}_i \in \mathbb{R}$, $k_i \in \mathbb{Z}$ are the ith components of $\boldsymbol{w}$, $\hat{\boldsymbol{\xi}}$, and $\boldsymbol{k}$, respectively. Note that $|w_i - \hat{\xi}_i| \leq \frac{\Delta}{2}$.

To prove the result of the lemma, it is enough to show that, if the bound in (6.5) holds, then there exists no 4-tuple $(\hat{\boldsymbol{\xi}}, \boldsymbol{w}_1, \boldsymbol{w}_2, \boldsymbol{\chi}_I)$, $\hat{\boldsymbol{\xi}} \in \mathbb{R}^n$, $\boldsymbol{w}_1, \boldsymbol{w}_2 \in \mathrm{K}(\mathbf{q}(\hat{\boldsymbol{\xi}}))$, and $\boldsymbol{\chi}_I \in \mathbb{R}^{n-n_\ell}$, such that the following equality applies:

$$\mathbf{0} = \begin{bmatrix} -k_P\left(\boldsymbol{D}\hat{\boldsymbol{\xi}} - \boldsymbol{A}\boldsymbol{w}_1\right) + \begin{bmatrix} \mathbf{1}_{n_\ell} \\ \boldsymbol{\chi}_I \end{bmatrix} \\ -k_I\boldsymbol{C}^\top\left(\boldsymbol{D}\hat{\boldsymbol{\xi}} - \boldsymbol{A}\boldsymbol{w}_2\right) \end{bmatrix}. \tag{B.55}$$

To prove this, we first consider the first n rows of equality (B.55) and multiply them on the left by $\boldsymbol{C}^\top$ to obtain

$$-k_P\boldsymbol{C}^\top\left(\boldsymbol{D}\hat{\boldsymbol{\xi}} - \boldsymbol{A}\boldsymbol{w}_1\right) + \boldsymbol{\chi}_I = \mathbf{0}\,.$$

Then, noting that the last $(n - n_\ell)$ rows of (B.55) imply that

$$\boldsymbol{C}^\top\left(\boldsymbol{D}\hat{\boldsymbol{\xi}} - \boldsymbol{A}\boldsymbol{w}_2\right) = \mathbf{0}\,,$$

it follows that equality (B.55) can be satisfied only if $\boldsymbol{\chi}_I = k_P\boldsymbol{C}^\top\boldsymbol{A}\,(\boldsymbol{w}_2 - \boldsymbol{w}_1)$. This result implies that the existence of a 4-tuple $(\hat{\boldsymbol{\xi}}, \boldsymbol{w}_1, \boldsymbol{w}_2, \boldsymbol{\chi}_I)$ satisfying (B.55) is equivalent to the existence of a triple $(\hat{\boldsymbol{\xi}}, \boldsymbol{w}_1, \boldsymbol{w}_2)$, $\hat{\boldsymbol{\xi}} \in \mathbb{R}^n$ and $\boldsymbol{w}_1, \boldsymbol{w}_2 \in \mathrm{K}(\mathbf{q}(\hat{\boldsymbol{\xi}}))$, such that the equality

$$\boldsymbol{L}\boldsymbol{w}_1 - \begin{bmatrix} \frac{1}{k_P}\mathbf{1}_{n_\ell} \\ \mathbf{0} \end{bmatrix} = \boldsymbol{D}\left(\boldsymbol{w}_1 - \hat{\boldsymbol{\xi}}\right) + \begin{bmatrix} \mathbf{0} \\ \boldsymbol{C}^\top\boldsymbol{A}\,(\boldsymbol{w}_2 - \boldsymbol{w}_1) \end{bmatrix} \tag{B.56}$$

is satisfied, where $\boldsymbol{L} \in \mathbb{R}^{n\times n}$ is the Laplacian of the network topology, while $\boldsymbol{D} \in \mathbb{R}^{n\times n}$ and $\boldsymbol{A} \in \mathbb{R}^{n\times n}$ are its degree and adjacency matrices, respectively. Recall that, in this lemma, we assume that the communication topology is static, and therefore $\boldsymbol{L}$, $\boldsymbol{D}$, and $\boldsymbol{A}$ are constant matrices.

The existence of vectors $\hat{\boldsymbol{\xi}}$, $\boldsymbol{w}_1$, and $\boldsymbol{w}_2$ such that equality (B.56) holds depends on the quantizer precision. For instance, if $\left\| \frac{1}{k_P} \boldsymbol{D}^{-1} \begin{bmatrix} \mathbf{1}_{n_\ell} \\ \mathbf{0} \end{bmatrix} \right\|_\infty < \frac{\Delta}{2}$, then the vectors

$$\hat{\boldsymbol{\xi}} = k\Delta \mathbf{1}_n + \frac{1}{k_P} \boldsymbol{D}^{-1} \begin{bmatrix} \mathbf{1}_{n_\ell} \\ \mathbf{0} \end{bmatrix}, \qquad \boldsymbol{w}_1 = \boldsymbol{w}_2 = k\Delta \mathbf{1}_n ,$$

verify equality (B.56) for any $k \in \mathbb{Z}$. On the contrary, if the bound in (6.5) holds, then there exist no vectors $\hat{\boldsymbol{\xi}}$, $\boldsymbol{w}_1$, and $\boldsymbol{w}_2$ such that equality (B.56) holds. To see this, consider the scalar equality

$$\frac{1}{k_P} n_\ell = \mathbf{1}_n^\top \boldsymbol{D} (\hat{\boldsymbol{\xi}} - \boldsymbol{w}_1) + \mathbf{1}_{n-n_\ell}^\top \boldsymbol{C}^\top \boldsymbol{A} (\boldsymbol{w}_1 - \boldsymbol{w}_2) , \tag{B.57}$$

which is obtained from (B.56) by multiplying on the left by $\mathbf{1}_n^\top$. Noting that $|w_{1i} - w_{2i}| \le \Delta$, the right-hand side of this equality can be bounded as

$$\left| \mathbf{1}_n^\top \boldsymbol{D} (\hat{\boldsymbol{\xi}} - \boldsymbol{w}_1) + \mathbf{1}_{n-n_\ell}^\top \boldsymbol{C}^\top \boldsymbol{A} (\boldsymbol{w}_1 - \boldsymbol{w}_2) \right| \le (3n - 2n_\ell)(n-1) \frac{\Delta}{2} .$$

If the step size of the quantizers is bounded as in (6.5), then we have

$$\left| \mathbf{1}_n^\top \boldsymbol{D} (\hat{\boldsymbol{\xi}} - \boldsymbol{w}_1) + \mathbf{1}_{n-n_\ell}^\top \boldsymbol{C}^\top \boldsymbol{A} (\boldsymbol{w}_1 - \boldsymbol{w}_2) \right| < \frac{1}{k_P} n_\ell ,$$

which implies that no vectors $\hat{\boldsymbol{\xi}}$, $\boldsymbol{w}_1$, and $\boldsymbol{w}_2$ satisfy (B.57), and thus (B.56). This, in turn, implies that there is no 4-tuple $(\hat{\boldsymbol{\xi}}, \boldsymbol{w}_1, \boldsymbol{w}_2, \boldsymbol{\chi}_I)$ such that equality (B.55) holds, and therefore the set Θ, defined in (6.4), is empty. □

B.2.10 Proof of Proposition 6.1

Let $\boldsymbol{\eta}(t) := [\boldsymbol{\eta}_1^\top(t), \, \boldsymbol{\eta}_2^\top(t)]^\top$ be defined as

$$\begin{aligned} \boldsymbol{\eta}_1(t) &:= \boldsymbol{\xi}(t) - \hat{\boldsymbol{\xi}} , \\ \boldsymbol{\eta}_2(t) &:= \boldsymbol{\chi}_I(t) - \hat{\boldsymbol{\chi}}_I , \end{aligned}$$

where $\hat{\boldsymbol{\xi}}$ and $\hat{\boldsymbol{\chi}}_I$ characterize the "zero-speed" equilibrium points introduced in (6.7). Since by assumption $\left\| \frac{1}{k_P} \boldsymbol{D}^{-1} \begin{bmatrix} \mathbf{1}_{n_\ell} \\ \mathbf{0} \end{bmatrix} \right\|_\infty < \frac{\Delta}{2}$, it follows that $\mathbf{q}(\hat{\boldsymbol{\xi}}) = k\Delta \mathbf{1}_n$, $k \in \mathbb{Z}$. Therefore, the closed-loop kinematic coordination dynamics (6.2) can be rewritten in terms of the states $\boldsymbol{\eta}_1(t)$ and $\boldsymbol{\eta}_2(t)$ as

$$\begin{aligned} \dot{\boldsymbol{\eta}}_1(t) &= -k_P \left(\boldsymbol{D}\boldsymbol{\eta}_1(t) - \boldsymbol{A}\,\mathbf{q}(\boldsymbol{\eta}_1(t) + \hat{\boldsymbol{\xi}}) \right) + \boldsymbol{C}\boldsymbol{\eta}_2(t) - k_P \boldsymbol{A} k \Delta \mathbf{1}_n , \\ \boldsymbol{\eta}_1(0) &= \boldsymbol{\xi}(0) - \hat{\boldsymbol{\xi}} , \\ \dot{\boldsymbol{\eta}}_2(t) &= -k_I \boldsymbol{C}^\top \left(\boldsymbol{D}\boldsymbol{\eta}_1(t) - \boldsymbol{A}\,\mathbf{q}(\boldsymbol{\eta}_1(t) + \hat{\boldsymbol{\xi}}) \right) - k_I \boldsymbol{C}^\top \boldsymbol{A} k \Delta \mathbf{1}_n , \\ \boldsymbol{\eta}_2(0) &= \boldsymbol{\chi}_I(0) - \hat{\boldsymbol{\chi}}_I . \end{aligned} \tag{B.58}$$

In a sufficiently small neighborhood of the origin $(\boldsymbol{\eta}_1, \boldsymbol{\eta}_2) = (\mathbf{0}, \mathbf{0})$ the nonlinear dynamics (B.58) evolve according to the linear equation

$$\dot{\boldsymbol{\eta}} = \boldsymbol{A}_\eta \, \boldsymbol{\eta} \, , \qquad \boldsymbol{A}_\eta := \begin{bmatrix} -k_P \boldsymbol{D} & \boldsymbol{C} \\ -k_I \boldsymbol{C}^\top \boldsymbol{D} & \mathbf{0} \end{bmatrix} .$$

The characteristic polynomial of $\boldsymbol{A}_\eta$ is given by

$$p_{A_\eta}(\lambda) = \det\left(\lambda \mathbb{I}_{2n-n_\ell} - \boldsymbol{A}_\eta\right) = \prod_{i=1}^{n_\ell} (\lambda + k_P d_i) \prod_{i=n_\ell+1}^{n} (\lambda^2 + k_P d_i \lambda + k_I d_i) \, ,$$

where d_i is the ith diagonal element of the degree matrix $\boldsymbol{D}$. Since the communication graph is assumed to be connected, it follows that $1 \le d_i \le n-1$, which implies that all of the eigenvalues of $\boldsymbol{A}_\eta$ have negative real part. Therefore, the equilibrium points (6.7) are locally asymptotically stable. □

B.2.11 Proof of Theorem 6.1

Let the function $\boldsymbol{\zeta}(t) = \begin{bmatrix} \boldsymbol{Q}\boldsymbol{\xi}(t) \\ \boldsymbol{\chi}_I(t) - \mathbf{1}_{n-n_\ell} \end{bmatrix}$ be a Krasovskii solution of (6.3) on $t \in I_t \subset \mathbb{R}$, that is, let $\boldsymbol{\zeta}(t)$ be absolutely continuous and satisfy the differential inclusion [1]

$$\dot{\boldsymbol{\zeta}} - \boldsymbol{A}_\zeta(t)\boldsymbol{\zeta} \in \mathrm{K}\left(\boldsymbol{f}_{\mathbf{q}}(t)\right)$$

for almost every $t \in I_t$. Then, letting $z(t) := \boldsymbol{S}_\zeta \boldsymbol{\zeta}(t)$, where $\boldsymbol{S}_\zeta$ was defined in (B.18), we have

$$\dot{z} - \boldsymbol{S}_\zeta \boldsymbol{A}_\zeta(t) \boldsymbol{S}_\zeta^{-1} z \in \boldsymbol{S}_\zeta \, \mathrm{K}\left(\boldsymbol{f}_{\mathbf{q}}(t)\right) \qquad \text{almost everywhere in } I_t \, .$$

Consider now the same Lyapunov function candidate (B.20) as in the proof of Lemma 4.1. Then, letting $\boldsymbol{w}_1(t), \boldsymbol{w}_2(t) \in \mathrm{K}(\mathbf{q}(\boldsymbol{\xi}(t)))$ and following the steps in the proof of Lemma 4.1, we have that, for the choice of parameters in (B.21), the following inequality holds:

$$\begin{aligned} \dot{V}_{cd} &\le -2\bar{\lambda}_{cd} V_{cd} + 2z^\top \boldsymbol{P}_{cd}(t) S_\zeta \begin{bmatrix} k_P \boldsymbol{Q}\boldsymbol{A}(t)(\boldsymbol{w}_1 - \boldsymbol{\xi}) \\ k_I \boldsymbol{C}^\top \boldsymbol{A}(t)(\boldsymbol{w}_2 - \boldsymbol{\xi}) \end{bmatrix} \\ &= -2\bar{\lambda}_{cd} V_{cd} + 2z^\top \begin{bmatrix} k_P \boldsymbol{P}_{cd_0}(t) \boldsymbol{Q}\boldsymbol{A}(t) \\ -\frac{k_P^3}{k_I^2} (\boldsymbol{C}^\top \boldsymbol{Q}^\top \boldsymbol{Q}\boldsymbol{C})^{-1} \boldsymbol{C}^\top \boldsymbol{Q}^\top \boldsymbol{Q}\boldsymbol{A}(t) \end{bmatrix} (\boldsymbol{w}_1 - \boldsymbol{\xi}) \\ &\quad + 2z^\top \begin{bmatrix} \mathbf{0} \\ \frac{k_P^3}{k_I^2} (\boldsymbol{C}^\top \boldsymbol{Q}^\top \boldsymbol{Q}\boldsymbol{C})^{-1} \boldsymbol{C}^\top \boldsymbol{A}(t) \end{bmatrix} (\boldsymbol{w}_2 - \boldsymbol{\xi}) \, . \end{aligned}$$

Noting that $\|\boldsymbol{w_1}(t) - \boldsymbol{\xi}(t)\| \le \sqrt{n}\frac{\Delta}{2}$, $\|\boldsymbol{w_2}(t) - \boldsymbol{\xi}(t)\| \le \sqrt{n}\frac{\Delta}{2}$, and also that

$$\left\| \begin{bmatrix} k_P \boldsymbol{P_{cd_0}}(t) \boldsymbol{Q} \boldsymbol{A}(t) \\ -\frac{k_P^3}{k_I^2} (\boldsymbol{C}^\top \boldsymbol{Q}^\top \boldsymbol{Q} \boldsymbol{C})^{-1} \boldsymbol{C}^\top \boldsymbol{Q}^\top \boldsymbol{Q} \boldsymbol{A}(t) \end{bmatrix} \right\| \le \sqrt{2} k_P (n-1) \max \left\{ \bar{c}_2 \,, \, \frac{k_P^2}{k_I^2} \frac{n}{n_\ell} \right\} =: \sigma_{B1} \,,$$

$$\left\| \begin{bmatrix} \boldsymbol{0} \\ \frac{k_P^3}{k_I^2} (\boldsymbol{C}^\top \boldsymbol{Q}^\top \boldsymbol{Q} \boldsymbol{C})^{-1} \boldsymbol{C}^\top \boldsymbol{A}(t) \end{bmatrix} \right\| \le k_P (n-1) \frac{k_P^2}{k_I^2} \frac{n}{n_\ell} =: \sigma_{B2} \,,$$

it follows that

$$\dot{V}_{cd} \le -2\bar{\lambda}_{cd} V_{cd} + \sqrt{n} \Delta (\sigma_{B1} + \sigma_{B2}) \|z\| \,.$$

We can now rewrite the above inequality as

$$\dot{V}_{cd} \le -2\bar{\lambda}_{cd}(1 - \theta'_\lambda) V_{cd} - 2\bar{\lambda}_{cd}\theta'_\lambda V_{cd} + \sqrt{n} \Delta (\sigma_{B1} + \sigma_{B2}) \|z\| \,,$$

where $0 < \theta'_\lambda < 1$. Then, for all $z(t)$ satisfying

$$-2\bar{\lambda}_{cd}\theta'_\lambda V_{cd}(t) + \sqrt{n} \Delta (\sigma_{B1} + \sigma_{B2}) \|z(t)\| \le 0 \,, \tag{B.59}$$

we have

$$\dot{V}_{cd} \le -2\bar{\lambda}_{cd}(1 - \theta'_\lambda) V_{cd} \,.$$

Inequality (B.59) holds outside the bounded set D_Δ defined as

$$D_\Delta := \left\{ z \in \mathbb{R}^{2n - n_\ell - 1} \; : \; \|z\| \le \frac{\sqrt{n} \Delta (\sigma_{B1} + \sigma_{B2})}{2\bar{\lambda}_{cd}\theta'_\lambda \min \left\{ \bar{c}_1, \frac{k_P^3}{k_I^3} \right\}} \right\} .$$

The set D_Δ is in the interior of the compact set Ω_Δ given by

$$\Omega_\Delta := \left\{ z \in \mathbb{R}^{2n - n_\ell - 1} \, : V_{cd}(t, z) \le \frac{n(\sigma_{B1} + \sigma_{B2})^2 \max \left\{ \bar{c}_2, \frac{k_P^3}{k_I^3} \frac{n}{n_\ell} \right\}}{4\bar{\lambda}_{cd}^2 \theta_\lambda'^2 \left(\min \left\{ \bar{c}_1, \frac{k_P^3}{k_I^3} \right\} \right)^2} \Delta^2 =: \kappa_V^2 \, \Delta^2 \right\} .$$

With this result and using a proof similar to that of Theorem 4.18 in [2], it can be shown that there is a time $T_b \ge 0$ such that

$$V_{cd}(t) \le V_{cd}(0, z(0)) \mathrm{e}^{-2\bar{\lambda}_{cd}(1-\theta'_\lambda)t} \,, \qquad \text{for all } 0 \le t < T_b \,,$$
$$V_{cd}(t) \le \kappa_V^2 \Delta^2 \,, \qquad \text{for all } t \ge T_b \,.$$

Then, the following inequalities

$$\min \left\{ \bar{c}_1, \frac{k_P^3}{k_I^3} \right\} \|z(t)\|^2 \le V_{cd}(t) \le \max \left\{ \bar{c}_2, \frac{k_P^3}{k_I^3} \frac{n}{n_\ell} \right\} \|z(t)\|^2 \,,$$

along with the similarity transformation in (B.18), yield

$$\|\boldsymbol{\zeta}(t)\| \leq \kappa'_{\zeta 0}\|\boldsymbol{\zeta}(0)\| \mathrm{e}^{-\bar{\lambda}_{cd}(1-\theta'_{\lambda})t}\,, \qquad \text{for all } 0 \leq t < T_b\,, \tag{B.60a}$$

$$\|\boldsymbol{\zeta}(t)\| \leq \kappa'_{\zeta 1}\Delta\,, \qquad \text{for all } t \geq T_b\,, \tag{B.60b}$$

where $\kappa'_{\zeta 0}$ and $\kappa'_{\zeta 1}$ are given by

$$\kappa'_{\zeta 0} := \|\boldsymbol{S}_{\zeta}^{-1}\| \left(\frac{\max\left\{\bar{c}_2, \frac{k_P^3}{k_I^3}\frac{n}{n_\ell}\right\}}{\min\left\{\bar{c}_1, \frac{k_P^3}{k_I^3}\right\}} \right)^{\frac{1}{2}} \|\boldsymbol{S}_{\zeta}\|\,,$$

$$\kappa'_{\zeta 1} := \|\boldsymbol{S}_{\zeta}^{-1}\| \left(\frac{1}{\min\left\{\bar{c}_1, \frac{k_P^3}{k_I^3}\right\}} \right)^{\frac{1}{2}} \kappa_V\,.$$

The bound in (6.8) follows immediately from the bounds in (B.60). □

B.2.12 Proof of Lemma 6.2

(*i*) To prove that $\boldsymbol{\zeta}_{\mathbf{eq}} = \mathbf{0}$ is an equilibrium point of the closed-loop kinematic coordination error dynamics (6.11), it is enough to notice that, for any admissible Laplacian $\boldsymbol{L}(t)$, the following inclusion holds for the pair $(\boldsymbol{\xi}_{\mathbf{eq}}(t), \boldsymbol{\chi}_{\boldsymbol{I},\mathbf{eq}}) = ((\xi_0 + t)\mathbf{1}_{\boldsymbol{n}}, \mathbf{1}_{\boldsymbol{n}-\boldsymbol{n}_{\ell}})$, $\xi_0 \in \mathbb{R}$:

$$\begin{bmatrix} \mathbf{1}_n \\ \mathbf{0} \end{bmatrix} \in \mathrm{K}\left(\begin{bmatrix} -k_P \boldsymbol{L}(t)\,\mathbf{q}(\boldsymbol{\xi}_{\mathbf{eq}}(t)) + \begin{bmatrix} \mathbf{1}_{n_\ell} \\ \boldsymbol{\chi}_{I,\mathbf{eq}} \end{bmatrix} \\ -k_I \boldsymbol{C}^\top \boldsymbol{L}(t)\,\mathbf{q}(\boldsymbol{\xi}_{\mathbf{eq}}(t)) \end{bmatrix}\right).$$

(*ii*) To prove this second result, it is enough to show that, if the bound in (6.12) holds, then there exists no 4-tuple $(\beta(t), \boldsymbol{w}_1(t), \boldsymbol{w}_2(t), \boldsymbol{\chi}_{\boldsymbol{I}})$, with $\beta(t) \in \mathbb{R}$, $\boldsymbol{w}_1(t), \boldsymbol{w}_2(t) \in \mathrm{K}(\mathbf{q}(\boldsymbol{\xi}(t)))$, and $\boldsymbol{\chi}_{\boldsymbol{I}} \in \mathbb{R}^{n-n_\ell}$, other than $(1, \mathbf{q}((\xi_0 + t)\mathbf{1}_{\boldsymbol{n}}), \mathbf{q}((\xi_0 + t)\mathbf{1}_{\boldsymbol{n}}), \mathbf{1}_{\boldsymbol{n}-\boldsymbol{n}_{\ell}})$, $\xi_0 \in \mathbb{R}$, such that the following equality holds:

$$\begin{bmatrix} \beta(t)\mathbf{1}_n \\ \mathbf{0} \end{bmatrix} = \begin{bmatrix} -k_P \boldsymbol{L}(t)\boldsymbol{w}_1(t) + \begin{bmatrix} \mathbf{1}_{n_\ell} \\ \boldsymbol{\chi}_I \end{bmatrix} \\ -k_I \boldsymbol{C}^\top \boldsymbol{L}(t)\boldsymbol{w}_2(t) \end{bmatrix}. \tag{B.61}$$

To show the above, in what follows we analyze separately the following cases: 1. $\beta(t) \equiv 0$ and $\xi_i \neq k_i\Delta - \frac{\Delta}{2}$ for all $i \in \{1, \ldots, n\}$; 2. $\beta(t) \equiv 0$ and $\xi_i = k_i\Delta - \frac{\Delta}{2}$ for (at least) one $i \in \{1, \ldots, n\}$; and 3. $\beta(t) \not\equiv 0$.

1. $\underline{\beta(t) \equiv 0 \text{ and } \xi_i \neq k_i\Delta - \frac{\Delta}{2} \text{ for all } i \in \{1, \ldots, n\}}$: In this case, the existence of an equilibrium point for the kinematic coordination error dynamics (6.11) is equivalent to the existence of a pair $(\boldsymbol{w}, \boldsymbol{\chi}_{\boldsymbol{I}})$, with $\boldsymbol{w} \in \mathrm{K}(\mathbf{q}(\boldsymbol{\xi}))$ and $\boldsymbol{\chi}_{\boldsymbol{I}} \in \mathbb{R}^{n-n_\ell}$, such that the following equality holds:

$$\mathbf{0} = \begin{bmatrix} -k_P \boldsymbol{L}(t)\boldsymbol{w} + \begin{bmatrix} \mathbf{1}_{n_\ell} \\ \boldsymbol{\chi}_I \end{bmatrix} \\ -k_I \boldsymbol{C}^\top \boldsymbol{L}(t)\boldsymbol{w} \end{bmatrix}. \tag{B.62}$$

Following similar derivations as in the proof of Lemma 6.1, it can be shown that the existence of a pair $(\boldsymbol{w}, \boldsymbol{\chi}_I)$ satisfying (B.62) is equivalent to the existence of a vector $\boldsymbol{w}$, $\boldsymbol{w} \in \mathrm{K}(\mathbf{q}(\boldsymbol{\xi}))$, such that the following equality is satisfied:

$$\boldsymbol{L}(t)\boldsymbol{w} - \begin{bmatrix} \frac{1}{k_P}\mathbf{1}_{n_\ell} \\ \mathbf{0} \end{bmatrix} = \mathbf{0}\,. \tag{B.63}$$

Since $\boldsymbol{L}(t)\mathbf{1}_n = \mathbf{0}$, $\boldsymbol{L}(t) = \boldsymbol{L}^\top(t)$, and $\boldsymbol{L}(t) \geq \mathbf{0}$, it follows that the vector $\begin{bmatrix} \mathbf{1}_{n_\ell} \\ \mathbf{0} \end{bmatrix} \in \mathbb{R}^n$ is not in the column space of any admissible $\boldsymbol{L}(t)$. Hence, equality (B.63) does not hold for any vector $\boldsymbol{w} \in \mathbb{R}^n$.

2. $\beta(t) \equiv 0$ and $\xi_i = k_i \Delta - \frac{\Delta}{2}$ for (at least) one $i \in \{1, \ldots, n\}$: In this case, the existence of an equilibrium point for the kinematic coordination error dynamics (6.11) is equivalent to the existence of a triple $(\boldsymbol{w}_1, \boldsymbol{w}_2, \boldsymbol{\chi}_I)$, with $\boldsymbol{w}_1, \boldsymbol{w}_2 \in \mathrm{K}(\mathbf{q}(\boldsymbol{\xi}))$ and $\boldsymbol{\chi}_I \in \mathbb{R}^{n-n_\ell}$, such that the following equality holds:

$$\mathbf{0} = \begin{bmatrix} -k_P \boldsymbol{L}(t)\boldsymbol{w}_1 + \begin{bmatrix} \mathbf{1}_{n_\ell} \\ \boldsymbol{\chi}_I \end{bmatrix} \\ -k_I \boldsymbol{C}^\top \boldsymbol{L}(t)\boldsymbol{w}_2 \end{bmatrix}. \tag{B.64}$$

We first consider the first n rows of equality (B.64) and multiply them on the left by $\boldsymbol{C}^\top$ to obtain $-k_P \boldsymbol{C}^\top \boldsymbol{L}(t)\boldsymbol{w}_1 + \boldsymbol{\chi}_I = \mathbf{0}$. Then, noting that the last $(n - n_\ell)$ rows of (B.64) imply that $\boldsymbol{C}^\top \boldsymbol{L}(t)\boldsymbol{w}_2 = \mathbf{0}$, equality (B.64) can be satisfied only if the following equality holds:

$$\boldsymbol{C}^\top \boldsymbol{L}(t)\,(\boldsymbol{w}_1 - \boldsymbol{w}_2) - \frac{1}{k_P}\,\boldsymbol{\chi}_I = \mathbf{0}\,. \tag{B.65}$$

We can now multiply equality (B.65) on the left by $\mathbf{1}_{n-n_\ell}^\top$ to obtain

$$\mathbf{1}_{n-n_\ell}^\top \boldsymbol{C}^\top \boldsymbol{L}(t)\,(\boldsymbol{w}_1 - \boldsymbol{w}_2) - \frac{1}{k_P}\,\mathbf{1}_{n-n_\ell}^\top \boldsymbol{\chi}_I = 0 \tag{B.66}$$

and, noting that from equality (B.64) it follows that $n_\ell + \mathbf{1}_{n-n_\ell}^\top \boldsymbol{\chi}_I = \mathbf{0}$ (which has been obtained by multiplying the first n rows of (B.64) on the left by $\mathbf{1}_n^\top$), we can rewrite equality (B.66) as

$$\frac{1}{k_P}\,n_\ell = \mathbf{1}_{n-n_\ell}^\top \boldsymbol{C}^\top \boldsymbol{L}(t)\,(\boldsymbol{w}_2 - \boldsymbol{w}_1)\;. \tag{B.67}$$

The right-hand side of this equality can be bounded as

$$\left|\mathbf{1}_{n-n_\ell}^\top \boldsymbol{C}^\top \boldsymbol{L}(t)\,(\boldsymbol{w}_2 - \boldsymbol{w}_1)\right| \leq 2n_\ell(n - n_\ell)\Delta\,.$$

If the step size of the quantizers is bounded as in (6.12), then we have

$$\left|\mathbf{1}_{n-n_\ell}^\top \boldsymbol{C}^\top \boldsymbol{L}(t)\,(\boldsymbol{w}_2 - \boldsymbol{w}_1)\right| < \frac{1}{k_P}\, n_\ell\,,$$

which implies that no vectors $\boldsymbol{w}_1$ and $\boldsymbol{w}_2$ satisfy equality (B.67). This, in turn, implies that, if the bound in (6.12) is satisfied, then there is no triple $(\boldsymbol{w}_1, \boldsymbol{w}_2, \boldsymbol{\chi}_I)$ such that equality (B.64) holds.

3. $\underline{\beta(t) \not\equiv 0}$: We start by noting that, in this case, the existence of an equilibrium point for the kinematic coordination error dynamics (6.11) requires that, at any time t' between "quantization jumps", there exist a triple $(\beta(t'), \boldsymbol{w}(t'), \boldsymbol{\chi}_I)$, with $\beta(t') \in \mathbb{R}$, $\boldsymbol{w}(t') \in \mathrm{K}(\mathbf{q}(\boldsymbol{\xi}(t')))$, and $\boldsymbol{\chi}_I \in \mathbb{R}^{n-n_\ell}$, such that the following equality holds:

$$\begin{bmatrix} \beta(t')\mathbf{1}_n \\ \mathbf{0} \end{bmatrix} = \begin{bmatrix} -k_P \boldsymbol{L}(t')\boldsymbol{w}(t') + \begin{bmatrix} \mathbf{1}_{n_\ell} \\ \boldsymbol{\chi}_I \end{bmatrix} \\ -k_I \boldsymbol{C}^\top \boldsymbol{L}(t')\boldsymbol{w}(t') \end{bmatrix}. \tag{B.68}$$

Following again similar derivations as in the proof of Lemma 6.1, it can be shown that the existence of triple $(\beta(t'), \boldsymbol{w}(t'), \boldsymbol{\chi}_I)$ satisfying equality (B.68) is equivalent to the existence of a pair $(\beta(t'), \boldsymbol{w}(t'))$, with $\beta(t') \in \mathbb{R}$ and $\boldsymbol{w}(t') \in \mathrm{K}(\mathbf{q}(\boldsymbol{\xi}(t')))$, such that the following equality is satisfied:

$$\boldsymbol{L}(t')\boldsymbol{w}(t') - \begin{bmatrix} \frac{\beta(t')-1}{k_P}\mathbf{1}_{n_\ell} \\ \mathbf{0} \end{bmatrix} = \mathbf{0}\,. \tag{B.69}$$

Since $\boldsymbol{L}(t)\mathbf{1}_n = \mathbf{0}$, $\boldsymbol{L}(t) = \boldsymbol{L}^\top(t)$, and $\boldsymbol{L}(t) \geq \mathbf{0}$, it follows that the vector $\begin{bmatrix} \mathbf{1}_{n_\ell} \\ \mathbf{0} \end{bmatrix} \in \mathbb{R}^n$ is not in the column space of any admissible $\boldsymbol{L}(t)$. Hence, equality (B.69) can hold only if $\beta(t') = 1$. Moreover, if the network topology is connected at all times, then the null space of $\boldsymbol{L}(t)$ is equal to the span of $\mathbf{1}_n$ for all $t \geq 0$, which implies that, in this case, equality (B.69) can hold only if $\boldsymbol{\xi}(t') \in \mathrm{span}\{\mathbf{1}_n\}$. This implies that, between "quantization jumps", the coordination state vector $\boldsymbol{\xi}(t)$ is required to evolve continuously according to $\boldsymbol{\xi}(t) \in \mathrm{span}\{\mathbf{1}_n\}$ and $\dot{\boldsymbol{\xi}}(t) = \mathbf{1}_n$. From equality (B.68), it further follows that $\boldsymbol{\xi}(t) \in \mathrm{span}\{\mathbf{1}_n\}$ and $\dot{\boldsymbol{\xi}}(t) = \mathbf{1}_n$ can hold simultaneously only if $\boldsymbol{\chi}_I = \mathbf{1}_{n-n_\ell}$. Finally, because the term $\boldsymbol{L}(t)\boldsymbol{w}(t)$ is bounded at the "quantization jumps", the coordination state vector $\boldsymbol{\xi}(t)$ is continuous for all $t \geq 0$, implying that equality (B.68) only holds if $\beta(t) = 1$ for almost every $t \geq 0$, $\boldsymbol{\xi}(t) = (\xi_0 + t)\mathbf{1}_n$ for some $\xi_0 \in \mathbb{R}$ and all $t \geq 0$, and $\boldsymbol{\chi}_I = \mathbf{1}_{n-n_\ell}$.

We can now conclude that, if the graph $\Gamma(t)$ is connected at all times and the step size of the quantizers is bounded as in (6.12), then $\boldsymbol{\zeta}_{\mathbf{eq}} = \mathbf{0}$ is the only equilibrium point of the closed-loop kinematic coordination error dynamics (6.11). □

B.3 Proofs and Derivations in Part III

B.3.1 Proof of Lemma 9.1

Consider the vector $\hat{\mathbf{b}}_{\mathbf{3D}}$ defined as follows:

$$\hat{\mathbf{b}}_{\mathbf{3D}} = \frac{(k_p + s_p)\boldsymbol{e_p} + (k_v + s_v)\boldsymbol{e_v} + mg\hat{\mathbf{e}}_3 + m\ddot{\boldsymbol{p}}_{\boldsymbol{d}}(\gamma)}{\|(k_p + s_p)\boldsymbol{e_p} + (k_v + s_v)\boldsymbol{e_v} + mg\hat{\mathbf{e}}_3 + m\ddot{\boldsymbol{p}}_{\boldsymbol{d}}(\gamma)\|}, \tag{B.70}$$

where $k_p, k_v > 1$, and

$$s_p = \begin{cases} \text{sign}(\boldsymbol{e_p}^\top \hat{\mathbf{e}}_3), & \text{if } \; \|k_p\boldsymbol{e_p} + k_v\boldsymbol{e_v} + mg\hat{\mathbf{e}}_3 + m\ddot{\boldsymbol{p}}_{\boldsymbol{d}}(\gamma)\| = 0 \\ 0 & \text{otherwise} \end{cases},$$

$$s_v = \begin{cases} \text{sign}(\boldsymbol{e_v}^\top \hat{\mathbf{e}}_3), & \text{if } \; \|k_p\boldsymbol{e_p} + k_v\boldsymbol{e_v} + mg\hat{\mathbf{e}}_3 + m\ddot{\boldsymbol{p}}_{\boldsymbol{d}}(\gamma)\| = 0 \\ 0 & \text{otherwise} \end{cases}.$$

The vector in (B.70) is not well defined if

$$(k_p + s_p)\boldsymbol{e_p} + (k_v + s_v)\boldsymbol{e_v} + mg\hat{\mathbf{e}}_3 + m\ddot{\boldsymbol{p}}_{\boldsymbol{d}}(\gamma) = \mathbf{0}. \tag{B.71}$$

Therefore, we need to show that Eq. (B.71) is never verified. To this end, we rewrite the previous vector equation in the form of the three following scalar equations:

$$\begin{cases} ((k_p + s_p)\boldsymbol{e_p} + (k_v + s_v)\boldsymbol{e_v} + m\ddot{\boldsymbol{p}}_{\boldsymbol{d}}(\gamma))^\top \hat{\mathbf{e}}_1 = 0 \\ ((k_p + s_p)\boldsymbol{e_p} + (k_v + s_v)\boldsymbol{e_v} + m\ddot{\boldsymbol{p}}_{\boldsymbol{d}}(\gamma))^\top \hat{\mathbf{e}}_2 = 0 \\ ((k_p + s_p)\boldsymbol{e_p} + (k_v + s_v)\boldsymbol{e_v} + mg\hat{\mathbf{e}}_3 + m\ddot{\boldsymbol{p}}_{\boldsymbol{d}}(\gamma))^\top \hat{\mathbf{e}}_3 = 0. \end{cases} \tag{B.72}$$

From the definition of s_p and s_v, and depending on the value of $\|k_p\boldsymbol{e_p} + k_v\boldsymbol{e_v} + mg\hat{\mathbf{e}}_3 + m\ddot{\boldsymbol{p}}_{\boldsymbol{d}}(\gamma)\|$, the last equation can be written as follows:

$$\begin{cases} |\boldsymbol{e_p}^\top \hat{\mathbf{e}}_3| + |\boldsymbol{e_v}^\top \hat{\mathbf{e}}_3| = 0, & \text{if } \|k_p\boldsymbol{e_p} + k_v\boldsymbol{e_v} + mg\hat{\mathbf{e}}_3 + m\ddot{\boldsymbol{p}}_{\boldsymbol{d}}(\gamma)\| = 0 \\ (k_p\boldsymbol{e_p} + k_v\boldsymbol{e_v} + mg\hat{\mathbf{e}}_3 + m\ddot{\boldsymbol{p}}_{\boldsymbol{d}}(\gamma))^\top \hat{\mathbf{e}}_3 = 0, & \\ & \text{if } \|k_p\boldsymbol{e_p} + k_v\boldsymbol{e_v} + mg\hat{\mathbf{e}}_3 + m\ddot{\boldsymbol{p}}_{\boldsymbol{d}}(\gamma)\| \neq 0. \end{cases}$$

Since we assumed that $\|\ddot{\boldsymbol{p}}_{\boldsymbol{d}}(\gamma)\| < g$, the first condition above is never verified. Moreover, the second condition cannot hold without violating either the first or second equation in (B.72). This completes the proof. □

B.3.2 Proof of Lemma 9.2

We start by noticing that in Ω_{pf} the following bounds hold:

$$\Psi(\tilde{\boldsymbol{R}}) \leq c^2 < \frac{1}{2}, \tag{B.73}$$

$$\|\boldsymbol{e}_{\tilde{\boldsymbol{R}}}\|^2 \leq \Psi(\tilde{\boldsymbol{R}}) \leq \frac{1}{1-c^2}\|\boldsymbol{e}_{\tilde{\boldsymbol{R}}}\|^2, \tag{B.74}$$

$$\|\boldsymbol{e}_{\boldsymbol{p}}\| \leq e_{p\,\max}. \tag{B.75}$$

We also notice that if (B.73) is verified, then

$$\|\boldsymbol{e}_{\tilde{\boldsymbol{R}}}\| \leq c^2. \tag{B.76}$$

We now choose the Lyapunov candidate function

$$V = \frac{k_p}{2}\|\boldsymbol{e}_{\boldsymbol{p}}\|^2 + \frac{m}{2}\|\boldsymbol{e}_{\boldsymbol{v}}\|^2 + \Psi(\tilde{\boldsymbol{R}}) + c_1 \boldsymbol{e}_{\boldsymbol{p}}^\top \boldsymbol{e}_{\boldsymbol{v}}, \tag{B.77}$$

with $c_1, k_p, m > 0$ as defined in Chapter 9. Using (B.74) it can be shown that

$$\boldsymbol{x}_{pf}^\top \boldsymbol{W}_1 \boldsymbol{x}_{pf} \leq V(\boldsymbol{x}_{pf}) \leq \boldsymbol{x}_{pf}^\top \boldsymbol{W}_2 \boldsymbol{x}_{pf}, \tag{B.78}$$

where $\boldsymbol{x}_{pf} = [\boldsymbol{e}_{\boldsymbol{p}}^\top, \boldsymbol{e}_{\boldsymbol{v}}^\top, \boldsymbol{e}_{\tilde{\boldsymbol{R}}}^\top]^\top$ and

$$\boldsymbol{W}_1 = \begin{bmatrix} \frac{k_p}{2} & -\frac{c_1}{2} & 0 \\ -\frac{c_1}{2} & \frac{m}{2} & 0 \\ 0 & 0 & 1 \end{bmatrix}, \quad \boldsymbol{W}_2 = \begin{bmatrix} \frac{k_p}{2} & -\frac{c_1}{2} & 0 \\ -\frac{c_1}{2} & \frac{m}{2} & 0 \\ 0 & 0 & \frac{1}{1-c^2} \end{bmatrix}.$$

Computing the derivative of the Lyapunov function yields

$$\dot{V} = k_p \boldsymbol{e}_{\boldsymbol{p}}^\top \dot{\boldsymbol{e}}_{\boldsymbol{p}} + m \boldsymbol{e}_{\boldsymbol{v}}^\top \dot{\boldsymbol{e}}_{\boldsymbol{v}} + \boldsymbol{e}_{\tilde{\boldsymbol{R}}}^\top \tilde{\omega} + c_1 \boldsymbol{e}_{\boldsymbol{v}}^\top \dot{\boldsymbol{e}}_{\boldsymbol{p}} + c_1 \boldsymbol{e}_{\boldsymbol{p}}^\top \dot{\boldsymbol{e}}_{\boldsymbol{v}}. \tag{B.79}$$

Next, consider the expression

$$m\dot{\boldsymbol{e}}_{\boldsymbol{v}} = m\ddot{\boldsymbol{p}}_d(\gamma) + mg\hat{\mathbf{e}}_3 - T\hat{\mathbf{b}}_3. \tag{B.80}$$

Adding and subtracting the term $\frac{T\hat{\mathbf{b}}_{3D}}{\hat{\mathbf{b}}_{3D}{}^\top \hat{\mathbf{b}}_3}$ on the right hand side gives

$$m\dot{\boldsymbol{e}}_{\boldsymbol{v}} = m\ddot{\boldsymbol{p}}_d(\gamma) + mg\hat{\mathbf{e}}_3 - \frac{T\hat{\mathbf{b}}_{3D}}{\hat{\mathbf{b}}_{3D}{}^\top \hat{\mathbf{b}}_3} - \frac{T}{\hat{\mathbf{b}}_{3D}{}^\top \hat{\mathbf{b}}_3}\left((\hat{\mathbf{b}}_{3D}{}^\top \hat{\mathbf{b}}_3)\hat{\mathbf{b}}_3 - \hat{\mathbf{b}}_{3D}\right). \tag{B.81}$$

Let $\boldsymbol{X} = \frac{T}{\hat{\mathbf{b}}_{3D}{}^\top \hat{\mathbf{b}}_3}\left((\hat{\mathbf{b}}_{3D}{}^\top \hat{\mathbf{b}}_3)\hat{\mathbf{b}}_3 - \hat{\mathbf{b}}_{3D}\right)$ and notice that $\boldsymbol{X}$ is bounded as follows [3]:

$$\|\boldsymbol{X}\| < ((k_p + s_p)\|\boldsymbol{e}_{\boldsymbol{p}}\| + (k_v + s_v)\|\boldsymbol{e}_{\boldsymbol{v}}\| + 2mg)\|\boldsymbol{e}_{\tilde{\boldsymbol{R}}}\|. \tag{B.82}$$

Define

$$A = -\left((k_p + s_p)\boldsymbol{e_p} + (k_v + s_v)\boldsymbol{e_v} + mg\hat{\mathbf{e}}_3 + m\ddot{\boldsymbol{p}}_d(\gamma)\right).$$

Then, from the definitions of $\hat{\mathbf{b}}_{3D}$ and T in (9.10) and (9.16), respectively, we note that $T = -A^\top \hat{\mathbf{b}}_3$, $\hat{\mathbf{b}}_{3D} = -\frac{A}{\|A\|}$ and $-A = \|A\|\hat{\mathbf{b}}_{3D}$. Therefore,

$$\frac{T\hat{\mathbf{b}}_{3D}}{\hat{\mathbf{b}}_{3D}^\top \hat{\mathbf{b}}_3} = \frac{(-A^\top \hat{\mathbf{b}}_3)\hat{\mathbf{b}}_{3D}}{\hat{\mathbf{b}}_{3D}^\top \hat{\mathbf{b}}_3} = \|A\|\hat{\mathbf{b}}_{3D} = -A\,,$$

and Eq. (B.81) can be rewritten as

$$m\dot{\boldsymbol{e}}_v = -(k_p + s_p)\boldsymbol{e_p} - (k_v + s_v)\boldsymbol{e_v} - X\,.$$

This allows us to rewrite the derivative of the Lyapunov function as

$$\begin{aligned}\dot{V} = &\Big(k_p \boldsymbol{e_x}^\top \boldsymbol{e_v} + \boldsymbol{e_v}^\top\left(-(k_p + s_p)\boldsymbol{e_p} - (k_v + s_v)\boldsymbol{e_v} - X\right) + \frac{1}{2}\boldsymbol{e}_{\tilde{R}}^\top \tilde{\omega} + c_1 \boldsymbol{e_v}^\top \boldsymbol{e_v}\\ &+ \frac{c_1}{m}\boldsymbol{e_p}^\top\left(-(k_p + s_p)\boldsymbol{e_p} - (k_v + s_v)\boldsymbol{e_v} - X\right)\Big)\,.\end{aligned}$$

Substituting the control law for the angular rate introduced in (9.17) in the above equation, straightforward computations lead to

$$\begin{aligned}\dot{V} \le &\Bigg(-\frac{c_1(k_p + s_p)}{m}\|\boldsymbol{e_p}\|^2 - (k_v + s_v - c_1)\|\boldsymbol{e_v}\|^2 - k_{\tilde{R}}\|\boldsymbol{e}_{\tilde{R}}\|^2\\ &+ \left(\frac{c_1(k_v + s_v)}{m} + s_p\right)\|\boldsymbol{e_p}\|\|\boldsymbol{e_v}\| + \|X\|\left(\frac{c_1}{m}\|\boldsymbol{e_p}\| + \|\boldsymbol{e_v}\|\right)\Bigg)\,.\end{aligned}$$

Finally, the bounds in (B.76) and (B.82) yield

$$\begin{aligned}\dot{V} \le &-\frac{c_1(k_p - 1)}{m}(1 - c^2)\|\boldsymbol{e_p}\|^2 - ((k_v - 1)(1 - c^2) - c_1)\|\boldsymbol{e_v}\|^2\\ &- k_{\tilde{R}}\|\boldsymbol{e}_{\tilde{R}}\|^2 + \left(\frac{c_1(k_v + 1)}{m}(1 + c^2) + 1\right)\|\boldsymbol{e_p}\|\|\boldsymbol{e_v}\|\\ &+ ((k_p + 1)e_{p\,\max} + 2mg)\|\boldsymbol{e}_{\tilde{R}}\|\|\boldsymbol{e_v}\| + 2gc_1\|\boldsymbol{e}_{\tilde{R}}\|\|\boldsymbol{e_p}\|\\ = &-\boldsymbol{x}_{pf}^\top W_3 \boldsymbol{x}_{pf}\,,\end{aligned} \tag{B.83}$$

where

$$W_3 = \begin{bmatrix} \frac{c_1(k_p-1)}{m}(1-c^2) & -\frac{1}{2}\left[\frac{c_1(k_v+1)}{m}(1+c^2)+1\right] & -gc_1\\ -\frac{1}{2}\left[\frac{c_1(k_v+1)}{m}(1+c^2)+1\right] & (k_v-1)(1-c^2)-c_1 & -\frac{k_p+1}{2}e_{p\,\max} - mg\\ -gc_1 & -\frac{k_p+1}{2}e_{p\,\max} - mg & k_{\tilde{R}} \end{bmatrix}. \tag{B.84}$$

By properly choosing k_p, k_v, and $k_{\tilde{R}}$, one can show that

$$\boldsymbol{W_3} - 2\lambda_{pf}\boldsymbol{W_2} \geq 0\,, \tag{B.85}$$

where λ_{pf} and $\boldsymbol{W_2}$ are defined in (9.19) and (9.22), respectively, and the matrix $\boldsymbol{W_3}$ is defined in (B.84). A proof of the existence of k_p, k_v, and $k_{\tilde{R}}$ such that (B.85) is satisfied is given at the end of this section.

We can now bound the derivative of the Lyapunov function from above to obtain

$$\dot{V}(t) \leq -2\lambda_{pf}V(t)\,.$$

Using the Comparison Lemma yields

$$V(t) \leq V(0)e^{-2\lambda_{pf}t}\,,$$

which implies that the inequality

$$\|\boldsymbol{x_{pf}}(t)\| \leq \sqrt{\frac{\lambda_{\max}(\boldsymbol{W_2})}{\lambda_{\min}(\boldsymbol{W_1})}}\|\boldsymbol{x_{pf}}(0)\|e^{-\lambda_{pf}t} \tag{B.86}$$

is verified for any $t \geq 0$, thus completing the proof. □

Proof of Inequality (B.85)

We now prove that the following inequality holds:

$$\boldsymbol{W_3} - 2\lambda_{pf}\boldsymbol{W_2} = \begin{bmatrix} \frac{c_1(k_p-1)}{m}(1-c^2)-\lambda_{pf}k_p & -\frac{c_1(k_v+1)}{2m}(1+c^2)-\frac{1}{2}-\lambda_{pf}c_1 & -gc_1 \\ -\frac{c_1(k_v+1)}{2m}(1+c^2)-\frac{1}{2}-\lambda_{pf}c_1 & (k_v-1)(1-c^2)-c_1-\lambda_{pf}m & -\frac{k_p+1}{2}e_{p\,\max}-mg \\ -gc_1 & -\frac{k_p+1}{2}e_{p\,\max}-mg & k_{\tilde{R}}-2\frac{\lambda_{pf}}{1-c^2} \end{bmatrix} \geq 0\,. \tag{B.87}$$

We notice that since $\lambda_{pf} < c_1(1-c^2)/m$, by letting $k_v > 1 + \frac{c_1+\lambda_{pf}m}{1-c^2}$, there exists k_p such that the upper left 2×2 corner of $(\boldsymbol{W_3} - 2\lambda_{pf}\boldsymbol{W_2})$ is positive definite. Since the only element of $\boldsymbol{W_3} - 2\lambda_{pf}\boldsymbol{W_2}$ that depends on $k_{\tilde{R}}$ is its (3, 3) entry, it can be shown that there exists $k_{\tilde{R}}$ such that $\det(\boldsymbol{W_3} - 2\lambda_{pf}\boldsymbol{W_2}) > 0$. □

B.3.3 Proof of Lemma 9.3

We start by considering the Lyapunov function in (B.77), with time derivative given by

$$\dot{V} = k_p\boldsymbol{e}_{\boldsymbol{p}}^\top\dot{\boldsymbol{e}}_{\boldsymbol{p}} + m\boldsymbol{e}_{\boldsymbol{v}}^\top\dot{\boldsymbol{e}}_{\boldsymbol{v}} + \boldsymbol{e}_{\tilde{\boldsymbol{R}}}^\top\tilde{\boldsymbol{\omega}} + c_1\boldsymbol{e}_{\boldsymbol{v}}^\top\dot{\boldsymbol{e}}_{\boldsymbol{p}} + c_1\boldsymbol{e}_{\boldsymbol{p}}^\top\dot{\boldsymbol{e}}_{\boldsymbol{v}}\,. \tag{B.88}$$

At this point, notice that

$$\boldsymbol{e}_{\tilde{\boldsymbol{R}}}^{\top}\tilde{\boldsymbol{\omega}} = \boldsymbol{e}_{\tilde{\boldsymbol{R}}}^{\top}\left(\begin{bmatrix} p \\ q \\ r \end{bmatrix} - \tilde{\boldsymbol{R}}^{\top}\{\boldsymbol{\omega}_{D/I}\}_D\right) = \boldsymbol{e}_{\tilde{\boldsymbol{R}}}^{\top}\left(\begin{bmatrix} p - p_c \\ q - q_c \\ r - r_c \end{bmatrix} + \begin{bmatrix} p_c \\ q_c \\ r_c \end{bmatrix} - \tilde{\boldsymbol{R}}^{\top}\{\boldsymbol{\omega}_{D/I}\}_D\right). \tag{B.89}$$

Letting p_c, q_c, r_c be governed by the control law given in (9.17), the previous equation becomes:

$$\boldsymbol{e}_{\tilde{\boldsymbol{R}}}^{\top}\tilde{\boldsymbol{\omega}} = \boldsymbol{e}_{\tilde{\boldsymbol{R}}}^{\top}\left(\begin{bmatrix} p - p_c \\ q - q_c \\ r - r_c \end{bmatrix} - 2k_{\tilde{R}}\boldsymbol{e}_{\tilde{\boldsymbol{R}}}\right).$$

Similarly, consider the term

$$\begin{aligned} m\dot{\boldsymbol{e}}_{\boldsymbol{v}} &= m\ddot{\boldsymbol{p}}_{\boldsymbol{d}}(\gamma) + mg\hat{\mathbf{e}}_{\mathbf{3}} - T\hat{\mathbf{b}}_{\mathbf{3}} + T_c\hat{\mathbf{b}}_{\mathbf{3}} - T_c\hat{\mathbf{b}}_{\mathbf{3}} \\ &= m\ddot{\boldsymbol{p}}_{\boldsymbol{d}}(\gamma) + mg\hat{\mathbf{e}}_{\mathbf{3}} - T_c\hat{\mathbf{b}}_{\mathbf{3}} - (T - T_c)\hat{\mathbf{b}}_{\mathbf{3}} \\ &= -(k_p + s_p)\boldsymbol{e}_{\boldsymbol{p}} - (k_v + s_v)\boldsymbol{e}_{\boldsymbol{v}} - \boldsymbol{X} - (T - T_c)\hat{\mathbf{b}}_{\mathbf{3}}. \end{aligned} \tag{B.90}$$

Substituting (B.89) and (B.90) in (B.88), and after some algebraic manipulations similar to the ones used in Appendices B.3.2 and B.86, we obtain the inequality

$$\dot{V} \leq -2\lambda_{pf}V + \gamma_\omega\|\boldsymbol{e}_{\tilde{\boldsymbol{R}}}\| + \gamma_T(\|\boldsymbol{e}_{\boldsymbol{v}}\| + \frac{c_1}{m}\|\boldsymbol{e}_{\boldsymbol{p}}\|),$$

which can be rewritten as

$$\begin{aligned} \dot{V} &\leq -2\lambda_{pf}(1-\delta_\lambda)V \\ &\quad - 2\lambda_{pf}\delta_\lambda\left(\frac{k_p}{2}\|\boldsymbol{e}_{\boldsymbol{p}}\|^2 + \frac{m}{2}\|\boldsymbol{e}_{\boldsymbol{v}}\|^2 + \frac{1}{1-c^2}\|\boldsymbol{e}_{\tilde{\boldsymbol{R}}}\|^2 + c_1\boldsymbol{e}_{\boldsymbol{p}}^{\top}\boldsymbol{e}_{\boldsymbol{v}}\right) \\ &\quad + \gamma_\omega\|\boldsymbol{e}_{\tilde{\boldsymbol{R}}}\| + \gamma_T\left(\|\boldsymbol{e}_{\boldsymbol{v}}\| + \frac{c_1}{m}\|\boldsymbol{e}_{\boldsymbol{p}}\|\right), \end{aligned}$$

where δ_λ satisfies $0 < \delta_\lambda < 1$. Therefore, for all $\boldsymbol{x}_{\boldsymbol{pf}}$ such that

$$\begin{aligned} &- 2\lambda_{pf}\delta_\lambda\left(\frac{k_p}{2}\|\boldsymbol{e}_{\boldsymbol{p}}\|^2 + \frac{m}{2}\|\boldsymbol{e}_{\boldsymbol{v}}\|^2 + \frac{1}{1-c^2}\|\boldsymbol{e}_{\tilde{\boldsymbol{R}}}\|^2 + c_1\boldsymbol{e}_{\boldsymbol{p}}^{\top}\boldsymbol{e}_{\boldsymbol{v}}\right) \\ &\quad + \gamma_\omega\|\boldsymbol{e}_{\tilde{\boldsymbol{R}}}\| + \gamma_T\left(\|\boldsymbol{e}_{\boldsymbol{v}}\| + \frac{c_1}{m}\|\boldsymbol{e}_{\boldsymbol{p}}\|\right) \leq 0, \end{aligned} \tag{B.91}$$

the derivative of the Lyapunov function satisfies the inequality

$$\dot{V} \leq -2\lambda_{pf}(1-\delta_\lambda)V.$$

Notice that (B.91), which can be rewritten as

$$2\lambda_{pf}\delta_\lambda \left(\frac{k_p}{2}\|\boldsymbol{e_p}\| \left(\|\boldsymbol{e_p}\| - \frac{\gamma_T c_1}{k_p \lambda_{pf}\delta_\lambda} \right) + \frac{m}{2}\|\boldsymbol{e_v}\| \left(\|\boldsymbol{e_v}\| - \frac{\gamma_T}{m\lambda_{pf}\delta_\lambda} \right) \right.$$
$$\left. + \|\boldsymbol{e_{\tilde{R}}}\| \left(\frac{1}{1-c^2}\|\boldsymbol{e_{\tilde{R}}}\| - \frac{\gamma_\omega}{2\lambda_{pf}\delta_\lambda} \right) + c_1 \boldsymbol{e_p}^\top \boldsymbol{e_v} \right) \geq 0\,,$$

is satisfied outside the closed set D defined by

$$D := \left\{ (\boldsymbol{x_{pf}} \in \mathbb{R}^9 \mid \|\boldsymbol{e_p}\| \leq \frac{\gamma_T c_1}{k_x \lambda_{pf}\delta_\lambda}, \quad \|\boldsymbol{e_v}\| \leq \frac{\gamma_T}{m\lambda_{pf}\delta_\lambda}, \quad \|\boldsymbol{e_{\tilde{R}}}\| \leq \frac{\gamma_\omega(1-c^2)}{2\lambda_{pf}\delta_\lambda} \right\},$$

which is contained inside the compact set

$$\Omega_b := \left\{ (\boldsymbol{x_{pf}} \in \mathbb{R}^9 \mid \|\boldsymbol{x_{pf}}\| \leq \frac{(c_1/m+1)\gamma_T + \gamma_\omega}{\lambda_{pf}\lambda_{\min}(\boldsymbol{W_2})\delta_\lambda} \right\}.$$

Thus, the design constraints for the performance bounds γ_ω and γ_T given in (9.23) imply that the set Ω_b is contained in Ω_{pf}. Finally, using a proof similar to that in [2, Theorem 4.18], it can be shown that for any initial state $\boldsymbol{x_{pf}}(0) \in \Omega_{pf}$, there is a time $T_b \geq 0$ such that the following bounds are satisfied:

$$\|\boldsymbol{x_{pf}}(t)\| \leq k_{pf}\|\boldsymbol{x_{pf}}(0)\|\mathrm{e}^{-\lambda_{pf}(1-\delta_\lambda)t}\,, \qquad \text{for all } 0 \leq t < T_b\,,$$
$$\|\boldsymbol{x_{pf}}(t)\| \leq \rho\,, \qquad \text{for all } t \geq T_b\,,$$

with

$$k_{pf} := \sqrt{\frac{\lambda_{\max}(\boldsymbol{W_2})}{\lambda_{\min}(\boldsymbol{W_1})}}\,,$$

and

$$\rho := \sqrt{\frac{\lambda_{\max}(\boldsymbol{W_2})}{\lambda_{\min}(\boldsymbol{W_1})}} \left(\frac{(c_1/m+1)\gamma_T + \gamma_\omega}{\lambda_{pf}\lambda_{\min}(\boldsymbol{W_2})\delta_\lambda} \right),$$

which completes the proof. □

B.3.4 Proof of Theorem 10.1

Consider the system

$$\dot{\boldsymbol{\phi}}(t) = -\frac{a}{b}\bar{\boldsymbol{L}}\boldsymbol{\phi}(t)\,, \tag{B.92}$$

where the matrix $\bar{\boldsymbol{L}}(t) = \boldsymbol{Q}\boldsymbol{L}(t)\boldsymbol{Q}^\top$ satisfies the (PE)-like condition in (2.8). Then, using the result reported in [4, Lemma 5], we conclude that the system in (B.92) is globally uniformly exponentially stable, and that the bound

$$\|\boldsymbol{\phi}(t)\| \leq k_\lambda \|\boldsymbol{\phi}(0)\| e^{-\gamma_\lambda t}$$

holds, with $k_\lambda = 1$ and $\gamma_\lambda \geq \bar{\lambda}_{cd} = \frac{a}{b}\frac{n\mu}{T(1+\frac{a}{b}nT)^2}$. This, together with [4, Lemma 1] or an argument similar to the one in [2, Theorem 4.14], implies that there exists a continuously differentiable, symmetric, positive definite matrix $\boldsymbol{P}(t)$ that satisfies the inequalities

$$\begin{aligned} 0 < \bar{c}_1\mathbb{I} := \frac{\bar{c}_3}{2n}\mathbb{I} \leq \boldsymbol{P}(t) \leq \frac{\bar{c}_4}{2\gamma_\lambda}\mathbb{I} =: \bar{c}_2\mathbb{I} \\ \dot{\boldsymbol{P}} - \frac{a}{b}\bar{\boldsymbol{L}}\boldsymbol{P} - \frac{a}{b}\boldsymbol{P}\boldsymbol{L} \leq -\bar{c}_3\mathbb{I}\,. \end{aligned} \tag{B.93}$$

Next, introducing the vector

$$\boldsymbol{\chi}(t) = b\boldsymbol{\zeta}_1(t) + \boldsymbol{Q}\boldsymbol{\zeta}_2(t)\,,$$

the auxiliary time-coordination state can be defined as $\bar{\boldsymbol{x}}_{TC} := [\boldsymbol{\chi}^\top, \boldsymbol{\zeta}_2^\top]^\top$, with dynamics given by

$$\begin{cases} \dot{\boldsymbol{\chi}} = -\frac{a}{b}\bar{\boldsymbol{L}}\boldsymbol{\chi} + \frac{a}{b}\boldsymbol{Q}\boldsymbol{L}\boldsymbol{\zeta}_2 - \boldsymbol{Q}\bar{\boldsymbol{\alpha}}_{pf}(\boldsymbol{x}_{PF}) \\ \dot{\boldsymbol{\zeta}}_2 = -(b\mathbb{I} - \frac{a}{b}\boldsymbol{L})\boldsymbol{\zeta}_2 - \frac{a}{b}\boldsymbol{L}\boldsymbol{Q}^\top\boldsymbol{\chi} - \bar{\alpha}_{pf}(\boldsymbol{x}_{pf})\,. \end{cases} \tag{B.94}$$

Consider the Lyapunov candidate function

$$V = \boldsymbol{\chi}^\top\boldsymbol{P}\boldsymbol{\chi} + \frac{\beta_1}{2}\|\boldsymbol{\zeta}_2\|^2 = \bar{\boldsymbol{x}}_{TC}^\top\boldsymbol{W}\bar{\boldsymbol{x}}_{TC}\,, \tag{B.95}$$

where $\beta_1 > 0$, $\boldsymbol{P}$ was introduced above, and

$$\boldsymbol{W} = \begin{bmatrix} \boldsymbol{P} & \boldsymbol{0} \\ \boldsymbol{0} & \frac{\beta_1}{2}\mathbb{I} \end{bmatrix}.$$

Using (B.94), the time-derivative of (B.95) can be computed to yield

$$\begin{aligned} \dot{V} = \boldsymbol{\chi}^\top\boldsymbol{P}\left(-\frac{a}{b}\bar{\boldsymbol{L}}\boldsymbol{\chi} + \frac{a}{b}\boldsymbol{Q}\boldsymbol{L}\boldsymbol{\zeta}_2 - \boldsymbol{Q}\bar{\boldsymbol{\alpha}}\right) + \left(-\frac{a}{b}\boldsymbol{\chi}^\top\bar{\boldsymbol{L}} + \frac{a}{b}\boldsymbol{\zeta}_2^\top\boldsymbol{L}\boldsymbol{Q}^\top - \bar{\boldsymbol{\alpha}}^\top\boldsymbol{Q}^\top\right)\boldsymbol{P}\boldsymbol{\chi} \\ + \boldsymbol{\chi}^\top\dot{\boldsymbol{P}}\boldsymbol{\chi} + \beta_1\boldsymbol{\zeta}_2^\top\left(-\left(b\mathbb{I} - \frac{a}{b}\boldsymbol{L}\right)\boldsymbol{\zeta}_2 - \frac{a}{b}\boldsymbol{L}\boldsymbol{Q}^\top\boldsymbol{\chi} - \bar{\boldsymbol{\alpha}}\right), \end{aligned}$$

leading to

$$\begin{aligned} \dot{V} \leq \boldsymbol{\chi}^\top\left(\dot{\boldsymbol{P}} - \frac{a}{b}\boldsymbol{P}\bar{\boldsymbol{L}} - \frac{a}{b}\bar{\boldsymbol{L}}\boldsymbol{P}\right)\boldsymbol{\chi} - \beta_1\boldsymbol{\zeta}_2^\top\left(b\mathbb{I} - \frac{a}{b}\boldsymbol{L}\right)\boldsymbol{\zeta}_2 \\ + 2\frac{a}{b}n\|\boldsymbol{P}\|\|\boldsymbol{\chi}\|\|\boldsymbol{\zeta}_2\| + 2\|\boldsymbol{P}\|\|\boldsymbol{\chi}\|\|\bar{\boldsymbol{\alpha}}\| \\ + \beta_1\frac{a}{b}n\|\boldsymbol{\zeta}_2\|\|\boldsymbol{\chi}\| + \beta_1\|\boldsymbol{\zeta}_2\|\|\bar{\boldsymbol{\alpha}}\|\,, \end{aligned}$$

where we used the fact that $\|\bar{\boldsymbol{L}}\| \leq n$. Using (B.93), straightforward computations show that

$$\dot{V} \leq -\bar{c}_3\|\boldsymbol{\chi}\|^2 - \beta_1\left(b - \frac{a}{b}n\right)\|\boldsymbol{\zeta}_2\|^2 + \left(2\frac{a}{b}n\bar{c}_2 + \beta_1\frac{a}{b}n\right)\|\boldsymbol{\zeta}_2\|\|\boldsymbol{\chi}\|$$
$$+ 2\left(2\bar{c}_2\|\boldsymbol{\chi}\| + \beta_1\|\boldsymbol{\zeta}_2\|\right)\frac{v_{\max}}{v_{\min}+\delta}\|\boldsymbol{x}_{pf}\|,$$

where $v_{\max} = \max_i\{v_{i,\max}\}$ and $v_{\min} = \min_i\{v_{i,\min}\}$.

Finally, using $\bar{c}_2 = \frac{\bar{c}_4}{2\gamma_\lambda}$, letting $\bar{c}_4 = \bar{c}_3$, and choosing $\delta > v_{\max} - v_{\min}$, we obtain

$$\dot{V} \leq -\bar{c}_3\|\boldsymbol{\chi}\|^2 - \beta_1\left(b - \frac{a}{b}n\right)\|\boldsymbol{\zeta}_2\|^2 + \left(\frac{a}{b}\frac{n\bar{c}_3}{\bar{\lambda}_{cd}} + \beta_1\frac{a}{b}n\right)\|\boldsymbol{\zeta}_2\|\|\boldsymbol{\chi}\|$$
$$+ 2\left(\frac{\bar{c}_3}{\bar{\lambda}_{cd}} + \beta_1\right)\|\bar{\boldsymbol{x}}_{TC}\|\|\boldsymbol{x}_{pf}\|.$$

The above inequality can be written in matrix form as

$$\dot{V} \leq -\bar{\boldsymbol{x}}_{TC}^\top \boldsymbol{M}\bar{\boldsymbol{x}}_{TC} + 2\left(\frac{\bar{c}_3}{\bar{\lambda}_{cd}} + \beta_1\right)\|\bar{\boldsymbol{x}}_{TC}\|\|\boldsymbol{x}_{pf}\|,$$

where

$$\boldsymbol{M} = \begin{bmatrix} \bar{c}_3 & -\left(\frac{a}{b}\frac{n\bar{c}_3}{\bar{\lambda}_{cd}} + \beta_1\frac{a}{b}n\right) \\ -\left(\frac{a}{b}\frac{n\bar{c}_3}{\bar{\lambda}_{cd}} + \beta_1\frac{a}{b}n\right) & \beta_1\left(b - \frac{a}{b}n\right) \end{bmatrix}.$$

Now, for any $\delta_{\bar{\lambda}}$ satisfying $0 < \delta_{\bar{\lambda}} < 1$, define $\lambda_{cd} := \delta_{\bar{\lambda}}\bar{\lambda}_{cd}$. Then, by choosing b large enough, the following matrix inequality holds:

$$\boldsymbol{M} - 2\lambda_{cd}\boldsymbol{W} \geq \begin{bmatrix} \bar{c}_3 - \frac{\bar{c}_3\lambda_{cd}}{\bar{\lambda}_{cd}} & -\left(\frac{a}{b}\frac{n\bar{c}_3}{\bar{\lambda}_{cd}} + \beta_1\frac{a}{b}n\right) \\ -\left(\frac{a}{b}\frac{n\bar{c}_3}{\bar{\lambda}_{cd}} + \beta_1\frac{a}{b}n\right) & \beta_1\left(b - \frac{a}{b}n\right) - \beta_1\lambda_{cd} \end{bmatrix} \geq 0. \tag{B.96}$$

As a consequence, the derivative of the Lyapunov function can be bounded as

$$\dot{V} \leq -2\lambda_{cd}V + 2\left(\frac{\bar{c}_3}{\bar{\lambda}_{cd}} + \beta_1\right)\|\bar{\boldsymbol{x}}_{TC}\|\|\boldsymbol{x}_{pf}\|.$$

Using [2, Lemma 4.6], one can conclude that the system (B.94) is input-to-state stable, with input $\boldsymbol{x}_{pf}$, and the following bound holds:

$$\|\bar{\boldsymbol{x}}_{TC}(t)\| \leq \sqrt{\frac{\max(\bar{c}_2, \beta_1/2)}{\min(\bar{c}_1, \beta_1/2)}}\|\bar{\boldsymbol{x}}_{TC}(0)\|e^{-\lambda_{cd}t}$$
$$+ \sqrt{\frac{\max(\bar{c}_2, \beta_1/2)}{\min(\bar{c}_1, \beta_1/2)}}\frac{\frac{\bar{c}_3}{\bar{\lambda}_{cd}} + \beta_1}{\lambda_{cd}\min(\bar{c}_1, \beta_1/2)}\sup_{t\geq 0}(\|\boldsymbol{x}_{pf}(t)\|). \tag{B.97}$$

Finally, from the definition of $\bar{x}_{TC}$ above, we can write

$$\bar{\boldsymbol{x}}_{TC} = \boldsymbol{S}\boldsymbol{x}_{cd}\,, \quad \boldsymbol{S} = \begin{bmatrix} b\mathbb{I}_{n-1} & \boldsymbol{Q} \\ 0 & \mathbb{I}_n \end{bmatrix},$$

and therefore we may conclude that

$$\|\boldsymbol{x}_{cd}(t)\| \leq \kappa_1 \|\boldsymbol{x}_{cd}(0)\| e^{-\lambda_{cd} t} + \kappa_2 \sup_{t\geq 0}(\|\boldsymbol{x}_{pf}(t)\|)\,, \tag{B.98}$$

with

$$\kappa_1 = \|\boldsymbol{S}^{-1}\| \sqrt{\frac{\max(\bar{c}_2, \beta_1/2)}{\min(\bar{c}_1, \beta_1/2)}} \|\boldsymbol{S}\| \tag{B.99}$$

and

$$\kappa_2 = \|\boldsymbol{S}^{-1}\| \sqrt{\frac{\max(\bar{c}_2, \beta_1/2)}{\min(\bar{c}_1, \beta_1/2)}} \frac{\frac{\bar{c}_3}{\lambda_{cd}} + \beta_1}{\lambda_{cd} \min(\bar{c}_1, \beta_1/2)}\,. \tag{B.100}$$

As a last step to complete the proof, we need to demonstrate that $\dot{\gamma}_i(t)$ and $\ddot{\gamma}_i(t)$, $i \in \{1 \ldots, n\}$, satisfy the bounds given in (9.6) and (9.7). To this end, notice that

$$\ddot{\gamma}_i \leq b\|\boldsymbol{\zeta}_2\| + an\|\boldsymbol{\zeta}_1\| + \|\boldsymbol{x}_{pf}\|\,.$$

For simplicity, let $b > an$. Using the bound in (B.98), and recalling the bound on the path-following error given in Lemma 9.3, the above inequality reduces to

$$\ddot{\gamma}_i \leq (b\kappa_1 + b\kappa_2 + 1)\max(\|\boldsymbol{x}_{cd}(0)\|, \rho)\,.$$

Moreover, using the fact that

$$\|\boldsymbol{\zeta}_2(t)\| \leq \kappa_1 \|\boldsymbol{x}_{cd}(0)\| e^{-\lambda_{cd} t} + \kappa_2 \sup_{t\geq 0}(\|\boldsymbol{x}_{pf}(t)\|)\,,$$

it can be proven that

$$\dot{\gamma}_i \leq 1 + (\kappa_1 + \kappa_2)\max(\|\boldsymbol{x}_{cd}(0)\|, \rho)\,,$$
$$\dot{\gamma}_i \geq 1 - (\kappa_1 + \kappa_2)\max(\|\boldsymbol{x}_{cd}(0)\|, \rho)\,.$$

Finally, since by assumption inequality (10.7) holds, then (9.6) and (9.7) are satisfied, and it can be shown that the bound in (B.98) holds for all $t \geq 0$. □

B.3.5 Proof of Corollary 10.1

Assume that the path-following algorithm satisfies

$$\|\boldsymbol{x}_{pf}(t)\| \leq k_{pf} \|\boldsymbol{x}_{pf}(0)\| e^{-\lambda_{pf} t} \tag{B.101}$$

and rewrite the inequality (B.98) as

$$\|\boldsymbol{x}_{cd}(t)\| \leq \kappa_1 \|\boldsymbol{x}_{cd}(s)\| e^{-\lambda_{cd}(t-s)} + \kappa_2 \sup_{s \leq \tau \leq t} (\|\boldsymbol{x}_{pf}(\tau)\|) , \tag{B.102}$$

where $t \geq s \geq 0$. Making $s = t/2$ in (B.102) yields

$$\|\boldsymbol{x}_{cd}(t)\| \leq \kappa_1 \|\boldsymbol{x}_{cd}(t/2)\| e^{-\lambda_{cd}(t/2)} + \kappa_2 \sup_{t/2 \leq \tau \leq t} (\|\boldsymbol{x}_{pf}(\tau)\|) . \tag{B.103}$$

Furthermore, making $s = 0$ and replacing t by $t/2$ in (B.102) allows for the computation of the estimate of $\boldsymbol{x}_{cd}(t/2)$ as

$$\|\boldsymbol{x}_{cd}(t/2)\| \leq \kappa_1 \|\boldsymbol{x}_{cd}(0)\| e^{-\lambda_{cd}(t/2)} + \kappa_2 \sup_{0 \leq \tau \leq t/2} (\|\boldsymbol{x}_{pf}(\tau)\|) . \tag{B.104}$$

Combining (B.103) and (B.104) yields

$$\begin{aligned} \|\boldsymbol{x}_{cd}(t)\| \leq \kappa_1 e^{-\lambda_{cd} t/2} & \left(\kappa_1 \|\boldsymbol{x}_{cd}(0)\| e^{-\lambda_{cd} t/2} + \kappa_2 \sup_{0 \leq \tau \leq t/2} (\|\boldsymbol{x}_{pf}(\tau)\|) \right) \\ & + \kappa_2 \sup_{t/2 \leq \tau \leq t} (\|\boldsymbol{x}_{pf}(\tau)\|) . \end{aligned} \tag{B.105}$$

Furthermore, using (B.101) we can write the inequalities

$$\sup_{0 \leq \tau \leq t/2} (\|\boldsymbol{x}_{pf}(\tau)\|) \leq k_{pf} \|\boldsymbol{x}_{pf}(0)\| ,$$

$$\sup_{t/2 \leq \tau \leq t} (\|\boldsymbol{x}_{pf}(\tau)\|) \leq k_{pf} \|\boldsymbol{x}_{pf}(0)\| e^{-\lambda_{pf} t/2} .$$

Finally, combining (B.105) with the previous two inequalities, and letting

$$\bar{\kappa}_1 := \kappa_1^2 , \quad \bar{\kappa}_2 := (1 + \kappa_1) \kappa_2 k_{pf} , \tag{B.106}$$

we obtain

$$\|\boldsymbol{x}_{cd}(t)\| \leq \bar{\kappa}_1 \|\boldsymbol{x}_{cd}(0)\| e^{-\lambda_{cd} t} + \bar{\kappa}_2 \|\boldsymbol{x}_{pf}(0)\| e^{-\frac{\lambda_{pf} + \lambda_{cd}}{2} t} ,$$

thus proving Corollary 10.1. □

References

[1] O. Hájek, Discontinuous differential equations, I, Journal of Differential Equations 32 (1979) 149–170.
[2] H.K. Khalil, Nonlinear Systems, Prentice Hall, Englewood Cliffs, NJ, 2002.
[3] T. Lee, M. Leok, N.H. McClamroch, Control of complex maneuvers for a quadrotor UAV using geometric methods on SE(3). Available online arXiv:1003.2005v4, 2011.

[4] A. Loría, E. Panteley, Uniform exponential stability of linear time-varying systems: revisited, Systems & Control Letters 47 (2002) 13–24.
[5] E. Panteley, A. Loría, Uniform exponential stability for families of linear time-varying systems, in: IEEE Conference on Decision and Control, Sydney, Australia, December 2000, pp. 1948–1953.

Index

A
Adjacency matrix, 105, 116, 198
Anti-windup, 74, 75
AR.Drone, 171–174
Arc length
  normalized, 62, 63, 68
  parameterization, 26, 51
Arrival margin, 28, 30
Arrival-time window, 8, 28, 95
Asymptotic stability
  definitions, 191, 193
  theorems, 191, 192, 194, 195
Auto-landing
  mission, 4, 8
  under low connectivity, 119
  under quantization, 111
  with absolute temporal constraints, 95, 99
  with relative temporal constraints, 78, 102
Autonomous system, 191
Autonomy, 184
Autopilot
  commercial off-the-shelf, 36, 52, 74, 139
  inner-loop, 35, 49, 71, 151, 162
  performance bound, 35, 36
  Piccolo Plus, 74, 129

B
Barbashin-Krasovskii theorem, 192
Bézier
  curve, *see* Bézier path
  path, 27, 156
  timing law, 27
Boundedness
  definition, 195
  theorem, 196

C
Characteristic distance, 44, 49, 55
Characteristic time, 118
Class $\mathcal{K}$ function, 193
Class $\mathcal{KL}$ function, 194
Communication
  constraints, 3, 32, 183
  graph, 33, 65, 197
  limitations, 3, 6, 27, 31, 115, 131
  link, 5, 118
  topology, 33, 64, 75, 115
Communications
  finite-rate, 13, 105
  logic-based, 115, 184
  range-based, 165–168
Communications network
  faulty, 31, 184
  static, 107
  switching, 12, 61, 159, 181, 182
  time-varying, 115, 159
Comparison lemma, 191
Complementary sensors, 6, 137
Connectivity
  any-to-any, 131
  integral, 33, 61
  low, 115, 183
  network, 33, 115, 117, 183, 198
  PE-like condition, 33, 66, 68, 91, 108, 116, 162
  pointwise in time, 67
Consensus
  practical, 34, 68, 108
  problem, 12, 32, 62, 87
  proportional-integral, 13, 65
  quantized, 13, 106
Constraints
  communication, *see* Communication constraints
  deconfliction, 27–29
  dynamic, 4, 25, 26–28, 145, 181, 183
  feasibility, 26–28
  geometric, 26–28
  spatial, *see* Spatial constraints
  temporal, *see* Temporal constraints
Control law
  path-following, 47, 51, 69, 149
  time-coordination, 64, 69, 88, 92, 105, 109, 117, 161

Convergence rate
path following, 48, 71, 149
time-coordination, 66, 67, 71, 91, 115, 162
Cooperative
mission, 3–9, 23, 87, 88, 129, 182
motion control, 3, 4, 181–183
path following, 4, 11, 23, 34, 69, 163
road search, *see* Road search
trajectory generation, 4, 24–31, 182
Coordination
closed-loop dynamics, 65, 67, 90, 106, 109, 161
control law, 64, 69, 88, 92, 105, 109, 117, 161
convergence rate, 66, 67, 71, 91, 115, 162
error dynamics, 66, 68, 90, 106, 110
error vector, 66, 90, 106, 160, 161
map, 33, 62, 76, 81
projection matrix, 66
state, 33, 62, 63, 87, 105, 115, 160
strategy, 8, 74
variable, *see* Coordination state

D
Deconfliction
spatial, *see* Spatial deconfliction
temporal, *see* Temporal deconfliction
Decoupling space and time, 11, 23, 181
Degree matrix, 105, 198
Digraph, 197–199
Disturbance rejection, 12, 13, 65, 67
Dwell time, 92, 93

E
Edge snapping, 118
Equilibrium point
Krasovskii, 107, 110
zero-speed, 107, 108
Error dynamics
coordination, 66, 68, 90, 106, 110
path-following, 31, 47, 148
Euler angles, 41
Existence and uniqueness
global theorem, 190
local theorem, 190
Exponential stability
definition, 194
global uniform, *see* GUES
theorem, 195

F
Feasible
path, 23, 31
speed profile, 23, 31, 61
trajectory, 26, 27
Field of view, *see* FoV
Flight tests
fixed-wing UAV, 129
multirotor UAV, 171
Follower vehicle, 65
FoV, 6–8, 131, 135, 137
Frame
body, 52, 144
desired, 44, 146
Frenet-Serret, *see* Frenet-Serret frame
inertial, 42, 144
parallel transport, *see* Parallel transport frame
velocity, 43
wind, 143
Frenet-Serret frame, 41, 43

G
Geo-referenced map, 138
Glide
path, 4, 8, 78, 94
slope, 4, 8, 78, 94
Graph
connected, 61, 65, 67, 107, 110, 198
directed, 117, 197, 198
disconnected, 33, 198, 199
undirected, 33, 65, 197–199
Graph theory
algebraic, 4, 33, 181, 198
spectral, 199
Ground station, 4, 130, 131, 171
GUES, 194

H
Hat map, 189

I
IceBridge program, 88, 89
Incidence matrix, 199
Input-to-state stability, *see* ISS
Inter-vehicle communications, 27, 34, 64, 131, 162, 182
Inter-vehicle schedule, 4, 27, 63, 92

Inter-vehicle separation, 28, 75, 84, 95, 133
Invariance principle, 192
ISS
  definition, 196
  theorems, 197

K
Kinematic
  coordination error dynamics, 66, 90, 106, 110
  model, 23, 52, 144
  path-following error dynamics, 31, 43, 47, 148
  state, 23, 32
Krasovskii
  equilibrium point, 107, 110
  operator, 106
  solution, 106

L
Laplacian matrix, 33, 65, 199
Leader vehicle, 65, 69, 88, 92, 133
Link weight
  binary, 121, 123
  dynamic, 87
  hybrid, 92, 118
  time-varying, 92, 117, 183
Lipschitz
  condition, 189
  globally, 190
  locally, 190
Local estimator, 115–117
Lyapunov function, 192, 197
Lyapunov stability
  autonomous systems, 191
  nonautonomous systems, 193
Lyapunov's stability theorem, 191

M
MANET, 129, 131, 138
Mission scenarios, 5–8, 74, 88, 129, 181, 184
Mobile Ad-hoc Network, *see* MANET
Motion-capture system, 171
Multirotor, 5–7, 12, 24, 32, 36, 143, 159

N
Neighboring vehicles, 24, 33, 64, 70, 88, 115, 117, 166, 198
Network
  evolving, 116, 182, 183
  extended, 116, 117, 182
  quality of service, *see* QoS
  synchronization, 12
  topology, 33, 64, 75, 78, 86, 99, 106, 116, 119, 182
Next Generation, *see* NextGen
NextGen, 8, 9, 185
Nonautonomous system, 193

O
Optimization algorithm
  centralized, 131
  distributed, 6, 131
Orbit coordination, 137, 171–175

P
Parallel transport frame, 43, 51
Path, *see* Spatial path
Path-following
  along-path error, 45
  approach shaping, 44, 52, 146
  characteristic distance, 44, 49, 55
  control law, 47, 51, 69, 149
  convergence rate, 48, 71, 149
  cross-track error, 44, 49, 134
  error dynamics, 31, 47, 148
  error vector, 31, 32, 47, 148
  strategy, 42, 51

Q
QoS, 3, 28, 33, 67, 78, 99, 115, 162, 165, 182, 183
Quadrotor, 163, 171–173, 182
Quality of service, *see* QoS
Quantization
  coarse, 107, 111, 113
  fine, 107, 109, 112, 114
  jumps, 106
Quantizer, 105

R
Rendezvous, 4, 75
Road search, 5–8, 129

S
Safety, 4, 8, 23, 74, 185
Sequential auto-landing, *see* Auto-landing

SES, 8, 9
SIG Rascal, 52, 129–131
Simultaneous arrival, 27, 28, 75, 131
Single European Sky, *see* SES
Single-point failure, 8, 65
SO(3), 41, 45, 143, 148, 181, 182
Solution
  Carathéodory, 13, 106
  existence and uniqueness, 189
  Krasovskii, 106
Spatial clearance, 25, 28, 70, 71, 183
Spatial constraints, 11, 23
Spatial deconfliction, 27–30, 70, 72
Spatial path
  collision-free, 6, 27, 131, 159
  desired, 4, 26, 32, 34, 42, 132, 143, 181
  framing, 51
  nominal, 4
  normal development, 43, 51
  polar coordinates, 43
Special Orthogonal Group, *see* SO(3)
Speed profile
  commanded, 61, 64, 69
  constant, 63, 68
  desired, 23, 26, 31, 33, 61–63, 68, 132, 145, 160
  nominal, 4
  path-dependent, 61, 181
Stability
  asymptotic, *see* Asymptotic stability
  exponential, *see* Exponential stability
  input-to-state, *see* ISS
  Lyapunov, *see* Lyapunov stability
  of equilibrium point, 191, 193
  uniform, 193–195
Swarming, 181

T

Temporal constraints
  absolute, 4, 8, 63, 87, 163
    loose, 87, 92, 95
    strict, 87, 88, 99
  relative, 4, 25, 33, 61–63, 75, 102, 131, 163
Temporal deconfliction, 27–30, 63, 71, 73
Terrain extraction, 138
Time-coordination, *see* Coordination
Time-critical
  cooperative path following, 23, 34, 163, 181
  coordination, *see* Coordination
  mission, 4, 12, 27, 31, 88, 182
Time-slot, 4, 8
Topology
  connected, 61, 65, 67, 107, 110, 198
  connected in mean, 61
  control, 4, 115, 116, 118, 119
  network, 33, 64, 75, 78, 86, 99, 106, 116, 119, 182
  switching, 5, 12, 61, 77, 98, 106, 120, 159, 181, 182
  unweighted, 119, 127
Tracking
  collective, 87
  target, 131, 136
  trajectory, 4, 5
Trajectory
  collision-free, 6, 26, 27, 131
  generation, 4, 24–31, 182
  nominal, 4
  replanning, 25, 51, 183
  sensor, 131
  spatially deconflicted, 27–30, 70, 72
  time-deconflicted, 27–30, 63, 71, 73

U

UAV
  AR.Drone, *see* AR.Drone
  avionics, 129, 130
  dynamics, 24, 35, 144
  fixed-wing, 4, 6, 24, 25, 27, 35, 41, 61, 181
  kinematics, 23, 52, 144
  multirotor, *see* Multirotor
  quadrotor, *see* Quadrotor
  SIG Rascal, *see* SIG Rascal
  tactical, 5, 129
Ultimate boundedness
  definition, 195
  theorem, 196
Unmanned Aerial Vehicle, *see* UAV

V

Vee map, 189
Vehicle
  air, 4, 8, 181, 185
  clock, 87, 88

ground, 10, 154, 155
marine, 4, 11, 185
neighboring, 24, 33, 64, 70, 88, 115, 117, 166, 198
virtual, 69, 88, 133, 135
Virtual
clock vehicle, 87, 88
leader vehicle, 65, 181
target vehicle, 10, 31, 41, 62, 144
time, 144, 145, 160
Vision sensor, 6, 8, 129–131, 133, 135
Vision-based control, 131, 136

W

Waypoint guidance, 139, 182

Z

Zero dynamics
unstable, 5

Printed in the United States
By Bookmasters